Making Capitalism Safe

THE WORKING CLASS
IN AMERICAN HISTORY

Editorial Advisors
James R. Barrett
Alice Kessler-Harris
Nelson Lichtenstein
David Montgomery

Making Capitalism Safe

Work Safety and Health
Regulation in America,
1880–1940

DONALD W. ROGERS

UNIVERSITY OF ILLINOIS PRESS

Urbana and Chicago

© 2009 by Donald W. Rogers
All rights reserved
Manufactured in the United States of America
C 5 4 3 2 1
♾ This book is printed on acid-free paper.

Library of Congress Cataloging-in-Publication Data
Rogers, Donald Wayne.
Making capitalism safe : work safety and health regulation in
America, 1880–1940 / Donald W. Rogers.
p.   cm. — (The working class in American history)
Includes bibliographical references and index.
ISBN 978-0-252-03482-4 (cloth : alk. paper)
1. Industrial welfare—Government policy—United States—
History. 2. Industrial safety—Law and legislation—United
States—History. 3. Industrial hygiene—Law and legislation—
United States—History. I. Title.
HD7654.R64   2009
363.110973'09041—dc22   2009026688

Contents

Acknowledgments

This book expands my 1983 study of Wisconsin's Industrial Commission into a comparative work on industrial safety and health regulation in six states. It aims to illuminate American occupational safety and health law's development not just in six different jurisdictions but also across the nation's decentralized federal system of government during the late nineteenth and twentieth centuries. The jump from a single-state case study to multistate analysis presented significant research and analytical challenges. I could not have met them without the kind assistance of numerous librarians, archivists, scholars, and colleagues.

When I started at the University of Wisconsin, Professors Morton Rothstein, Stanley Schultz, Stewart Macaulay, and James Willard Hurst, numerous colleagues, and especially esteemed legal historian Stanley Kutler helped me to launch this project. Simultaneously, Wisconsin's Department of Industry, Labor, and Human Relations, Wausau Insurance Companies, and *The Capital Times* opened key archives, while a grant from the U.S. Department of Labor's National Council on Employment Policy funded my initial research. The Wausau firm graciously granted me permission to cite its collection in this study. When I broadened this work later, the University of Hartford gave me release time and travel money, and then a long series of part-time employers, including Central Connecticut State University (CCSU), Wesleyan University, Housatonic Community College, and Stop-and-Shop Supermarket in Vernon, Connecticut, kept me in academic life and afloat financially. A small CCSU-AAUP faculty development grant aided my completion of this project.

Among scholars, Harry Scheiber and Joseph Tripp provided excellent insights on my first conference paper and journal article. Later, David Courtwright, Robert Asher, Patrick Reagan, and Clarence Wunderlin astutely commented on a book prospectus. Victoria Hattam broadened my knowledge of administrative law scholarship. Morton J. Horwitz received me into his office to make suggestions and share chapters of his (then) forthcoming second volume of *The Transformation of American Law*. William Nelson and the New York University Legal History Colloquium gave early chapters of my revised manuscript a thorough critique. Daniel Ernst patiently tolerated my feisty defense of this work and made studious suggestions for improvement. R. Kent Newmyer, Norton Mezvinsky, and Tony Ball encouraged my project as I brought it along. Most important, Ronald Schatz stood as a superb resource on labor history, and steadfastly supported my study to its end. Inspired by our discussions about labor economist John R. Commons, Ron suggested this book's title, *Making Capitalism Safe*.

Librarians and archivists gave me consistently courteous attention, beginning with staff at Wisconsin's State Historical Society, which generously opened its superb public records archive and granted me permission to cite documents from a restricted collection and utilize photographs from its iconography department. Later, Elizabeth Barnaby Keeney (then a Harvard instructor) arranged my access to Harvard University Libraries, whose staff graciously opened their collections of old state labor department records. Equally supportive were the Cornell University Labor-Management Documentation Center, the Connecticut State Law Library, the University of Connecticut School of Law Library and Homer Babbidge Library, Wesleyan University's Olin Library, Yale University's Sterling and Mudd Libraries, and CCSU's Burritt Library interlibrary loan office. Pete Cannon from Wisconsin's Legislative Reference Library secured a publication, and *The Capital Times* editor Phil Haslanger documented an old news story.

When my manuscript reached the University of Illinois Press, law professor John Fabian Witt, another anonymous reviewer, and series editor David Montgomery offered superb suggestions for my book's improvement, while senior acquisitions editor Laurie Matheson, assistant acquisitions editor Breanne Ertmer, assistant managing editor Angela Burton, copy editor Annette Wenda, and other staff members gracefully guided my book through the publication process.

Finally, I could not have long persisted through this project without the examples provided by my late loving parents, Ethan and Mary Rogers. Their separate struggles with the Depression, war, illness, and family dislocations taught me how to overcome obstacles and live life constructively.

Introduction

During a June 1911 Milwaukee political speech, former president Theodore Roosevelt praised a measure then pending before the Wisconsin legislature—a bill to create a state "industrial commission" to regulate work safety and health conditions.[1] At that moment of immense economic growth, when industrial accidents and labor disputes ran rampant, and when court judgments and factory legislation failed to contain them, Roosevelt commended Wisconsin's new commission for ensuring "stronger enforcement of factory laws than has been possible under the complicated and ambiguous labor laws of the past." Two years later, journalist Ida Tarbell chimed in by heralding Wisconsin's new agency as a model for the nation. "We in the East," she told *The Milwaukee Journal,* "look upon Wisconsin as the Germany of the United States, in the same sense that it has blazed the trail for many of the labor laws which are now being taken up nationally." She added, "Your industrial commission is unique. . . . I look forward to great things from it."[2]

Roosevelt's and Tarbell's acclamations dramatized the Wisconsin Industrial Commission's rising fame as a milestone in American labor law history. Indeed, for cynical twenty-first-century Americans wary of governmental action today, Roosevelt's and Tarbell's comments revive a bygone image of triumphant regulatory governance. Just as Tarbell saw Wisconsin's commission to resemble pathbreaking social legislation emanating out of Bismarckian Germany, subsequent observers regarded it as a "pioneering" agency that sprang from Progressive Era Wisconsin's "laboratory of reform." As reputation has it, the commission was a "powerful advocate for working people" that epitomized Wisconsin Progressives' passion for curbing business abuses and promoting social justice. It also exemplified the "Wisconsin Idea's" affinity for "quasi-independent commissions of experts," and fulfilled Wisconsin labor economist John R. Commons's plan to institutionalize state mechanisms for resolving industrial disputes.[3]

More specifically, commentators credited Wisconsin's Industrial Commission for two legal and administrative innovations. One directed the agency to make safety and health regulations in the form of administrative rules, rather than perpetuate the legislative system of enacting piecemeal factory laws. That new arrangement, commission secretary Arthur Altmeyer later said, was "nothing less than a work of genius." The other change abetted the commission's practice of assembling employers, workers, and experts on to "representative" advisory committees to formulate safety and health codes, a method of gathering expertise and empowering interest groups in administrative proceedings. According to reputation, these two devices produced "effective," "scientific," and "state-of-the-art" safety and health rules that both business and labor could support.[4]

Most important, observers lauded the Wisconsin commission for its supposed national influence. Use of an administrative agency to prescribe and enforce safety rules, John Commons's biographer, Lafayette Harter, asserted in 1962, "was an innovation of great importance" that "had a considerable impact on the national scene." Sure enough, in 1931, Commons's student Elizabeth Brandeis boasted that nineteen states had emulated the Wisconsin agency's safety code powers. Other commentators saw the commission's advisory committees as the archetype for the New Deal–era labor policy of "industrial pluralism," a program long associated with Commons's "Wisconsin School" of labor economics that proposed "tripartite" councils of business, labor, and public representatives to formulate industrial policy. As U.S. labor secretary Frances Perkins told Wisconsin's legislature in 1933, "The procedure of taking into an advisory committee all parties in interest before a code is adopted or rule put into effect" was a new approach to industrial regulation "worked out in Wisconsin and New York" that "can now be applied throughout the country."[5]

Historical orthodoxy thus treats the Wisconsin Industrial Commission as a founding institution of modern American labor law administration, "a guide for future State and Federal regulation of occupational safety and health." At a critical juncture in the nation's industrial growth, settled opinion says, the agency modeled administrative methods for improving work safety and health conditions and easing worker-employer conflict. To paraphrase John Commons, the commission's legal-administrative innovations "save[d] Capitalism by making it good."[6]

Surprisingly, Wisconsin's commission has largely escaped the condemnatory post–World War II historical assessment of Progressive regulatory reform. Since the 1950s, scholars have contended that external political forces and business influence dominated Progressive programs, marking that era's new regulatory agencies as either "captives" of corporate enterprise,[7] components of the rising "organizational society's" interlocking business and government bureaucracies,[8] or compromises resulting from "pluralistic" interest-group competition.[9] Likewise, postwar scholars have depicted Progressive child labor, women's protective, factory inspection, mine safety, and workers' compensation laws as merely interest-group bargains that placated business, neglected occupational disease, and remained too "piecemeal" and "spare" to provide much relief.[10] Hence, recent studies regard corporate executives and engineers as the spearheads of early-twentieth-century safety reform, not state regulators.[11] In 1970, consequently, Congress created what reformers hoped would be a better national program, the Occupational Safety and Health Administration (OSHA).[12]

Yet a few studies have described even the Wisconsin agency as a minion of business interests. One writer portrayed Wisconsin's commission as beholden to Milwaukee manufacturers when it hired a corporate safety engineer and accommodated corporate labor practices. Another ascribed the agency's supposed success to its support of big employers' pursuit of manufacturing efficiency and workforce stability.[13] Other commentators have faulted the "industrial pluralism" policy that supposedly originated in Wisconsin advisory panels. Reviewing that arrangement in New Deal collective-bargaining institutions, critical labor law historians have found it not to promote "industrial democracy" with equal business and union power but to subordinate workers to managerial and governmental dictates for industrial efficiency and peace.[14]

The clashing heroic and political images of the Wisconsin Industrial Commission and contemporaneous state labor departments raise real uncertainties about the character, internal workings, and impact of Progressive labor law programs. Did most of those plans adopt Wisconsin's commission model? What were their powers? Did they equalize business and labor influence in policy proceedings, or succumb to business "capture"? Did their new administrative capacities improve working conditions and alleviate accident and disease rates? Already, since the 1970s, diverse commentators have advocated "bringing the state [and the law] back in" to the study of such Progressive institutions. These analysts maintain variously that legal ideas guided Progressive social programs, that America's federal constitutional structure shaped reform, that Progressive agencies' bureaucratic and technical proclivities affected regulatory outcomes, and that some agencies became "autonomous" "state actors" pursuing reform themselves.[15] In such a spirit, this book reappraises Progressive safety and health regulation as a multistate system. The book traces the development of Wisconsin's administrative plan from 1880 to 1940, comparing it to labor law programs in Ohio, California, New York, Illinois, and Alabama. Especially, this study emphasizes labor law in action. It begins with state labor departments' heritage in nineteenth-century common law and factory legislation (Chapter 1), examines their foundation in Progressive labor law politics and public utility law reform (Chapter 2), and then considers their educational, safety code–making, and enforcement functions in the 1910s (Chapters 3–5) and the 1920s and 1930s (Chapters 6–8), finishing with their troublesome occupational disease work (Chapter 9) and their demise in the OSHA era (Epilogue).

The Wisconsin Industrial Commission itself marked an important change in substantive work safety law. Historians have long regarded Progressive workers' compensation statutes as a watershed in work accident law that ended harsh nineteenth-century common-law rules relegating most accident liability to workers, and then instituted a new legal regime stressing employer responsibility for workplace injuries instead.[16] This book argues, however, that "safe place statutes" in Wisconsin, Ohio, and California enlarged "enterprise responsibilities" that already existed under common law. Though interpreted narrowly to encourage entrepreneurial activity, the common-law "due care" doctrine affirmed employers' obligation to exercise ordinary prudence in providing safe work. Safe place statutes raised that duty to a higher technological plain. While retaining the developmental judicial view that regulation had to be "reasonable," safe place laws required employers to install the best protection economically and technologically feasible.

The Wisconsin Industrial Commission was, of course, better known for improving the administration of safety standards. In 1911, amid a complex situation that featured socialist agitation, electoral unrest, the ascendancy of the political ideology known as the "Wisconsin Idea," the triumph of cooperative "commonwealth" politics, growing interest in European social insurance, and the coalescence of support behind workers' compensation programs, Wisconsin lawmakers embraced administrative reform and a Progressive new governor convened a special committee headed by Charles McCarthy and John R. Commons to draw up legislation. More concerned with the implementation of safety laws than with "industrial pluralism," McCarthy and Commons proposed a new industrial commission based on the model of public utility regulation. Their plan eliminated old factory laws and empowered the new agency to formulate safety rules, just as other commissions prescribed railroad and utility rates. According to reformer Frederic Howe, the Industrial Commission constituted "an arm of the legislature in permanent session" to carry out the legislature's mandate. And according to Elizabeth Brandeis, Wisconsin's commission finished work safety law's evolution from "pre-enforcement" (judicial) to "enforcement" (factory law) and then to "administrative" arrangements. Wisconsin's agency thus exhibited Progressivism's "transformation of governance" and "new American administrative state," as well as the shift from nineteenth-century "Classical" rule-based, court-dominated law to twentieth-century "Progressive" fact-based administrative law.[17] Indeed, the Wisconsin Industrial Commission accentuated Progressive reformers' effort to transfer the work accident issue from the quintessentially "American" arena of litigation and rights-based torts to a "European" system of administrative decision making.[18]

The alleged virtue of Progressive commissions was their flexibility and technical capacity, but Wisconsin's Industrial Commission offered something more: a complete new macroeconomic regulatory scheme to improve safety across industry. Safe place duties raised safety standards, rule-making proceedings actualized those standards in administrative codes, advisory panels incorporated interest groups into code-making processes, workers' compensation largely eliminated adversarial tort claims, compensation insurance rates and penalties induced employer compliance with safety codes, and service-oriented inspectors helped employers to install protective devices. Despite reputation, however, Wisconsin's commission plan rarely received complete implementation elsewhere. California and Ohio erected similar agencies, but New York established a balkanized bureaucracy without safe place legislation, Illinois belatedly enacted factory laws and a separate compensation statute, and Alabama meekly adopted compensation legislation and a few child labor, textile, and mining statutes. In coal-mining regions, meanwhile, states established various mine inspection programs, but only a few, such as Pennsylvania and Utah, erected commissions to regulate mine safety. Progressive politics sustained a patchwork of local safety law institutions, not a uniform regulatory model nationwide.

American industry's uneven evolution from "proprietary" to "corporate" capitalism during the early 1900s complicated new state labor departments' functions. In Wisconsin and elsewhere, big rationalized corporations overpowered small traditional companies, spreading systematic engineering, scientific management, and in-house welfare programs throughout industry. After the mid-1910s, corporate authority congealed in national firms, business associations, "technical systems" of experts, and industry-sponsored safety campaigns. Unfolding differently from state to state, subsequently, such industrial modernization provoked structural economic conflicts that all safety regulators had to face. Corporate managerial strategies clashed not only with union work rules and organizing efforts but also with local employers' traditional "shop culture."[19]

For its part, Wisconsin's program sided with corporate modernizers. Based on the safe place statute, the state's industrial commissioners adopted "prevailing good practice" as their guide for safety law administration, thereby favoring advanced industrial techniques. In its educational work, Wisconsin's agency launched a "Safety First" campaign that not only broadcast technical information but also embraced big industry's paternalistic antiunion welfare practices. In code making, Wisconsin's commission used the "representation of interests" principle to assemble safety code advisory committees, but skewed committee membership more toward specialists and corporate representatives than toward small employers and union delegates. Consequently, Wisconsin committees stressed engineering, medical, and statistical analysis over interest-group bargaining. And in safety law enforcement, Wisconsin officials implemented a distinctive system of compensation insurance regulation, deploying insurance rates, merit plans, and penalty-clause actions to induce employer "good practices" more than they utilized "command-and-control" processes such as fines and prosecution.

The Wisconsin Industrial Commission, therefore, cultivated a unique technomanagerial safety policy that aligned with modernizing big industry. The agency was not "captured," however. Nor was its work simply a product of "pluralistic" business-versus-labor lobbying. Instead, like Progressive Wisconsin's celebrated "service state," and like Theodore Roosevelt's New Nationalism, Wisconsin's commission leaned toward the politics of "left corporate liberalism," even "state corporatism."[20] With its safe place statute, safety code powers, and insurance regulatory scheme, Wisconsin's commission extended remarkable state influence over the local economy to satisfy public demands for industrial efficiency, justice, and stability. Simultaneously, the Wisconsin agency collaborated with manufacturers, insurers, and "technical networks" to formulate scientifically up-to-date safety and health regulations. Then it engaged business, labor, and other local interest groups on advisory committees, not to adopt the "pluralist" agenda for "industrial democracy" but to gather expertise and secure interest-group cooperation with state regulatory goals. The commission thus stressed state guardianship over workplaces rather than union empowerment. And under separate legislation reflecting contemporary gender attitudes, Wisconsin's commission established a woman and child labor subdivision that enforced women's work hour and wage laws, though it rarely if ever included women in safety code proceedings.

Like the union movement nationwide, organized labor in Wisconsin staunchly supported strong state workplace protection, but wavered over its practical operation. Labor officials vacillated between abiding the Industrial Commission's protective activities and guarding their own interests as trade unionists. As elsewhere, Wisconsin labor leaders became indefatigable legislative lobbyists who defended the agency's mandate and sometimes tried to expand its powers. They also sought participation on safety code advisory committees, though they seldom enjoyed equal representation with business and experts. As elsewhere, too, Wisconsin unions sometimes faulted the state commission for inaction, rejected regulations threatening craft skills or job security, and resisted rules slighting full protection. Wisconsin unionists especially evaded paternalistic government welfare schemes copied from industry, though they rarely advanced plans such as employer–union shop safety committees and union-sponsored educational programs that labor groups advocated elsewhere.

Across other states, federalism perpetuated a checkerboard of safety law programs, not a duplication of Wisconsin's "left corporate liberal" approach. California's Industrial Accident Commission set up competitive state compensation insurance and technocratic safety code procedures but pluralistically accommodated business, labor, and insurers on advisory committees, while a separate agency handled woman and child labor issues. The Ohio Industrial Commission, by comparison, administered an exclusive state compensation fund but spared local corporations from safety code requirements until labor agitation compelled action in the 1920s. New York's labor department, however, was fragmented into uncoordinated compensation, safety code, and inspection divisions that catered to the state's array of manufacturer, labor, reform, and women's groups. Illinois, by contrast, adopted a broad compensation law and extended adversarial litigation rights to occupational disease but erected an old-fashioned, partisan factory inspection bureau to implement the state's anachronistic factory laws. Simultaneously, Alabama belatedly enacted compensation legislation, though its characteristically southern antilabor politics blocked most protective laws and the creation of a state labor department until the 1930s.

Wisconsin's and other states' labor departments certainly were diverse, but contrary to reputation, they were not invariably "weak" or "spare" institutions, even under the probusiness "associative state" of the 1920s.[21] Wisconsin, California, New York, and Ohio agencies sponsored lively educational campaigns and promulgated dozens of safety codes. Even Illinois's antiquated labor bureau performed thousands of inspections, whereas Wisconsin's commission freely invoked a 15 percent compensation penalty clause to enforce safety codes, and New York's department coerced employers with "calendar hearings" and "complaint inspections." Regulation was so heavy that one disgruntled employer condemned Wisconsin's commission as "Soviet Government." From the 1910s through the 1920s, state action likely improved workplace engineering and lowered machine-injury rates, particularly in marginal firms.

Even in the occupational disease field, state agencies took more action than historians recognize. Admittedly, industrial illness remained largely beyond work injury lawsuits and outside of most compensation laws. Nonetheless, labor agitation caused states such as New York and Ohio to schedule some illnesses for compensation and to establish industrial hygiene plans. Moreover, Wisconsin and California adopted comprehensive disease compensation laws and then incorporated occupational health into their technocratic regulatory systems. In fact, while reformers elsewhere struggled to extend compensation coverage to diseases such as lead poisoning, radium poisoning, black lung, and silicosis, Wisconsin's program precociously formulated dusts-and-fumes regulations, compensated silicotic workers, and largely suppressed silicosis with preventive measures.

Whatever "decline" that Progressive work safety and health agencies experienced after 1920 came not simply from inherent weaknesses but from the way in which their distinctive institutional arrangements fell out of step with industrial, political, and legal change. Modern manufacturers shifted from engineering controls to personnel management to control workplace accidents, labor organizers fought mainly for collective-bargaining rights rather than for stronger protective legislation, employers pursued the reform of administrative procedures to curtail labor departments' power, and employers and trade unionists turned workers' compensation into an adversarial process by contesting worker claims. These developments weakened state programs, but politics and law through the 1930s still discouraged national involvement in industrial working conditions. Not until 1970 did the "new social regulation" fully extend national authority over occupational safety and health. By establishing OSHA then (as well as the Mining Enforcement and Safety Administration in 1973), reformers finally nationalized work safety and health standards and dramatically toughened procedures for their enforcement.

OSHA proved to be disappointing, but its creation left a negative impression of state safety and health programs that still obscures their historical significance today. Bracketed historically by the nineteenth-century common-law system and the late-twentieth-century OSHA plan, state programs constituted a crucial intermediate early-twentieth-century Progressive stage of work safety and health law development. In this phase, aptly captured by historian Jon Teaford's expression "the rise of the states," state governments broadened employer responsibility for work safety and health, though they still followed the nineteenth-century "federal" pattern of establishing varied state-to-state institutions to enforce that duty without embracing a particular regulatory model.[22] Among the states, Wisconsin erected an extraordinary administrative plan, and pieces of it were duplicated elsewhere. Though unevenly during the Progressive phase of occupational safety and health regulation, Wisconsin and other states expanded their administrative capacities over industrial workplaces with significant effects on working conditions and on the relationship between workers and employers.

1

From Common Law to Factory Laws

Work safety and health law first developed in the United States during the mid-nineteenth century, when railroads, mining operations, and mechanized factories expanded industry and multiplied the number of work accidents. Arising mainly on the state level, work safety and health law began in the courts and moved haltingly into legislatures. Starting in the 1840s, judges developed common-law rules for work injury cases that defined employers' and workers' respective duties for work safety, and their respective liabilities for accidents. When the pace of industrialization and accidental injury quickened in the late 1800s, workers in Wisconsin and other states began agitating for protective labor legislation, yielding an uneven patchwork of local statutes that modified employer liability rules, set up state bureaus of labor, instituted safety and sanitation requirements, and created state factory inspection staffs.

Though fragmented and incomplete nationally, this transition from common law to factory laws exhibits an important change in work safety and health policy in Wisconsin and the nation. As legal historian James Willard Hurst once observed, nineteenth-century legal development shifted public policy from the business-friendly principle of "releasing individual creative energy" to the public-spirited idea of "social cost accounting."[1] The move from common-law work accident rules to factory legislation, along with modifications in common law itself, seemingly embodied that shift by increasing governmental supervision of employers' safety and health efforts. Yet jurisprudence, politics, and bureaucracy retarded work safety policy reform until the early 1900s.

The limitations of nineteenth-century work safety and health law emanated out of the earliest court decisions. In South Carolina and Massachusetts cases, especially *Farwell v. The Boston and Worcester Rail Road Corporation* (1842), judges prescribed "assumption of risk" and "fellow servant" rules that placed much of the burden for work safety on wage earners. Dismissing railroad liability for an accident caused by a coworker in *Farwell,* Massachusetts chief justice Lemuel Shaw denied applicability of the *respondeat superior* doctrine that held employers responsible when employee negligence caused injuries to third parties. In job accident cases, he ruled, workers assumed the "natural" risks of employment, including those created by fellow laborers. Indeed, Shaw said that public policy obligated workers themselves, not employers, to assume the risk of coworker negligence, because workers were better positioned to observe and police coworkers' behavior. Northern state courts rapidly embraced this rule, though some state judges made exceptions for negligent coworkers who were supervisors or worked in separate departments. Pre–Civil War southern courts initially modified the rule to suit the South's hierarchical slave economy, but they quickly adopted northern analysis after the war.[2]

This jurisprudence served business enterprise at workers' expense, according to most modern accounts. Early scholars attributed such pro-employer inclinations to "instrumental" judicial efforts to "subsidize" business with lower liability costs, and subsequent commentators grounded them in the complex heritage of medieval English master-and-servant law, although a few recent analysts have seen mid-nineteenth-century judges balancing moral and community values against business interests in accident cases. Nowadays most historians agree that nineteenth-century work accident law harshly ignored the real hazards of modern business. Infused with "Classical" moral and economic thinking, most scholars now contend, early work accident law adopted contractual language that treated employers and workers as free and equal economic actors in small-scale enterprises, wherein workers controlled the flow of work and assumed most accident risks. Applied to emerging industry, however, the law's fellow-servant and assumption-of-risk rules effectively "immunized" employers from accident liability without acknowledging that big industrial operations reduced worker control and increased accident hazards. Despite industrial change, work accident law sustained workers' primary responsibility for safety until compensation statutes changed that policy in the 1910s.[3]

Courts in Wisconsin and other states absorbed this jurisprudence but cultivated a growing body of employer-responsibility law at the same time.

As early as *Farwell*, Lemuel Shaw recognized "implied warranties" that railroads might owe their workers, including "a sufficient engine, a proper rail road track, a well constructed switch, and a person of suitable skill and experience to attend it." Using the language of emerging negligence law, judges in Wisconsin and elsewhere transformed this dictum into the rule that employers owed workers a duty to exercise "due care" or "ordinary care" in providing safe tools, machinery, and equipment along with competent coworkers, in effect holding employers accountable for safe workplace engineering and management.[4]

Though "due care" established employer responsibility, the "Classical" nineteenth-century "rule of law" judicial ideology constricted its impact. Drawn from moral philosophy, Classical economics, and republican political thought, legal "Classicism" stressed uniform judicial adherence to abstract principles, particularly those protecting individual will and private property from state interference. Classical jurisprudence thus presumed a developmental economic policy that treated entrepreneurial technological freedom as the best route to public welfare and social progress.[5] Although the rising tide of industrial injury lawsuits may have eventually caused judges to qualify the fellow-servant rule, enlarge the list of employer duties, and affirm employer negligence, the Classical ideology and developmental policy caused Wisconsin's and other states' judges to interpret the employer "due care" standard narrowly to protect business initiative in the market, much as they did with restrictive rulings toward union activities and wages and hours laws.

Indeed, the narrow Classical vision of employer responsibility persisted even when work safety and health law shifted into the legislative arena after the 1860s. It was not just, as historians suggest, that legislative modifications of the fellow-servant rule had questionable effect on employer liability, or that factory safety and sanitation measures were "weak," "poorly written," and badly enforced,[6] but that factory statutes in Wisconsin and other states possessed basic limitations. Politically, factory laws rested on the rising labor movement's slim political base, resulting in only a spotty patchwork of protective legislation from state to state. Administratively, factory laws foundered upon a rudimentary criminal law model of policing that denied factory inspectors the power and technical expertise needed to enforce statutory requirements. Legally, factory laws did not secure total judicial acceptance of new legislatively formulated standards of employer care. In work injury cases, courts accepted factory safety legislation as a proper exercise of states' police powers, but still treated factory laws as mere elaborations of the judge-made due care standard.

In very uneven fashion, hence, court cases and factory legislation in Wisconsin, New York, Ohio, Illinois, California, Alabama, and other states expanded employer responsibility for work safety during the late 1800s and early 1900s, but they did so without fully breaking the judiciary's ideological dominance over work safety policy. Litigation and factory laws notwithstanding, most courts perpetuated the Classical developmental economic outlook that acknowledged employers' duty to exercise due care in providing safe work, but restrained that duty to avoid stifling business initiative. Only after 1900 did factory legislation gather enough public support to challenge judicial policy's persistence.

Work Accidents and the Courts

In 1880 Wisconsin was a relatively small industrial state. Out of a total working population of 417,455, almost half of its laborers remained on farms, whereas only 57,109 were factory workers, most in tiny carriage and wagon, footwear, cooperage, and blacksmith shops. The state's factory labor force consequently was larger than those of agricultural states like Alabama and California, but much smaller than the diverse manufacturing workforces that already ran into the hundreds of thousands in Ohio, Illinois, and New York. Aside from Milwaukee foundries and upstate lumber mills, Wisconsin was still a state of farms and small enterprises.[7]

In the 1860s and 1870s, however, railroads brought modern industry and work accidents to Wisconsin, leading the state's supreme court to adopt the emerging common law of industrial torts, along with its underlying developmental economic policy. At first, in 1860, the Wisconsin Supreme Court waved the fellow-servant doctrine aside on the grounds that a railroad owed the public and workers alike a broad duty to conduct its business with "proper care and skill." In *Moseley v. Chamberlain* (1861), however, it acceded to the fellow-servant rule. Because the court's initial ruling stood alone against English and American precedents, Chief Justice Luther Dixon explained, it had to be reversed. After *Moseley*, the Wisconsin Supreme Court "fell into line" with fellow-servant jurisprudence, though not without exceptions. In 1877, the court adopted the so-called vice-principal rule exempting accidents caused by supervisor negligence from the fellow-servant rule's operation.[8]

By accepting fellow-servant analysis, Wisconsin's supreme court accommodated Classical mid-nineteenth-century legal mores that required judges to "put a premium on harmony and consistency across jurisdictions." It also embraced the judiciary's prevailing developmental economic philosophy.

As commentators of the time rationalized, holding employers responsible for coworker negligence would ruin business by encouraging lax worker habits, raising employers' accident liability costs, and impeding technological innovation. The fellow-servant rule removed this burden and unleashed entrepreneurial initiative and economic progress.[9]

Wisconsin's court achieved similar results when it turned to employers' "implied warranties" for safety, though it stressed employer-employee relations rather than public good. In *Wedgwood v. The Chicago & Northwestern Railway Company* (1877), brakeman Wedgwood complained that the railroad negligently caused his injury by allowing a brake-beam bolt to protrude hazardously. Relying on *Clarke v. Holmes* (1862), one of the few early English cases to favor injured workers, Associate Justice Orsamus Cole agreed that the railroad owed Wedgwood a duty to exercise "due care" and use "all reasonable means" to guard against such defective equipment. That duty arose from a presumed employment contract, wherein the company promised to exercise due care to avoid aggravating employment dangers beyond what Wedgwood could have expected. Through defective design and failure to inspect, Cole decided, the railroad broke its promise and left Wedgwood's work "unnecessarily hazardous and unsafe." In the subsequent case of *Smith v. The Chicago, Milwaukee & St. Paul Railway Company* (1877), however, he cautioned that "due care" did not guarantee safety. Citing a New York precedent, Cole observed that due care required employers to take reasonable steps to supply adequate safety appliances, materials, and helpers. Once that was done, workers assumed "the risks of employment, and of a failure of machinery from any latent defects not discovered by practical tests."[10]

Central to accident litigation, the "due care" doctrine reflected mid-nineteenth-century judges' application of the rising common law of "negligence" to employer-worker relations. Negligence presumed that a cause for legal action lay in employer "fault," specifically the breach of a legal duty that "proximately caused" a harm. In work injury cases, that duty was "due care," sometimes called "reasonable care" or "ordinary care." Generally, it referred to that level of diligence that ordinarily prudent persons exercised in a given situation. Consequently, Wisconsin courts did not require firms to exercise extraordinary or "utmost" care, exhibit the highest level of diligence possible, or even use the latest equipment, just to do what was ordinarily prudent. And other courts North and South ruled the same way. New York courts required "reasonable care, not the highest efficiency," whereas Alabama judges demanded "such care as men of reasonable and ordinary prudence exercise," and Illinois courts mandated "reasonable and ordinary care and diligence."[11]

The ordinary-prudence standard limited due care. In *Smith,* for instance, the Wisconsin Supreme Court said that due care required railroads "to adopt all reasonable and usual tests" to discover defects in brake rods, one of which had broken and caused Smith's injury, but that it was "impracticable" and of "doubtful utility" to require the company to do more. Still, employers had to do more than just what was customary. In *Dorsey v. The Phillips & Colby Construction Company* (1877) involving a cattle chute that knocked Dorsey off a moving freight car, Chief Justice Edward G. Ryan ruled that customary use of dangerous equipment could not satisfy employers' duty to exercise due care. The public interest, he declared, "cannot tolerate a practice which unnecessarily exposes workers to danger."[12]

Dorsey's reference to the public interest was unusual, because Wisconsin's and other states' courts typically treated due care just as a private duty. Although it may be true that the fellow-servant rule absorbed the post–Civil War "free labor" ideology celebrating workers' independent control over work,[13] judges viewed due care as an obligation emanating out of implied reciprocal employment contracts. For increased wages, they reasoned, laborers bore the risk of accidents resulting from usual work hazards, while employers assumed the burden of those resulting from their failure to exercise proper care. Judges postulated that such agreements formed freely in the private economic market would facilitate "fair" distribution of the benefits and costs of economic growth.[14]

This legal image of economic fairness was dubious. Legal historians have hotly disputed whether nineteenth-century contract and property law treated commercial interests more generously than farmers, consumers, and other groups, but they overwhelmingly agree that work accident law favored business over workers. Not only did employer defenses such as the fellow-servant and assumption-of-risk rules "shelter" industrialists from liability, but also the courts' contractual language sustained the fiction that employers and employees stood as equally powerful bargaining partners in huge industrial operations actually dominated by employers. Even in meritorious injury cases, it was questionable whether comparatively weak worker litigants could recover damages.[15]

More important, Classical nineteenth-century legal ideology constrained judges from shifting more responsibility for job safety to employers. Work safety rulings did not, as some legal historians suggest, emerge as an "exception" to the fellow-servant rule that created an expansive new duty for employers to provide "safe workplaces."[16] Actually, they amplified the original suggestion propounded in Lemuel Shaw's *Farwell* ruling that employer "warranties" were part of the same employment agreements that created the fellow-servant principle. Nonetheless, courts rarely interpreted those "warranties" under the due care doctrine to impose an absolute duty on employers to provide "safe workplaces" but instead followed the logic of a *Clarke v. Holmes* concurrence that "the master is neither . . . at liberty to neglect all care, nor . . . is he to insure safety, but he is to use due and reasonable care."[17]

American courts faithfully adhered to this middle-ground policy. Although they sometimes stated that employers had a duty to provide "safe and suitable machinery," judges construed such language only to require employer due care, not to guarantee a "safe" workplace. As such, the due care principle developed side by side the fellow-servant rule, and both rules together perpetuated the Classical vision of progress and economic growth.

The Judicial Response to Industrialization

From 1870 to 1910, industrial development transformed the economic environment in which Wisconsin judges applied common-law work accident doctrines. By 1910, the number of Wisconsin's manufacturing workers had jumped to 182,583, about one-fifth of the state's total labor force, compared to one-third still in agriculture. This ranked Wisconsin ahead of Alabama's 72,148 and California's 115,296 factory workers, but well behind Ohio's 446,934, Illinois's 465,764, and New York's whopping 1,003,981 wage earners.[18] Wisconsin industry simultaneously grew more dangerous. Big integrated factories replaced craftsmen's workshops. Unguarded ripsaws, planers, and punch presses driven by steam- and water-powered belt and pulley systems replaced hand tools. With cheap energy readily available, employers accelerated machinery at safety's expense, while shop foremen and workers alike sustained the work culture of individual responsibility and "manly" independence.[19]

Compared to other states, however, Wisconsin's industrial development was distinctive. The state's furniture, railroad car construction, leather, and paper industries grew, but lumber and foundry products predominated. After 1900, upstate lumber mills peaked, while southeastern Wisconsin steel- and foundry works matured into "highly mechanized mass production" operations. By contrast, Alabama remained a "colonial economy" led by textile and lumber industries, despite some coal mining and pig-iron production. As in Wisconsin, lumber topped California industries, but that state's manufacturing was scattered among small railroad car construction, foundry, and food preservation firms, while wheat, citrus products, and wine generated the most wealth. With rich natural resources, meantime, Ohio and Illinois grew rapidly toward diversified economies involving huge basic industries such as steel, mining, and oil, and large peripheral businesses in meatpacking, rubber, and glass. In 1910, New York remained the nation's premier industrial state with diverse manufacturers spread among Buffalo, Elmira, and of course New York City.[20]

Despite economic differences, the amount of accident litigation grew everywhere. In Wisconsin, work accident appeals to the state supreme court climbed from 19 cases before 1881 to 44 during 1881–1890, 153 from 1891 to 1900, and 167 from 1901 to 1910. Similarly, in New York City's trial courts, work injury suits jumped from 0 in 1870 to 24 in 1890 and 160 in 1910, reflecting a "tort explosion" exceeding even the growing accident rate.[21] The substance of work injury suits changed accordingly. Early Wisconsin accident cases mainly involved railroads, plus a few in building construction, shipping, and lumber, but after 1890, accidents in paper mills, lumber mills, and factories predominated. And whereas many post-1890 suits involved falls, objects striking victims, cuts, asphyxiation, electrical shock, or explosions, the single largest nonrailroad category (25 percent) entailed the crushing of limbs or fingers in gearing, pulleys, elevators, and other machinery.[22] Historians disagree as to whether rapidly increasing work injury suits changed legal doctrine. Some scholars contend that new safe place and safe tool doctrines nearly overthrew the fellow-servant rule, and that a growing body of judge-made rules "heightened demands" on employers to provide safe work. Yet others maintain that judges hewed consistently to traditional standards of employer care, despite pressure from plaintiffs and juries to elevate employer responsibilities.[23]

Wisconsin's late-nineteenth-century supreme court certainly expanded rules to govern employers' duties in the workplace, but it did so firmly within the orthodox common-law due care standard. Assuming that doctrine to be "well settled," the Wisconsin court reiterated that due care did not require employers to guarantee safety or eliminate all hazards, but only to use the level of care, judgment, and skill of businessmen in like circumstances. The court thus ruled that manufacturers had to purchase safe and suitable machinery, use it for intended purposes, service it often enough to keep it safe and operational, warn laborers of machinery's special hazards, and promulgate regulations where work was complicated and dangerous.[24] Rather than "heighten demands," Wisconsin's court, like other states' tribunals, merely inventoried changes in what employers themselves found to be prudent in light of industrial change.

Remarkably, the Wisconsin Supreme Court's fidelity to orthodox due care did not deviate in women's work accident suits, and only partially in child labor injury actions. Studies of non-work-related accidents to women on railroads and other conveyances have shown that courts "bifurcated" accident law along gender lines, applying different standards of care to women's cases based on presumed physical debilities, "ladylike" behavior, and subordinate status. Yet in work accident cases, Wisconsin's court cited no gender-based disabilities, nor did it require employers to raise standards of care for female workers, even when hair fashion, physical strength, height, or familiarity with machinery was involved. The court simply recognized some female laborers' "tender age" and status as minors. As for child laborers, Wisconsin's court relaxed the assumption-of-risk rule based on the obviousness of dangers and the extent of children's experience and intelligence, while elevating employer duties to give warnings and instructions for dangerous jobs. Even here, Wisconsin's court still adhered to orthodox employer due care regarding machinery, tools, and coworkers.[25]

Other states' courts followed similar logic in female and child labor cases. California decisions gave female workers no special treatment, though rulings stipulated that due care required employers to take precautions for these typically teenage laborers' "youth and inexperience." Illinois's supreme court held women to the same standard of "due and ordinary care for his or her own personal safety" as it did for all adult workers, while demanding that employers exercise consistent due care to provide "reasonably safe work." Alabama's high court ruled that employers had a constant duty to warn and instruct "inexperienced and immature" female child laborers, but found a twelve-year-old girl in one case to be guilty of contributory negligence for cleaning a textile machine's unguarded cogs while being "fully aware of the danger." In New York, the Court of Appeals not only sustained the principles that employers owed female laborers "reasonably safe work" and that the women themselves assumed the risk of open and obvious hazards but also affirmed women workers' "free agency" and "right to manage their own affairs." Only one rare New York decision hinted that female laborers deserved different treatment. "If the master hires younger children, especially girls with long hair," it stated, "it is his duty to guard [dangerous machinery] and . . . take into consideration their age, inexperience, and lack of care and discretion."[26]

Comparatively, the Wisconsin Supreme Court's doctrinal reasoning was very stable in a contentious area of law. True, the court was not immune to the "wobbly" state of employer defenses during the late 1800s and early 1900s, as it developed exceptions such as the vice-principal doctrine and accommodated employer liability statutes that delimited the fellow-servant rule and other employer exemptions.[27] Yet when it came to defining employers' standard of care, the Wisconsin court retained a steady Classical or "formalistic" style in work injury cases. Of the 364 industrial accident cases that it heard on appeal between 1881 and 1910, the Wisconsin Supreme Court reversed 175 of them (136 in favor of employers and only 39 for workers), not only by upholding the fellow-servant, contributory negligence, and assumption-of-risk rules but also by nullifying trial court evidence, judges' rulings, and jury verdicts that tended to enlarge employer duties and worker protection above due care standards.[28] Wisconsin's high court especially resisted the inclination of a few other states' courts to endorse the apparent absolute requirement that employers provide "a reasonably safe place to work in." Wisconsin's supreme court avoided changes in standards. As one insurance executive observed in 1909, "The Wisconsin courts have been holding a very consistent position, and we do not fight a case as much as you would in some other states."[29]

The Wisconsin Supreme Court's *Knudsen v. The La Crosse Stone Company* (1911) decision illustrates its persisting Classical style and doctrinal consistency. No matter how great the human tragedy, Associate Justice Roujet D. Marshall lamented when he invoked the fellow-servant rule in this case, "courts cannot shape their decrees to meet their personal ideas or merely satisfy human sensibilities to human sorrow and suffering." Common law "commanded" the court as much as the statutes did. Though an active supporter of workers' compensation reform, Marshall insisted that only the legislature, not the courts, could change the law. Associate Justice William Timlin dissented vigorously, but only on the grounds that the fellow-servant rule did not apply in this case.[30] Conditions changed, but the spirit of Luther Dixon's 1861 *Moseley* opinion stressing conformity to prevailing common-law rules lived on in the 1911 court of Marshall and Timlin.

The Wisconsin court's adherence to the due care principle and fellow-servant rule perpetuated nineteenth-century law's developmental economic policy. As Associate Justice Alfred Newman explained in *Guinard v. Knapp-Stout & Co.* (1897), the law aimed to be "practical." Because the expense of extraordinary care might "defeat the success of the enterprise," he said, the law would require no person to use greater care than was exercised by the mass of mankind.[31] Extraordinary care, he implied, would impede economic progress. Just at that moment, however, the growing industrial accident toll generated political agitation that would challenge this policy and the courts' control over it.

Invoking the Police Power

During the late nineteenth and early twentieth centuries, organized labor emerged all over the country as a catalyst for work safety legislation, though in a rough way. Before the 1890s, as Morton Keller observed, political activity exhibited a confusing "uncertainty of purpose," suspended as it was between the fading mid-nineteenth-century values of individualism, localism, and limited government and the rising needs of industrial society. Consequently, early labor groups in Wisconsin and elsewhere lobbied state legislatures for protective laws only in a piecemeal, haphazard way, sometimes to have those laws struck down in court. After the mid-1890s, however, old political values gave way to a broad new public outlook encouraged by economic pressure groups, including state federations of labor. Organized largely as legislative lobbyists, local labor federations united with middle-class reformers to obtain an outpouring of protective factory laws, despite business opposition.[32]

In fact, organized labor played a bigger role in the early-twentieth-century safety movement than historians generally recognize. Conventional history holds that labor injunctions, state police action, and judicial nullification of labor laws so thoroughly suppressed labor agitation in the 1880–1910 era that the American Federation of Labor and other labor groups largely abandoned governmental protection and pursued a self-help policy of "voluntarism" to secure their goals through collective action instead.[33] Some historians thus dismiss organized labor as a factor in the industrial safety movement, saying that the movement came "from the top down" from business managers and engineers.[34] Yet other scholars have shown the significance of *local* turn-of-the-century labor organizations in reform. Though "ambivalent" toward government, those groups did enter politics to accumulate a wide, if piecemeal, array of fire safety, machine safeguard, mining, and workshop sanitation measures.[35] Agitating "from below," local labor organizations increasingly demanded state action to place more responsibility for accident prevention on employers.

And so it was in Wisconsin. Before 1890, short-lived labor groups addressed factory safety only sporadically, but produced important early results. Momentarily flirting with political action, the Wisconsin Knights of Labor teamed up with Milwaukee labor leaders, several newspapers, and Republican governor Jeremiah Rusk to establish the state's Bureau of Labor Statistics in 1883 over rural objections. The new bureau provided for a labor commissioner who collected employment statistics, issued biennial reports, and enforced laws regarding "the employment of children, minors, and women" and "the protection of the health and lives of operatives." In 1885, lawmakers added a deputy, a clerk, and a factory inspector. Rural opposition continued, but the labor bureau became a fixture of Wisconsin government like boards set up to regulate railroads, public health, insurance, pharmacists, and illuminating oils. Indeed, eight other states simultaneously created their own bureaus of labor due to agitation by the Knights of Labor, trade unions, municipal trade assemblies, and city labor councils, including Ohio (1877), Illinois (1879), New York (1883), and California (1883), but not Alabama, where the South's weak array of Knights, trade unions, and organized coal miners faced racial divisions, employer dominance, and anti–labor law sentiment.[36]

The Wisconsin labor bureau's first factory inspector initially targeted child labor law enforcement, but labor unrest caused lawmakers to redirect that officer toward factory safety. In 1887, Wisconsin's legislature enacted the state's first machine safety law, requiring businesses to guard or fence dangerous belting, shafting, gearing, hoists, flywheels, and elevators, as well as stationary vats or pans holding molten metal or hot liquids. In a companion measure, legislators instructed the factory inspector to enter factories and workshops in order to examine elevators and machinery for dangerous conditions, condemn and post notices on unsafe elevators, and order the installation of guards or protection on machinery, elevator wells, and stairways.[37]

Wisconsin's 1887 machine safety law signaled a nationwide shift of work safety policy out of the courts and into legislatures. That shift occurred unevenly from state to state. It began in coal mining states, when mine disasters and worker agitation—first in the East and Midwest (Pennsylvania, Illinois, Ohio, and West Virginia from 1869 to 1883) and then in the Rocky Mountain West (Colorado, Wyoming, New Mexico, Montana, and Utah from 1883 to 1896)—inspired an array of state mine safety laws that instigated rudimentary regulation and inspection to prevent explosions and roof collapses. Almost simultaneously, Massachusetts enacted America's first factory act in 1877, with Ohio following in 1884, and New York in 1887. Yet Illinois passed no machine safety legislation until 1909, despite its important 1893 sweatshop act. Neither did California until 1913, although it enacted shop sanitation requirements in 1889 that included seats and separate wash facilities for women. And neither did Alabama, though it passed an isolated law requiring seats for salesgirls in 1889 and another establishing a cotton-mill sanitation inspector in 1907. A dozen states enacted factory laws by 1900, but coverage varied. Kansas's supreme court rightly complained about American factory laws' "confused condition."[38]

Late-nineteenth-century factory law administration impaired what legislation that states did enact. Illinois, for example, founded a factory inspection force in 1893 to implement the child labor and women's hours-of-work sections of the state's new sweatshop act, but law and politics disrupted that program's operation. At the start, prolabor Democratic governor John Peter Altgeld named sweatshop reformer Florence Kelley to head this staff of ten inspectors, including five women and several trade unionists. Kelley, with a European socialist's sense of governmental power, mobilized inspectors efficiently and prosecuted delinquent employers aggressively. Employers fought back, however, leading to the state's *Ritchie v. People* (1895) decision striking down the law's hours provisions. Then Altgeld's electoral defeat in 1896 turned Kelley's regime out of office. Illinois's inspectorate remained intact and eventually received authority over work safety and health, but rapidly lost force as it became entangled in partisan politics.[39]

Labor law administration elsewhere had internal problems. It was not just underfunding, understaffing, political corruption, or lack of will that caused trouble but also the way in which state labor bureaus worked as institutions. Those bodies were, for one thing, initially headed by partisans, not professionals. Wisconsin's first labor commissioner, Frank Flower, was a journalist and political ally of Republican governor Rusk. Succeeding commissioners Henry M. Stark (Democrat, 1889–1890) and Halford Erickson (Republican, 1895–1904) both were former railroad accountants with political connections. Only Joseph D. Beck, who joined Wisconsin's labor bureau as a clerk in 1901 and then became commissioner in 1905, could claim expertise. Like New York labor commissioner P. Tecumseh Sherman's arrival in 1905, Beck's appointment reflected movement toward professionalism.[40]

Labor bureaus' factory inspection regimes were, for another thing, clumsy. Originally established to enforce child labor laws, Wisconsin's tiny inspectorate initially adopted the "reactive" criminal law approach of police officers who patrolled their beats looking for law violations rather than the "proactive" strategy of securing safety improvements before accidents happened. Frank Flower told inspector Henry Siebers in 1887 simply to "enforce the laws strictly and impartially and let it offend whomever it may." Flower's successor, Halford Erickson, by contrast, inaugurated an "educational" approach, counseling inspectors to obtain employer compliance with frank and open talks while pressuring them with frequent visits. "Factory inspectors [are] like policemen," Erickson remarked. "[They] must call early and often, in order to serve their purpose." Neither Flower nor Erickson followed inspections, however, with vigorous prosecution of delinquent employers.[41]

Factory legislation's ambiguity, for one final thing, left inspectors without clear instructions or legal authority. Factory laws specified only that machinery be "guarded" or "fenced." Meanwhile, Wisconsin's labor bureau provided inspectors with blank forms, asking them to ascertain, "Do you consider floors, walls, etc., safe?" or "Is any workroom so poorly ventilated or overcrowded as to endanger health and safety of employees?" Inspectors reacted to this broad language in different ways, still acting like police officers who issued on-the-spot "curbside justice" as they saw fit. Some inspectors simply noted dangers without issuing orders, whereas others adopted a "reformatory" posture, telling employers about dangerous conditions and suggesting remedies. Occasionally, inspectors issued drastic orders. Without explicit statutory authorization, for example, factory inspector John Zwaska ordered one firm to replace the cable on its elevator, directed another to provide self-closing elevator gates on each floor of a building, and asked a final company to cut three windows in a wall and install an exhaust fan with hooded pipes to improve ventilation.[42]

The results were mixed. Employers initially resisted factory laws, and occasionally tried to bribe factory inspectors to avoid compliance, but commissioners and inspectors alike reported eventual cooperation. Labor commissioner Henry Stark wrote in 1890 that most employers complied with inspectors' orders "without murmur," whereas Erickson indicated in 1896 that employers readily obeyed most inspector directives. After ordering the Neillsville Manufacturing Company in 1892 to fix its elevator, inspector S. L. Van Etten reported that "my suggestion was complied with at once." Yet there was evasion. When Erickson wrote to employers in 1901 to verify their adherence to orders, nearly all affirmed compliance. Surprise visits, however, revealed many employer reports to be "misleading." Only frequent reinspection, Erickson decided, would keep some employers honest.[43]

Then, in the decade after 1898, factory legislation changed and expanded. Rising state labor federations turned work safety into a major political issue, paralleling business managers' and engineers' growing concern with that problem. The Wisconsin State Federation of Labor (WSFL), for example, began lobbying for safety measures in 1899, and the state legislature responded with laws governing cigar manufacturing, sanitary conditions, emery-wheel hoods, and separate dressing rooms and seats for women workers. In the wake of growing female factory employment in Wisconsin—from 6,670 in 1890 to 16,266 in 1900—these latter requirements codified "maternalist" notions promoted by women reformers that female workers were potential (or actual) mothers who faced special risks in industry and needed state protection. In other states as well, labor lobbyists prompted the expansion of machine guarding, ventilation, and sanitation laws, as well as women's work hour and protective measures, and a new round of mine safety laws. Just after the Ohio State Federation of Labor (OSFL) was founded in 1897, for instance, Ohio lawmakers strengthened laws regarding dust-producing apparatus, seats for women, and machine dangers. Simultaneously, the Workingmen's Federation of New York secured improvement of protective legislation. In 1901, the California State Federation of Labor (CSFL) got started with a legislative agenda.[44]

The early 1900s also brought increases in the size of inspection forces. Wisconsin lawmakers agreed to labor commissioner Halford Erickson's request to improve child labor law enforcement by raising the number of factory inspectors from two to seven, a change that helped safety inspection. Meanwhile, New York's legislature boosted its inspection staff from twenty-six to fifty, and Illinois expanded its sweatshop inspectorate from ten to eighteen.[45]

Despite these legislative achievements, factory laws remained controversial. Industrial managers and engineers may have started to rethink work accidents as a managerial responsibility, and they may have initiated in-house relief and prevention programs, but business associations and individual employers continued fiercely to oppose factory legislation until 1908 or 1909.[46] Labor leaders, meanwhile, hardly abandoned such measures. In Wisconsin, all the way down to 1911, with individual firms and the Merchants and Manufacturers Association of Milwaukee (MMAM) opposed, the WSFL legislative committee sought additional laws covering machinery, electrical wiring, sanitation, ventilation, building construction, and emery wheels, as well as more inspectors.[47]

Legislative action expanded factory inspection during the first decade of the twentieth century. By 1910, Wisconsin lawmakers had enacted more than a dozen factory laws and had gradually enlarged the inspection force, adding the first female inspector in 1901 and four new male inspectors in 1905. At the behest of women's groups, New York, Illinois, and other states simultaneously added female investigators to enforce woman and child labor laws. The frequency of Wisconsin factory inspections consequently jumped from 9,976 in 1905 to 21,701 in 1906. Inspectors visited every factory in the state at least twice during 1906, some up to six times.[48] By the early 1900s, hence, Wisconsin's bureau of labor was hardly just an "informational" agency or a lethargic bureaucracy but a dynamic institution. It promised to revolutionize work safety policy by imposing a legislatively sponsored factory law discipline on employers in place of litigation-based rules of negligence.

Nonetheless, administrative and legal problems held that revolution back. In an era of government expansion, state legislatures piled new responsibilities on labor bureaus without giving them additional clerks and inspectors to do the work, much less keep up with growing industries. Wisconsin commissioner Joseph Beck complained that his understaffed bureau was "up against a stone wall" in meeting new and old obligations. At the same time, New York commissioner Tecumseh Sherman bemoaned extraneous duties' "constant drain" on his bureau's resources and factory inspection operation.[49]

States' failure to collect adequate accident statistics further impeded labor bureaus' work. In the 1880s, Wisconsin and New York began tabulating work injuries based on voluntary employer information, but aside from Ohio's 1888 law requiring manufacturers to report accidents, states did not mandate accident accounting by physicians and manufacturers until the early 1900s. Even then, state labor law officials did not immediately heed New York labor commissioner Sherman's advice that factory inspectors utilize "a knowledge of conditions which can only be derived from practically complete statistics of the causes and results of all important accidents." When Commissioner Beck issued Wisconsin's comprehensive accident study in 1909, he used that report to advocate workers' compensation legislation, not better factory inspection.[50] Only after Wisconsin created its Industrial Commission in 1911 would administrators use statistics for accident-prevention work.

Factory legislation's technical imprecision continued to hinder labor bureaus' work, as well. Factory laws remained vague, requiring "proper" safeguards around "dangerous" machinery. Enforcing such generalities strained factory inspectors' ingenuity and mechanical skill. As Wisconsin commissioner Erickson explained in 1898, "Much of the machinery or many of the objects included are so constructed that no safety appliance yet known can offer effective protection without lowering their usefulness for the purposes intended. What steps to take in such cases are always more or less puzzling. Often the matter has to be left where found." Without statutes technically precise enough to meet such perplexities, factory law enforcement became a battleground between inspectors and employers over what "proper" safeguards were. The uncertainty discouraged labor departments from prosecuting resistive employers.[51]

Lack of statutory uniformity from state to state aggravated this problem. Factory laws in different states covered different hazards and fixed different standards of care, varying from "properly guarded" or "securely guarded" to simply "enclosed," "covered," or "guarded." Deep South states such as Alabama generally enacted no machine-guarding laws, whereas other states such as Illinois and California did not establish them until the early 1900s. American factory legislation was consequently "quite inferior" to European measures, according to the International Association of Factory Inspectors. In America's federated government, that group argued in 1905, factory laws' national strength was "no greater than that of its weakest link." Employers could flee from enlightened states to weak states, putting pressure on lawmakers everywhere to slacken statutory requirements. The patchwork quality of turn-of-the-century factory legislation evinced a growing conflict between a nationalizing economy and the U.S. government's decentralized federal structure.[52]

Labor bureaus' relations to local labor organizations also complicated their enforcement work. Sometimes, labor groups supported bureau activities. In Wisconsin, local unions assisted inspectors by applying pressure to noncompliant companies, while the WSFL and Milwaukee Federated Trades Council encouraged union locals to help bureau officials to collect employment and accident statistics. At other times, labor organizations disrupted bureau operations. In New York, unions harassed employers in labor disputes by invoking a statute requiring immediate state inspection of work sites when workers submitted complaints. Unions distracted inspectors with so many complaints that they could barely do their regular jobs.[53]

Judicial treatment of factory laws constituted one final impediment to labor bureau work. Although turn-of-the-century judges notoriously struck down a wide range of labor legislation, they generally upheld safety and sanitation measures as permissible applications of state police powers, just as they approved statutes regulating banks and pharmacies. The judicial question lay in determining factory laws' scope. As legal authority Ernst Freund noted, factory laws presented the quandary of balancing worker protection with property rights. "The peculiar difficulty of safety and health legislation," he asserted, "is that the possible causes of injury to person and property are extremely numerous and practically ubiquitous, . . . and that while the danger is often slight and remote, the measure devised to combat it may profoundly affect economic interests, favoring one set of interests and prejudicing another." Concluding that legislation best resolved this dilemma, Freund urged courts to decide only whether factory laws utilized police powers reasonably.[54]

In practice, however, many courts perpetuated traditional legal standards when they applied factory law requirements to work injury lawsuits. In this era when factory law enforcement was spotty, work injury litigation was an important arena for implementing factory legislation's new criteria for employer safety activity. Like other state courts, Wisconsin's supreme court ruled that violation of safety statutes was prima facie evidence of employers' failure to exercise traditional due care. This holding appeared to accept legislative efforts to raise employers' standard of safety, but it contained some large interpretive loopholes. Wisconsin's 1887 law, for instance, required machinery to be guarded or fenced when it was "so located as to be dangerous to employees when engaged in their ordinary duties." The Wisconsin Supreme Court ascertained that "ordinary duties" referred to work proximate to or directly on machines, leaving juries to decide whether machinery was "so located as to be dangerous" enough to require the guards specified by law. The court virtually allowed jurors to determine when statutes applied. In addition, the court ruled that the statutes did not impose specific methods of safeguarding but permitted employers to determine the manner of compliance with the law, even to forego safeguards where they interfered with machine operations or were unnecessary for worker protection.[55]

Most important, the Wisconsin Supreme Court ruled that factory laws did not replace common-law negligence standards with strict liability. In *West v. Bayfield Mill Company* (1910), Associate Justice John Barnes said that common law required employers only to exercise ordinary care in providing guards or fences mandated by statutes, the measure being whether the employer used a similar safeguard, fastened in a similar way, to those utilized by other employers under similar circumstances. Justice William Timlin loudly dissented that Barnes's ruling practically annulled factory legislation. The decision, he said, gave employers "the right to determine the diligence of its own members by comparison among themselves, and thus to decide whether or not they will comply with the statute." The Wisconsin court subsequently reaffirmed its majority position, but Timlin had a point. If factory legislation aimed to compel better business practices, then the Wisconsin court's majority undermined that purpose by perpetuating ordinary business prudence. Legislation in 1911 corrected this problem by declaring that "the exercise of ordinary care . . . shall not be deemed a compliance" with factory laws in work injury lawsuits.[56]

New York's court of appeals rendered similar decisions. In *Glen Falls P. C. Co. v. Travelers' Insurance Co.* (1900), it ruled that factory laws did not establish a new absolute standard of care for work accident litigation that would require manufacturers to guard "every piece of machinery in a large building," but only "gave more force" to the common-law rule that manufacturers should provide employees with reasonably safe workplaces. Employers were "required only to guard against such dangers as would occur to a reasonably prudent man as liable to happen." Reviewing this and subsequent cases in

1910, the Kansas Supreme Court condemned its New York counterpart for failing to accommodate new legislative standards. By interpreting the law "as if the statute did not exist," the Kansas jurists stormed, New Yorkers "simply fritter[ed] away serious efforts on the part of the legislature to secure factory workers against the barbarities of an industrial system which has been conducted with prodigality for human life and limb."[57]

Like the Kansans, other early-twentieth-century courts accommodated new statutory requirements in work injury litigation. In "factory act cases," Washington State's supreme court held such laws to require all "necessary" safeguards to prevent accidents. In an 1899 federal appeals court case, Judge William Howard Taft declared that an Ohio enactment requiring railways to block guardrails changed liability for employers. To rule otherwise, he argued, would turn that statute into "a dead letter," even with the availability of criminal enforcement sanctions. Later U.S. Supreme Court cases affirmed that the 1893 federal Safety Appliance Act replaced common-law reasonable care with an "absolute duty" to use safety devices. In 1914, Ohio's supreme court held a state statute requiring covers on bolts and set screws to replace ordinary care with an absolute duty.[58] In some courts, hence, common-law rules yielded to new legislative standards in work injury cases. Yet other courts, such as Wisconsin's and New York's, retained common-law precepts. In safety law, 1910 was an awkward legal moment.

The Unfinished Legal Transformation

At the start of the twentieth century, America's courts earned a reputation for inflexible conservatism. To many reform-minded Americans, the courts did not respond creatively to industrialization, restricted as they seemed to be by a "mechanical" style of reasoning that ignored social reality and rested rigidly on legal principle. By striking down social legislation, especially, the courts seemed to erect a barrier to Progressive reform. Americans "are irritated and constrained by a legal system that was developed in a different civilization," journalist Walter Lippmann complained in 1914, "and they find the courts, as Professor Roscoe Pound says, 'doing nothing and obstructing everything.'"[59]

Modern scholars debate whether the turn-of-the-century courts deserve this reputation. Recent studies demonstrate that both the U.S. Supreme Court and state appellate courts upheld the vast majority of regulatory laws. Yet political scientist Paul Kens has suggested that courts still utilized standing legal principles to restrict legislatures' exercise of state police powers.

Lawmakers, he contends, adapted regulatory laws to the courts' thinking. Thus, the U.S. Supreme Court's affirmation of reform statutes, such as those modifying employers' common-law defenses in accident cases, reflected not judicial liberalization but lawmakers' success in fitting reform measures into the Court's narrow guidelines.[60]

When it came to work safety law, state courts reacted to industrial change unevenly during the early 1900s. In Wisconsin and New York, judges certainly comprehended industrialism's harsh realities, as evidenced by Wisconsin justice Roujet Marshall's agonized *Knudsen* opinion, but their Classical rule-of-law philosophy sustained the due care standard, even though they built a substantial body of work safety rules, and generally approved factory statutes. Wisconsin and New York courts, at least, perpetuated the nineteenth-century developmental economic policy and continued to limit employers' responsibility for work safety to ordinarily prudent business practices.

The rapid expansion of factory legislation in Wisconsin and other states after 1899 challenged both judicial supremacy over workplace accidents and the economic policy that courts perpetrated. Factory laws increased legislative and executive intrusion into business operations and enlarged employers' legal responsibilities for workplace dangers. Yet they fell short of reshaping the legal environment. Even after 1900, factory laws were still special-interest legislation widely promoted by labor and a few middle-class reformers but generally lacking business and public support, as well as effective enforcement methods. Various courts, moreover, watered down the stringency of factory laws' statutory requirements in work injury lawsuits.

As late as 1910, then, the Classical judicial vision of enterprise and work safety still survived, though it was much under assault. Some state courts perpetuated the old Classical outlook, while factory legislation as a system failed to overthrow the court-dominated Classical regime. Change was imminent, however. Beginning in 1911, reformers in Wisconsin and other states began establishing powerful new institutions that ended the Classical judge-dominated era of work safety law and inaugurated the new age of administrative regulation.

2

The Administrative Transformation
of Work Safety and Health Law

In May and June 1911, Wisconsin governor Francis E. McGovern approved two statutes that transformed work safety policy. On May 3 he signed Wisconsin's Workmen's Compensation Act, replacing the old work injury litigation process with a new administrative system of industrial-accident compensation. Then on June 30 the governor signed Wisconsin's Industrial Commission Act. This measure repealed the state's old factory laws and then consolidated workers' compensation, safety regulation, women's and child labor protection, industrial arbitration, and state employment services all under control of a single new industrial commission, authorizing that body to fix work safety and health requirements in the form of administrative rules.

Together, these two statutes erected a powerful administrative regime that exemplified the Progressive Era "transformation of governance." More than Wisconsin's and other states' old factory laws had done, the statutes shifted power from local courts and the legislature to an executive body, thereby inflating government's administrative capacity to intervene in the economy. Consequently, the two acts curtailed the power of the allegedly reactionary courts more than contemporary judicial reform efforts did, substituting fact-oriented "Progressive" administrative decision making for Classical nineteenth-century rule-based judging.[1]

Moreover, Wisconsin's workers' compensation and industrial commission statutes revolutionized work accident law. By itself, Wisconsin's compensation law exhibited the "profound paradigm shift" taking place in accident law across the country. As in other states, Wisconsin's compensation measure abandoned the common-law assumption that accident liability followed from worker or employer negligence, and embraced the new "presumption of managerial responsibility" that required employers automatically to remunerate injured workers regardless of fault. Like other states, also, Wisconsin's act made compensation the "exclusive" remedy for work accidents (with exceptions), removing them from the individualized litigation process and "socializing" them under state administration instead.[2] Wisconsin's Industrial Commission Act took work accident law even further. Responding to defects in nineteenth-century common-law and factory-law regimes, this act replaced the common-law due care principle with a new safe place statute that raised employers' standard of care. To implement that new standard, the act authorized the new Industrial Commission to issue and enforce safety and health rules in the place of piecemeal factory laws. Finally, Wisconsin's law deployed compensation's financial inducements to secure employer compliance with safety rules rather than relying on "command-and-control" enforcement sanctions.

A unique sequence of local political events brought this novel administrative program about. At first, as elsewhere, the burgeoning accident problem inspired Wisconsin labor leaders, industrialists, reformers, and politicians to coalesce around workers' compensation legislation, though Milwaukee's rising socialist movement impelled this coalition toward an unusually strong state administrative plan. Whereas negotiations between businessmen, laborers, and reformers continued to shape labor laws in New York, Illinois, and other states, electoral politics in Wisconsin, Ohio, and California diverted subsequent safety law development to the executive arena. In these latter states, agitated voters elected strong Progressive governors and reform-minded legislatures in 1910 and 1912, creating a favorable environment for executive-sponsored legislative action. This situation especially encouraged the governmental-academic partnership celebrated in the "Wisconsin Idea." Aiming largely to preempt the socialists, newly elected Wisconsin governor McGovern assembled a special committee headed by University of Wisconsin labor economist John R. Commons and state Legislative Reference Library (LRL) chief Charles McCarthy to write a new work safety measure rather than leave it to legislators.

This Commons-McCarthy committee formulated Wisconsin's vaunted industrial commission statute. Synthesizing law, social science theory, practical politics, and European precedents, the committee emulated "strong" railroad and utility commissions by proposing a safe place statute to prescribe a legislative standard of employer care (mimicking requirements for reasonable railroad rates), and then by establishing the Industrial Commission to make rules fulfilling that standard (just as railroad commissions fixed rates). As such, the Commons-McCarthy bill embraced the emerging American law of public utility regulation, while it left such regulation subject to Classical judge-made law's "delegation doctrine" and substantive due-process "reasonableness" principle. Moving beyond public utility law, moreover, the Commons-McCarthy bill imaginatively provided for administrative use of unpaid advisers, the seed of Wisconsin's famous advisory committee system, and for coordinating workers' compensation and safety regulation under a single agency's control. Commons and McCarthy thus projected an extraordinarily powerful administrative system.

The political complexion of Wisconsin's administrative work safety and health plan varied from politics often attributed to Progressive regulation. Scholars have debated whether Progressive reform served the public interest, reflected pluralistic compromises between interest groups, or succumbed to "capture" by business interests. Although it embodied some of these patterns, Wisconsin's program more centrally reflected its executive-branch origins. Capitalizing on the consensus that formed around compensation legislation, and on its broad 1910 electoral mandate, Wisconsin's McGovern administration designed the Industrial Commission's special apparatus largely to secure business and labor cooperation with executive leaders' pursuit of industrial and political stability. It was a plan that veered uniquely toward "left corporate liberalism" and "state corporatism," and was not widely imitated in its entirety.[3]

Administrative Consolidation in Workers' Compensation Reform

After 1900, industrial change propelled Wisconsin into the national workers' compensation movement, but with important local permutations. As everywhere, escalating accidents disrupted the work injury lawsuit system, leading Wisconsin manufacturers, labor leaders, reformers, and politicians eventually to unite behind compensation reform. Also as elsewhere, Wisconsin lawmakers passed a compensation measure that was not "class legislation" imposed by business or by labor but a law backed by a coalition of interests.[4] Yet because socialist agitation made class harmony an especially urgent issue, Wisconsin's compensation legislation addressed industrial stability as much as it grappled with work injury law reform.

A new phase of industrialism occasioned Wisconsin's distinctive compensation plan. Like business nationally, Wisconsin industry expanded rapidly from

1900 to 1910, boosting output by two-thirds and enlarging the total manufacturing workforce by 28 percent and the female labor force by 46 percent. Simultaneously, the nationwide "factory revolution" reorganized Wisconsin industry. While mechanized upstate paper mills prospered, the state's economic center of gravity shifted southeastward to the Milwaukee area, where steel plants, foundries, and other giant firms established mechanized, rationalized, vertically integrated, and horizontally incorporated mass-production operations typical of advanced corporate industrialism. Such restructuring and growth inflated work accidents to about seven thousand a year, including a hundred or more women. It also ignited worker-employer conflict.[5]

In Wisconsin, labor-business strife moved in a political direction that heavily affected work accident law reform. As happened all over the nation, craft unionism thrived during Wisconsin's prosperous postdepression years of 1898–1903, but then employers' "open-shop" campaign crushed strikes and stymied union organizers until the 1910s. Unique to Wisconsin, however, skilled German workers rallied newly arrived Poles and eastern Europeans in the Milwaukee-based Wisconsin State Federation of Labor (WSFL) and Social-Democratic Party (SDP), aligning the state's labor movement with European political socialism. Socialists elected state legislators, a congressman, and a Milwaukee mayor, and simultaneously adopted a practical legislative platform.[6]

Building on late-nineteenth-century legislation promoted by railroad brotherhoods, WSFL and SDP leaders sought employer liability reform. In 1905, they obtained a statute removing the assumption-of-risk defense in accident cases involving employer failure to safeguard machinery. Still, working-class hostility deepened. Angry workers condemned employers and courts alike for the "hellish and damnable" fellow-servant and assumption-of-risk rules that allowed judges to "skin injured workmen out of their rights." Business owners and engineers may have begun to recognize managerial responsibility for work safety, but Wisconsin workers were already clamoring to hold employers more legally accountable. In fact, the WSFL endorsed workers' compensation as early as 1905.[7]

Working-class ferment and liability law's problems changed Wisconsin manufacturers' outlook around 1908–1909. Like most industrialists around the country, large Wisconsin companies led by the Merchants and Manufacturers Association of Milwaukee (MMAM) initially met labor unrest by quelling unionization campaigns, opposing labor laws, and adopting in-house safety and welfare programs. Yet unfavorable court decisions, rising litigation levels, and growing evidence of accident litigation's injustices persuaded Wisconsin industrialists to embrace workers' compensation. Milwaukee socialism's growing strength likely hastened manufacturers' conversion. When voters elected Socialist Emil Seidel as Milwaukee mayor in 1910, MMAM employers held out the olive branch of cooperation, denouncing "class hatred" and beckoning workers to embrace "peace and good order" for community progress.[8]

After industrialists came around, political interests coalesced in Wisconsin's legislatively sponsored Industrial Insurance Committee (IIC), wherein interest-group bargaining achieved a compromise over compensation legislation, much in the spirit of "social harmony and efficiency" that unified manufacturers and labor federationists behind compensation reform in Ohio. Elsewhere, compromise came less easily. In Illinois, the moderate State Labor Federation clashed with militant Chicago unionists over compensation versus employers' liability reform, while the powerful Illinois Manufacturers Association (IMA) resisted state intervention. Illinois consequently enacted a nonadministrative voluntary compensation statute in 1911. In New York, meanwhile, the state's notorious *Ives v. South Buffalo Railway Company* (1911) decision struck down an early compulsory law, but then in late 1913, after Democratic Party infighting impeached Governor William Sulzer, and after voters ratified a constitutional enabling amendment, newly elevated governor Martin Glynn secured compromise legislation amid intense lobbying by industrialists, casualty insurers, labor leaders, and American Association for Labor Legislation (AALL) reformers. The result was a strong administrative compensation program that made employer participation compulsory but the mode of insurance voluntary. In California, Progressive Republican governor Hiram Johnson accommodated the politically surging state labor movement by backing an elective compensation law in 1911, and then a compulsory statute in 1913, but in Alabama, industrialists and even reformers resisted labor law initiatives, and finally accepted a nonadministrative voluntary compensation plan during Governor Thomas Kilby's reform administration in 1919.[9]

Nonetheless, Wisconsin's compensation statute had much in common with other states' laws. As enacted in April 1911, and upheld by the state supreme court, Wisconsin's Workmen's Compensation Act duplicated other states' measures by establishing a no-fault system of compensating employees (not "casual workers") for injuries "proximately caused by accident" at companies electing the program. And Wisconsin's law was generous. Financial provisions altruistically allowed injured workers to get 65 percent of lost wages up to three thousand dollars (wholly from employer contributions), a benefit less than New York's initial stipend but more than those in California, Ohio, Illinois, and Alabama.[10]

Otherwise, Wisconsin law provided compensation coverage in distinctive ways. For one thing, Wisconsin lawmakers deferred to employers (and to constitutional doubts about compulsory plans raised by New York's *Ives* ruling) by instituting a voluntary system that manipulated the mode of election to induce employer participation. Wisconsin's original statute encouraged employer involvement by abolishing the common-law fellow-servant and assumption-of-risk defenses in accident litigation, except when workers themselves rejected compensation. Subsequent amendments enrolled employers with four or more employees (later three) automatically in compensation, unless they explicitly withdrew. This approach left some, mainly small, firms subject to litigation. In New York, California, and Ohio, by contrast, labor leaders and reformers secured constitutional amendments and legislation to make compensation compulsory, with some exemptions. Wisconsin's WSFL-SDP partnership lacked the statewide appeal to get such a compulsory plan.[11]

In addition, Wisconsin's compensation law did not, as WSFL and SDP leaders wanted, allow workers to choose litigation or compensation after accidents happened. Yielding again to manufacturers (and perhaps *Ives*), lawmakers made compensation the "exclusive" remedy for workers covered by the plan, joining states like California by preserving workers' right to sue only in cases of "willful" employer misconduct. Wisconsin law consequently did not match Illinois's 1911 legislation (repealed in 1913) or Ohio's 1911 law, both of which permitted injured workers to sue employers who violated safety regulations. Those latter two states perpetuated employer liability litigation to encourage employer safety work, but Wisconsin and California soon shifted such inducements to compensation proceedings.[12]

Indeed, Wisconsin's compensation program assigned resolution of injury claims to a new Industrial Accident Board, not to the courts. IIC members hoped that such a body would stop profit-seeking insurance firms from devaluing worker benefits, while Wisconsin employers expected the board to issue impartial and conservative decisions, as most state commissions did. Other states also established compensation commissions, but some enforced compensation judicially. Fearing that protective labor laws would weaken their state's competitive position, Alabamans enacted compensation legislation in 1919 that left local courts in control of contested claims. Opposed to state administration, big manufacturers ensured that Illinois's original 1911 act also allowed only judicial oversight over compensation disputes.[13]

Meanwhile, political struggle seethed over compensation insurance. According to historians, different states chose between allowing employers to insure their risks privately, as the British plan did, and establishing state insurance as a mild alternative to compulsory German social insurance and quasi-public employer-risk pools. In fact, states adopted a spectrum of programs. In the antistatist political climate of Illinois and Alabama, casualty firms operated in the open market regulated only by basic insurance laws. In California, workers and employers compromised on competitive state insurance that operated alongside casualty firms and self-insurance arrangements. In New York, employers secured a similar competitive state plan over labor's demand for monopoly state insurance. In Ohio, workers and big manufacturers united against private insurers to establish a monopolistic state fund.[14]

Wisconsin reformers, by contrast, decided to regulate compensation insurance. During IIC hearings, speakers roundly condemned private casualty firms but resisted state insurance, either competitive or monopolistic. Unlike Ohio and New York labor federations, the WSFL flirted only briefly with state insurance, doubting such a plan's efficacy and constitutionality. Eyeing Continental European examples, John Commons and insurance commissioner Herman L. Ekern then recommended a private mutual insurance company regulated by the state, but their proposal died for lack of support. After Wisconsin's compensation act passed, however, upstate Marathon County lumbermen led by W. A. Fricke founded Employers Mutual Liability Insurance Company, a firm that later collaborated with state regulators. Then, to overcome stock insurance companies' wild pricing tactics, Wisconsin employers and regulators secured amendments to the state's compensation law in 1913 and 1917 that required all insurance carriers to join the quasi-public Rating and Inspection Bureau to formulate uniform insurance rates. Avoiding "socialistic" state insurance, Wisconsin normalized compensation insurance costs by regulating them.[15]

Although it was a highly centralized administrative system that typified Wisconsin's "service state" and "general welfare state," Wisconsin's compensation program retained conservative features. Like all compensation measures, Wisconsin's plan abandoned common-law negligence rules that held individual workers and employers accountable for accidents, and embraced instead the new actuarial view that ascribed work injuries intrinsically to modern business. Otherwise, as historian Robert Asher contends, it was a "conservative reform" that neither redistributed wealth nor guaranteed full worker income but bolstered capitalist institutions. The plan promised injured workers automatic financial relief but regularized employer expenses and preserved employers' private managerial "prerogatives."[16]

Still, Wisconsin's compensation plan had exceptional administrative features, reflecting its distinctive local origins. Labor unrest and socialist activism threatened Progressive Republican dominance, especially after Progressives lost legislative seats to Social-Democrats in 1910 elections. Progressives then sought a strong administrative version of compensation reform to mitigate labor problems. Remarkably, the socialists acquiesced. More radical than Ohio's business unionists, but less militant than New York labor activists, Wisconsin WSFL and SDP leaders were practical men who collaborated with Progressives to secure laws improving workers' lives. Meanwhile, unlike ultraconservative, antistatist big employers in Illinois and Alabama, Wisconsin industrialists accommodated the administrative approach to compensation reform. Wary of labor unrest but cautiously tolerant of state regulation, big manufacturers accepted the state's administrative compensation plan as a means of advancing their own agenda of rationalizing production and stabilizing labor relations.[17] Wisconsin's 1911 compensation law was thus not just an interest-group compromise over the costs and benefits of compensation reform but also a political bargain over administrative means to secure "social peace." This bargain dominated the next phase of reform.

The Wisconsin Industrial Commission Act

After meeting with manufacturers in 1908 to discuss compensation reform, Wisconsin labor commissioner Joseph Beck bemoaned reformers' indifference to accident prevention. "It is more important to prevent accidents than to compensate the injured," he wrote. Employers had to be "compelled to provide safe appliances and ways." Modern historians might speculate that Beck was talking both about ineffective factory legislation and about the persisting nineteenth-century "commonsense" view that work accidents were the result of individual employers' fault. Not until the workers' compensation movement and tragedies like New York City's 1911 Triangle Shirtwaist Factory fire, recent scholars say, did lawmakers and the public adopt a new "common sense" that employers were generally responsible for job injuries and work safety.[18]

Yet Beck's observation actually revealed growing nationwide support for enlarging employer safety requirements. Nineteenth-century law's "common sense" included employer responsibility for accident prevention in the common-law "due care" doctrine and in factory laws. Workers and their leaders now sought to expand that duty. From 1903 to 1911, workers filed an increasing number of work injury lawsuits. Meanwhile, prolabor legislators secured new safety and health acts, such as Wisconsin laws governing emery wheels, ventilation, building construction, corn shredders, and machine guards. Women's groups simultaneously pursued women's protective measures and more female factory inspectors. And in coal mining states, lawmakers toughened requirements for mine operators to prevent explosions and roof collapses, although the mining industry's special characteristics placed many of the new safety demands on miners.[19]

In Wisconsin and a few other states, moreover, political developments diverted safety law reform in a new administrative direction. During the winter of 1910–1911, labor commissioner Beck convened manufacturers, labor leaders, insurance officials, and state factory inspectors in a series of conferences to formulate uniform inspection standards. Far from abandoning state factory inspection, Beck was trying to improve it through interest-group cooperation. In fact, use of collaborative bodies like the National Civic Federation (NCF), Wisconsin's own IIC, the Massachusetts Board of Boiler Rules, and Illinois's labor code commission to resolve labor problems had become commonplace. At Beck's Wisconsin conferences, big firms like International Harvester presented company safety and sanitation manuals as models, and participants reworked those private standards as statewide inspection guidelines.[20] The conferences were pathbreaking. They revealed business's accommodation to state regulation, a new method of labor law making, and a new partnership between employers, labor leaders, and government.

Electoral, not interest-group, politics soon institutionalized this new kind of labor law. From 1910 to 1912, voters in Wisconsin and other industrial states rejected the conservative Republicanism associated with President William Howard Taft by electing Progressive Republican or reform-minded Democratic governors and legislators, and (in some states) socialist lawmakers. Wisconsin Republican Francis McGovern, California Republican Hiram Johnson, and Ohio Democrat James Cox triumphed as gubernatorial candidates then. Popular discontent over rising consumer prices partly accounted for this upsurge, but so too did middle-class worries about predatory corporate power and growing labor unrest.[21] Contemporary magazines captured middle-class unease when they condemned America's "industrial juggernaut" for its awful accident toll, and exhorted readers to support compensation and protective legislation.[22] Popular alarm about industrial problems intensified.

Newly elected Progressive governors addressed such sentiment with strong executive leadership. Cox of Ohio advocated an end to industrialism's "waste" and class animosity, reminding Ohioans of "the element of interdependence . . . between social entities." Wisconsin's McGovern demanded cooperation and social justice, calling on government to protect "the weak, the unselfish, and the man of average ability." Simultaneously, by comparison, interest-group struggle dominated labor law development in Illinois and New York, while an antilabor coalition obstructed safety legislation in Alabama. In Illinois, where labor groups temporarily united despite chronic disagreements and persistent rivalry with machine politicians, and where the IMA continued to oppose legislative action, state officials secured the appointment of a joint commission of unionists, manufacturers, and academic experts to update labor laws. This led Illinois lawmakers formally to establish the Department of Factory Inspection in 1907 and enact the state's first factory safety law in 1909, but they did not create a modern regulatory commission. In New York, where manufacturers, unionists, and AALL reformers jockeyed to shape protective labor laws after the 1911 Triangle Shirtwaist Factory fire, Democratic lawmakers consolidated labor law administration (including statistics, factory inspection, and new safety code powers) under a Department of Labor headed by a single labor commissioner in 1913, but then in 1915 a conservative Republican legislature reorganized those functions (now including workers' compensation) under an "expert" industrial commission instead. In Alabama, manufacturers, politicians, and even reformers avoided labor legislation, except for weak child labor, textile factory, and mining laws, all to sustain business dominance and the state's competitively low labor costs. Yet in Ohio, Wisconsin, and California, Governors Cox, McGovern, and Johnson emerged as "leaders of a policy crusade," as historian Jon Teaford puts it. Each appealed to voters across class lines by promoting labor legislation in a sympathetic but paternalistic way without directly collaborating with labor groups.[23] Remarkable administrative reforms resulted.

In Wisconsin, moreover, 1910 elections reinforced the political tradition known as the "Wisconsin Idea." Emanating from the Social Gospel, German historical economics, and other intellectual forces, the Wisconsin Idea affirmed that state government led by a strong executive and university-trained experts should act as a moral steward to harmonize conflicting interest groups in a cooperative "commonwealth" and serve the public welfare. Robert La Follette instituted the tradition as governor (1901 to 1906), enlisting dozens of University of Wisconsin faculty members to address public policy problems, and Francis McGovern continued the practice with a twist. In his Saturday Lunch Club, McGovern drew on academic advice, but he rejected La Follette's populistic, small-government brand of Progressivism for a program similar to Theodore Roosevelt's New Nationalism—a strong, efficient regulatory state guided by expert administrators. As Charles McCarthy's 1912 book, *The Wisconsin Idea,* explained, expert commissions could best implement public policy when complex industrial problems required state action.[24]

In his 1911 inaugural message, consequently, McGovern promised to "reconstruct" government. Much like Cox in Ohio, Johnson in California, and Charles Evans Hughes in New York, as well as presidential candidate Theodore Roosevelt, McGovern proposed to strengthen government's capacity to protect ordinary citizens from business abuse. As an urban Milwaukee reformer, McGovern especially proffered government action for workers who might vote socialist, urging not only compensation reform but also a consolidated new labor department.[25] In the Wisconsin Idea's spirit, he turned to "experts" in McCarthy's LRL to draft legislation. A small LRL committee consisting of McCarthy, labor commissioner Beck, Professor John Commons, and his student Francis Bird drew up the measure. Young McGovern loyalist Thomas J. Mahon then introduced it in the state assembly.[26]

As the bill-drafting committee's intellectual forces, Commons and McCarthy exhibited "professional experts'" special function as technical advisers in Progressive reform, unlike interest-group lobbyists' political role. After a controversial career, Commons especially prepared for the impartial expert's service when he joined other social scientists in the AALL to promote rational design of social and labor legislation.[27] Although recent scholarship has stressed Commons's "nonstatist" views of trade unionism, he joined McCarthy on this occasion to advocate state administrative expansion. When McCarthy advocated flexible public commissions, Commons added ideas from contemporary academic discussion. From labor economics Commons gathered views about the state's moral power to guide progress. From political science he adopted the notion of bypassing political parties and three-branch government to create a separate realm of "administration" to perform daily governmental duties. And from his early writings, reform experience, and teaching, he embraced the European model of "functional" interest-group representation, a potentially "corporatist" system for managing industrial conflict.[28] Ultimately, Commons and McCarthy chose domestic public utility law to synthesize these ideas. Having studied utility regulation for the NCF, and having taught university courses on the subject, Commons had helped to establish Wisconsin's commission method of regulating utilities in the state's Public Utilities Act of 1907.[29]

A moderate alternative to European-style public ownership, public utility law involved state regulation of business through powers legislatively delegated to independent commissions. It started with the U.S. Supreme Court's *Munn v. Illinois* (1877) decision that upheld government's authority to regulate businesses "affected with a public interest," an exception to Classical law's defense of property rights under the Fourteenth Amendment due-process clause. Railroads were the most important publicly interested business, and railroad rates emerged as the central regulatory problem. Antebellum state lawmakers had enacted statutes to govern railroad rates and services, but post–Civil War legislators established regulatory commissions, some as "strong" agencies empowered to fix maximum rates, but most as "weak" investigatory bodies. After 1900, lawmakers revived "strong" regulation by giving agencies like the Interstate Commerce Commission and Wisconsin's Railroad Commission qualified power to prescribe rates. The key legal qualification, originating in *Munn*, was that rates be "reasonable," a question that *Chicago, Milwaukee & St. Paul Railway v. Minnesota* (1890) said belonged to the courts.[30]

In a bill closely resembling Wisconsin's Railroad and Public Utility Acts of 1905 and 1907, the Commons-McCarthy committee utilized the "strong" commission model to create a new Wisconsin labor department. They proposed an "Industrial Commission of Wisconsin" consisting of three appointed members to replace the old Bureau of Labor and Industrial Statistics. The new agency would assume control over all state labor laws, including the newly enacted Workmen's Compensation Act, thereby replacing the Industrial Accident Board. Women's and child protective acts were covered too, not relegated to a separate agency, as in California.[31]

Imitating statutes that required railroads and public utilities to "furnish reasonably adequate services and facilities," the proposed law replaced factory legislation with a general requirement called the "safe place statute." This measure expanded employers' legal responsibility. Whereas common-law rules and factory laws governed specific aspects of employment—safe appliances, adequate materials, and competent helpers—the proposed safe place statute demanded reasonably safe conditions in "place[s] of employment" and, after amendment, in "employment," as well. Granted, the proposed commission act also forbade workers from removing safety devices, disrupting safety procedures, or neglecting coworkers' safety. Yet the proposed safe place statute enlarged employers' duty to make the whole of employment reasonably safe, not just its physical apparatus.[32]

The bill drafters hoped that the safe place statute would raise safety standards, but McCarthy warned that the proposal's requirement for "reasonably safe" places of employment remained vulnerable to the old judicial conception of ordinary care. Francis Bird then suggested language redefining "safety" as "such freedom from danger . . . as the nature of the employment will reasonably permit," a true legal breakthrough. It surpassed the due care standard, though it lagged behind the modern Occupational Safety and Health

Act's stipulation that employers provide employment "free from recognized hazards." The new language linked safety to advancing technology, requiring employers to install the best new safeguards proven feasible. According to John Commons, "If any new methods of safety have been devised and are practicable, and therefore reasonable, [then] the commission can require all employers to put them in."[33]

Wisconsin's safe place statute would ultimately create new grounds for liability lawsuits against firms exempted from workers' compensation, but bill drafters saw it as only the basis for regulation. To specify how manufacturers were to meet safe place statute requirements, the proposed Wisconsin law authorized the new Industrial Commission to prescribe safeguards in the same way that the state Railroad Commission fixed rates and schedules. The bill empowered the new agency to issue general safety orders for all companies, occupations, and workplaces uniformly and to promulgate special orders for unique situations. "A work of genius," as future commission secretary Arthur Altmeyer subsequently called it, order-making authority relieved the cumbersome legislature of adjusting factory laws to meet evolving industrial conditions and assigned that task to a "flexible" administrative body instead.[34]

This change dramatically expanded administrative power. The Industrial Commission's order-making authority exceeded the state Railroad Commission's power, because the new agency could prescribe safety codes on its own volition. Simultaneously, judges helped by relaxing Classical law's "delegation doctrine" that forbade legislatures from vesting their own functions in executive agencies and allowed administrative bodies only to "fill up the details" of legislative policy. After 1910, courts accommodated broad legislative grants of administrative jurisdiction, a trend confirmed by judicial decisions upholding the Wisconsin Railroad Commission's authority. Judicial reaction to New Deal programs would later revive the delegation debate, but back in the Progressive Era, the doctrine's erosion afforded Wisconsin's Industrial Commission exceptional order-making discretion.[35]

As to how Wisconsin's commission would formulate safety and health orders, the proposed law did not create an "expert" administrative staff or allow paid outside consultants, as the Railroad Commission did, but authorized the new agency "to appoint [unpaid] advisors who shall . . . assist the industrial commission in the execution of its duties." Without saying so, this provision sanctioned Wisconsin's famous committee system under which the Industrial Commission assembled "advisers" from business, labor, government, and the professions to draft safety and health regulations. Imitating the

Massachusetts Board of Boiler Rules and labor commissioner Beck's 1910–1911 conferences, and drawing on Commons's knowledge of European interest representation, the committees promised to blend workers' and employers' practical experience with technicians' expertise, and thus produce safety rules more effective than old-fashioned factory laws.[36]

Wisconsin advisory committees were a novelty in American law. In the early 1900s, courts treated administrative bodies as agents of legislative will whose inner workings were to be constrained, if at all, by trial-type procedures. Not until the 1970s would courts see agencies as representative bodies, the validity of whose action depended on fair interest-group participation.[37] Back in 1911, hence, neither statutes nor judge-made rules governed how committees worked: which interests were represented, who chose them, what decisions they made or how. Committee proceedings were a matter of agency discretion.

The industrial commission bill's limitations on judicial review further augmented agency power. Like Wisconsin's Railroad and Public Utility Acts, the proposed law required aggrieved employers to bring questionable orders first before the new commission, and then to appeal commission rulings to the Dane County Circuit Court to ascertain whether orders should be vacated as "unlawful" or "unreasonable." No court could review safety orders in any other proceeding, including prosecution actions, wherein orders were presumed "prima facie" lawful. Combined with the delegation doctrine's temporary decline, these restrictions shifted authority over economic affairs from courts to agencies, and from adversarial judicial proceedings to technical administrative decision making. The courts' role narrowed to mere policing of the administrative process.[38]

Nonetheless, remnants of Classical judicial authority persisted. In a 1908 precedent involving Wisconsin's Railroad Commission, state supreme court justice William H. Timlin recognized new restrictions on judicial review but still left room for judicial intervention. A state circuit court, he ruled, could not specify just and reasonable rates itself—that was the Railroad Commission's prerogative—but it could determine when a commission order was "*un*reasonable," fixing administrative action's outer boundary, just as U.S. Supreme Court decisions had done. Indeed, as Stephen Skowronek suggests, the reasonableness doctrine was American courts' modus vivendi with the rising administrative state. Courts tolerated legislative delegation of rate-making authority to administrative agencies but reserved power to intervene in "proper cases."[39] Bodies like the Industrial Commission expanded rule-making authority but were limited by judicial ideas of reasonableness.

Wisconsin's proposed commission law also strengthened administrative authority by restructuring the office of factory inspector. Whereas factory acts had allowed inspectors broad discretion to tell employers how to satisfy state statutes, the new law created "deputies" who were simply to secure compliance with protective techniques already mandated by safety orders. Deputies were just technicians who implemented safety codes. The power to design safety devices stayed in the Industrial Commission that formulated regulations.[40]

An effort to exempt deputies from state civil service laws provoked the commission bill's only controversy. Fearing that civil service requirements might discourage qualified experts from undertaking government work, the Commons-McCarthy committee proposed a special system like the Railroad Commission's, whereby the Industrial and Civil Service Commissions would jointly evaluate applicants' fitness as deputies. Civil service advocates, however, wanted no "pets" appointed to the new agency. After a heated floor debate, legislators voted to bring commission staff fully under civil service rules.[41]

This controversy illustrates the gap between the industrial commission law's legal contents and the politics of its enactment. Public utility law and academic theory may have shaped the bill, but political considerations determined its legislative fate. In public appearances and magazine articles, for example, Wisconsin assemblyman Thomas Mahon tapped Progressivism's conservation and efficiency ethics. Condemning industrial accidents' enormous waste, and criticizing factory laws' chronic obsolescence and unenforceabilty, Mahon lauded the proposed Industrial Commission's "scientific" investigatory powers and flexible procedures for adjusting laws to changing economic conditions. In *The Survey,* moreover, Francis Bird reassured employers that the new commission would be fair. Its deputies would not harass but provide technical assistance. Its "labor reference library" would help employers develop safety devices. Altogether, Bird promised, the new agency would promote cooperation among employers, workers, and factory inspectors, not impose "drastic labor laws to which employers are opposed."[42]

With such endorsements, Wisconsin's industrial commission bill rolled easily to enactment. In Progressive spirit, newspapers noted former president Theodore Roosevelt's support and commended the proposed agency's promise to standardize safeguards, use taxpayer dollars efficiently, and improve work safety conditions. In final legislative action, Democrats opposed the bill, but all Social-Democrats and most Republicans backed it, reflecting the Progressive Republican and Social-Democratic alliance that passed many new-fashioned economic welfare and labor laws in 1911. Governor McGovern signed the bill on June 30, 1911, and then nominated former labor commissioner Joseph Beck, Professor John Commons, and attorney Charles Crownhart (members of the now displaced Industrial Accident Board) as the first three industrial commissioners.[43]

The Wisconsin Industrial Commission Act's easy passage obscured its immense legal and administrative importance. Although not entirely repudiating nineteenth-century Classical ideas, the measure dramatically revised work safety law. For one thing, the act's safe place statute hardened the new legal "presumption of managerial responsibility" for accident prevention and enlarged legal standards of employer care. In addition, the commission act replaced factory laws with administrative rules to implement new safety standards. Finally, the commission law harnessed workers' compensation administration to enforce safety requirements, mostly eliminating work injury litigation and factory-law sanctions as methods of holding employers accountable.

This remarkable measure built on the political bargain underlying Wisconsin's compensation act. Having endorsed state administration when they backed compensation, Wisconsin manufacturers easily supported the commission statute. Adhering to the "corporate liberal" policy of using governmental means to secure corporate ends, industrialists likely saw the new agency as a way both to extend ameliorative in-house labor and welfare programs into the public sphere and to save money.[44] Meanwhile, Social-Democrats voted unanimously for the Industrial Commission Act, as they had for workers' compensation. Skeptical of commission government, socialists normally preferred direct legislation like factory laws. Yet after many fruitless legislative sessions, WSFL and SDP leaders joined the 1911 legislature's governing coalition in a practical effort to enact favorable labor laws, including the compensation and industrial commission acts.[45]

Most important were politicians and reformers in Francis McGovern's governorship. Capitalizing on the 1910 electoral upsurge and the compensation law's political settlement, McGovern and the Commons-McCarthy bill-drafting committee bypassed regular legislative bargaining and devised the commission law as an executive-branch solution to work accidents and industrial unrest. For McGovern, the plan parried socialists' challenge to Progressive Republicans, while for reformers, the program actualized professionals' faith in the power of organized knowledge and bureaucratic management to improve industry.[46] Executive politics created Wisconsin's Industrial Commission and established its new safety policy.

Administrative Labor Law:
A New Vision of Modern Government

Wisconsin's Industrial Commission quickly became a symbol of administrative reform, although that agency's champions soon shifted the terms of their support from familiar Progressive arguments to the new academically inspired theory of "administrative labor law." Initially, newly appointed Wisconsin industrial commissioners Charles Crownhart and John Commons trumpeted their agency as a "fourth branch of government" that would deal better with modern industry than the judicial and legislative branches had done—that is, better than the nineteenth-century governmental system that Stephen Skowronek labels as the "state of courts and parties." On the one hand, Crownhart predicted, the new commission would use "progressive" legal science to investigate factory conditions and make decisions on the facts rather than relying on secondhand testimony and static legal principles, as the courts did. On the other hand, Crownhart added, the new agency's flexible safety code process would accommodate technological change better than the politically entangled state legislature could. If new dangers arose and new safety devices were invented, "then the commission [could] investigate the fact and issu[e] the proper order."[47]

Commons joined Crownhart, moreover, to reassure legal conservatives that Wisconsin's new commission satisfied the "delegation" and "reasonableness" doctrines. Safety order making, they explained, was only a fact-finding function, not a usurpation of legislative authority. The Industrial Commission merely investigated working conditions and issued orders fulfilling legislative intent. Indeed, unlike European programs that allowed scientific "experts" to impose "ideal" regulations on manufacturers, Wisconsin advisory committees would produce practical and reasonable orders that would easily pass judicial review. Because Wisconsin committees utilized "constructive investigation" to secure "practical ends under existing conditions," Commons asserted, their method was constitutionally sounder than Europe's "scientific investigation." Though meeting constitutional tests, Commons and Crownhart stated further, Wisconsin's program would remain free from judicial interference, because the industrial commission law directed courts to treat safety orders as conclusive and binding findings of fact, except in specified procedures. Such restrictions, Commons affirmed, protected the new agency's special ability to make technical decisions.[48]

After Commons and Crownhart initially defended Wisconsin's Industrial Commission in constitutional terms, Commons and his students John B.

Andrews, Elizabeth Brandeis, and Arthur Altmeyer introduced a bolder academic argument. In Commons and Andrews's textbook, *Principles of Labor Legislation,* and other works, Commons and students made the Wisconsin commission the centerpiece of "administrative labor law," their vision of the administrative state's role in industrial society.[49] That vision began with labor economists' view of the "state" as a moral force "above" politics and the private economy, an idea common to turn-of-the-century reformers, including police power authority Ernst Freund. Commons and students retained Classical economists' faith in the capitalist market, but they contended that government should guide economic growth in a socially equitable direction, "sav[ing] Capitalism by making it good," as Commons said.[50]

To facilitate government's guiding moral influence, Commons and students supported institutional changes proposed by political scientists Frank Goodnow and Woodrow Wilson. Sidestepping American government's three-branch constitutional structure, Goodnow and Wilson focused on two distinct "functions"—politics and administration—and they recommended expanding the administrative realm to improve governmental efficiency. Wisconsin's Industrial Commission embodied this idea. "Modern industrial conditions have become so complex, and the laws deal with such a variety of facts," *Principles of Labor Legislation* reported, "that a fourth branch of government is emerging whose purpose is primarily investigation. This is administration." Through "constructive investigation," administrative bodies like Wisconsin's commission could apply policy to industrial realities better than courts, legislatures, and executives could.[51]

In this regard, Commons and students seemed to anticipate New Deal–era defenses of administrative government, but there was a crucial difference. New Dealers like James Landis would contend that administrators needed relief from three-branch American government to apply independent technical judgment to modern problems. "Expertise," Landis said, would guide administrative decisions without danger of arbitrary governance. In *Principles of Labor Legislation,* however, Commons and Andrews rejected pure "expert" authority, and sanctioned administrative labor law on the grounds of representativeness. Denouncing bureaucrats who "formulate their own rules as they please," they endorsed safety code advisory committees that gave "due weight" to employer and worker viewpoints.[52]

Representing interests in administrative decision making had two public benefits, Commons and Andrews said. First of all, it improved the private economic market's operation. "Competition," they explained, "tends to drag down all employers to the level of the worst." Administrative labor law, however, would reverse this market process. "Instead of permitting the worst [employers] to set the standard for the best," Commons and Andrews declared, the new laws "can assist the best in setting standards for the worst." And by "best," they meant large, technologically advanced manufacturers. As Commons later observed, "Obviously I wanted all employers to be forced by law to follow the lead of [firms like U.S.] Steel Corporation, and to make a profit by doing good, instead of . . . neglecting the safety of their employees."[53]

The second benefit of interest representation was class stabilization. If civil service reform handled "political antagonism" between rival political parties, Commons and students argued, then labor law resolved "industrial antagonism" between employers and workers. To Commons and Andrews, safety and health code advisory committees properly adjusted business-labor relations. Like collective bargaining, in which "neither the union, the employer, nor the politician dominates," advisory committees were "cooperative" bodies that brought employers and workers together to devise mutually beneficial safety rules.[54]

Representative advisory committees became a hallmark of "Wisconsin School" labor policy, but their modern reputation obscures their original purpose. Because Commons and students later cited Wisconsin committees as a model of "tripartite committees" of business, labor, and the public for use in industrial relations policy, historians have seen Wisconsin's plan as a forerunner of New Deal "industrial pluralism," a "nonstatist" policy in which government established forums to encourage private agreements between business and worker organizations, typically in collective bargaining. Yet in administrative labor law, Commons and students portrayed the committee process more in "statist" and "corporatist" terms. True, the committees' "tripartite" composition resembled industrial pluralism. Nonetheless, the committee process envisioned in administrative labor law verged on "state corporatism" in the way that it fostered business-labor cooperation with state-mandated goals. State officials granted employers and workers a voice on advisory committees but expected their compliance with state-sanctioned safety standards in return. Such a quid pro quo promoted the state policy of industrial stability.[55]

Administrative labor law's "statism" also appeared in Commons and Andrews's view of workers' compensation. In capitalism, they argued, labor law had to recognize that employers responded only to "inducement of financial gain." Accordingly, compensation's financial incentives called employers to "public service." Embracing the European social insurance idea, Commons and Andrews explained that compensation treated each worker as a "public utility" and made "an injury to one the concern of all." Compensation thus induced employers to protect not only their own employees, but also all workers as well.[56]

In administrative labor law, then, Commons and associates advanced a theory of work safety and health law that went beyond the conventional views of regulation. Their "fourth branch of government" proposed a new state administrative apparatus to guide market activity, humanize industrial production, and stabilize industrial relations, all to steer industrial capitalism toward a better future. It was a bold vision, but one that had only a modest national impact.

The Political Limits of Administrative Reform

Reputation has long credited Wisconsin's Industrial Commission for transforming labor law administration nationwide. Led by John Commons's student John Andrews in the 1910s, the AALL championed the Wisconsin agency as a legislative model for other states. Commons's students subsequently claimed that nearly two dozen states followed Wisconsin's example. In 1962, hence, Commons's biographer, Lafayette Harter, celebrated Wisconsin's safety code process as an innovation of national importance.[57]

Actually, the Wisconsin Industrial Commission's influence was spotty. Few other states embraced the whole Wisconsin system, whose true genius was to integrate administrative powers, not just to institute safety code making. The crux of Wisconsin's program lay in how its safe place statute elevated employers' legal duty to provide safe and healthful employment, and then arranged commission functions to enforce that duty through safety code making, advisory committees, technical factory inspection, and workers' compensation administration. Soon to include power to fix women's work hours, this was an extraordinarily comprehensive administrative program.

By comparison, only Ohio and California duplicated Wisconsin's whole plan. In 1913, under strong Progressive gubernatorial leadership, those two states adopted safe place statutes and created powerful commissions—the Ohio Industrial Commission and the California Industrial Accident Commission. Both of these agencies had authority to issue safety orders, assemble advisory panels, and utilize workers' compensation to enforce safety codes. Indeed, Ohio went further by setting up a monopolistic state compensation fund, while California created competitive state compensation insurance, strengthened order-making procedures, and founded a special Industrial Welfare Commission to enforce women's labor laws.[58] New York, Pennsylvania, and Massachusetts also established commissions with safety code authority, but deviated from Wisconsin's model in important ways. In lieu of safe place statutes, New York and Pennsylvania codified common-law standards requiring employers only to install "reasonable and adequate" safeguards. None of these states coordinated administrative functions as Wisconsin did, with New York in particular erecting separate compensation, safety code, and inspection bureaucracies.[59]

Beyond these states, administrative safety law reform advanced slowly. During the early 1900s, chief mine inspectors in West Virginia, Ohio, and Pennsylvania unilaterally issued safety rules, but their action lacked statutory authority and set no precedent for administrative regulation. By 1930, only fourteen states beyond the original six had legislated administrative procedures, mainly in the upper South, upper New England, the western mountain states, and the Northwest. Big industrial states were not among them: Illinois established factory laws and an old-fashioned factory inspection department without creating safety code authority, while Michigan and New Jersey set up agencies that issued only factory inspection guidelines. Alabama, like other Deep South states, spurned administrative safety regulation.[60]

Variegated state-to-state politics slowed administrative labor law's advance, just as it hindered administrative expansion generally.[61] In most states, interest-group politics impeded centralized safety and health programs. In New York, political discord among manufacturers, labor unions, and AALL-led social Progressives fractured the state's safety law bureaucracy into separate departments, with business seeking a friendly compensation bureau, unions promoting a strong inspection division, and reformers advancing an efficient safety code commission. In Illinois, party politics, labor movement divisions, and employer opposition impeded factory safety law reform until 1909, and even then little support for centralized state administrative control emerged. In Alabama, weak unionism and antilabor business attitudes blocked protective measures except for its modest child labor, textile factory, and mining laws and its rudimentary compensation act.[62]

In the three strong commission states—Wisconsin, Ohio, and California—the unusual 1910–1912 electoral upsurge momentarily eclipsed interest-group struggle. Voters elected reform-minded lawmakers and countenanced strong governors, releasing a flood of executive-sponsored legislation in what Kevin Starr, writing about California, aptly calls a Progressive "Camelot." In this, governors with advisers like John Commons became "state actors" who advanced state-centered administrative schemes to serve their own interest in industrial and political harmony.[63] How well their new administrative programs would actually improve working conditions and industrial relations depended on how new commissions operated in practice.

3

Selling the Safety Spirit

Established in 1911, Wisconsin's Industrial Commission inaugurated its new work safety and health policy in a half-changed industrial culture. Expanding on the nineteenth-century due care doctrine and factory laws, the commission's founding legislation codified reformers' emerging consensus that business was mainly responsible for stopping accidents and that government had to enforce that duty.[1] Yet most Wisconsin manufacturers and laborers still sustained nineteenth-century attitudes toward accident causation, workplace prerogatives, and government power. Many employers still resisted state regulation and treated accidents as inevitable events to be endured primarily by employees. Shop foremen and skilled laborers still defended traditional workplace liberties against managerial and state intrusion. Before implementing new laws, consequently, Wisconsin's new industrial commissioners tried to break these old outlooks. With evangelical zeal, the commissioners instigated a spirited "safety first" campaign to persuade local employers and workers to adopt safe practices and accept the state's vastly expanded involvement in their activities. In particular, the commissioners organized safety workshops, dispatched lecturers to business meetings, built a traveling safety exhibit, presented safety movies, distributed technical bulletins, and posted safety banners.

This state-sponsored campaign drew heavily on early-twentieth-century business practices as industry evolved from "proprietary" to "corporate" phases of development. To address rising competition, big firms rationalized their operations by mechanizing production processes, "systematizing" machine use and "scientifically managing" workers. Simultaneously, a few large steel, railroad, business machine, and coal mining companies adopted a variety of injury relief programs, in-house safety systems, employee representation plans, and "welfare capitalism" schemes to cut accident rates, improve working conditions, reduce labor turnover, mollify worker unrest, discourage unionism, meet the cost of newly enacted compensation measures, and generally offer a business alternative to Progressivism's "embryonic welfare state." Among manufacturers, especially, employers cultivated worker loyalty by designing welfare plans to imitate the gendered interdependence of Victorian families. That is, they treated business owners as paternalistic supervisors, male workers as responsible breadwinners, and female workers as vulnerable potential mothers needing special protection. After Wisconsin and other states initiated compensation laws in 1911, big firms launched industry's safety first movement and founded the National Safety Council (NSC) to spread safety practices nationwide.[2]

In its own safety first campaign, Wisconsin's Industrial Commission reproduced many corporate practices, but for state purposes. The commission hired International Harvester safety director Charles W. "C. W." Price to run its safety campaign, and it appropriated technical information and educational methods from the NSC, the National Association of Manufacturers (NAM), and the Merchants and Manufacturers Association of Milwaukee (MMAM). Nonetheless, Wisconsin's commission was not just an industry pawn. As Daniel Carpenter observes about national bureaus, some Progressive agencies cultivated impressive political "autonomy," whereby they "sustained patterns of action consistent with their own wishes" and became policy initiators in their own right, "chang[ing] terms of legislative delegation" and altering public preferences "through experimentation, rhetoric and coalition building." As Jon Teaford contends, moreover, such bureaucratic initiative developed in states as well as national government.[3] In its safety first campaign, Wisconsin's Industrial Commission cultivated such "autonomy" on the state level. The agency enlisted corporate techniques to establish itself as the state's moral leader and technical authority on industrial safety, a guide for local employers to follow.

The Industrial Commission's safety first campaign rested, however, on transitory political circumstances. From 1911 to 1913, popular favor, gubernatorial support, generous legislative funding, and solid business and labor backing all emboldened the commission's educational work. Then, in 1914, a split in Progressive ranks and the election of Stalwart Republican Emanuel Philipp as governor restrained commission activities. A spokesman for conservative industrialists, lumbermen, and insurance companies, Philipp named political moderates to head the agency, while stingy legislative budgets stymied its educational work. Thereafter, the commission's safety campaign withered.

Yet few other state labor departments matched the Wisconsin agency's educational activities. California's Industrial Accident Commission developed a similar program, but the Ohio Industrial Commission's campaign deferred to large manufacturers, whereas the New York labor department's extensive educational work avoided business-oriented rhetoric. Illinois's Factory Inspection Department neglected safety publicity, and Alabama (lacking a labor department) ignored it altogether. Federalism sustained a diverse patchwork of state educational programs. Through the 1910s, however, Wisconsin's plan remained the national leader.

The Politics of Commission Leadership

During Progressivism's high tide from 1911 to 1913, politics afforded Wisconsin's Industrial Commission far more autonomy than most other state labor departments enjoyed. The electoral mandate that chose Progressive Republican Francis McGovern as Wisconsin governor and produced the state's workers' compensation and industrial commission laws also encouraged the appointment of reform-minded administrators and the appropriation of ample budgets. To head Wisconsin's new commission, McGovern named Professor John Commons, Robert La Follette's ally Charles H. Crownhart, and former labor commissioner Joseph Beck—an academic, lawyer, and bureaucrat, not one a business or labor representative. Notably absent were women. Meanwhile, lawmakers generously funded the new agency at seventy-five thousand dollars per year.[4]

Elsewhere, governors paid closer attention to interest groups. In New York, 1913 legislation created a Department of Labor separate from the state's Workmen's Compensation Commission to govern safety code making and factory inspection. Appealing to trade unionists, Democratic governor Martin Glynn named former International Typographical Union president James M. Lynch as the department's first commissioner. When a 1915 law consolidated the labor and compensation departments under a five-member industrial commission, however, a new Republican governor appointed two business leaders and an attorney to the new body to join Lynch and former mine workers president and compensation commission chair John Mitchell. The 1915 statute also required the governor to appoint a ten-member industrial council evenly divided between business and workers. Bypassing women's group leaders, New York labor law administration catered heavily to business and labor interests.[5]

In Ohio, Democratic governor James Cox placated reformers as well as business and labor representatives. To the state's new Industrial Commission he named former Ohio State Federation of Labor (OSFL) and pottery union leader Thomas Duffy, probusiness attorney Wallace Yapple, and Ohio State University Professor Matthew Bray Hammond. Again, women's groups were not a concern. When Hammond resigned, Cox replaced him with industrialist Morris Woodhull, tilting Ohio's commission in industry's direction.[6]

In California, Progressive Republican governor Hiram Johnson mollified antiunion Los Angeles employers, San Francisco labor leaders, and the California State Federation of Labor with appointments to the state's new Industrial Accident Commission in 1913. He named Progressive editor A. J. Pillsbury as chairman, along with militant San Francisco Labor Council president Will J. French and Progressive San Francisco merchant Harris Weinstock, who had befriended Los Angeles employers by proposing a controversial public utility labor arbitration bill. With women's issues covered by a separate Industrial Welfare Commission, Johnson's nominations both accommodated northern and southern California regional interests and assuaged business and labor groups.[7]

Back in Wisconsin, politics moved the Industrial Commission's leadership away from its initially strong reform orientation. When John Commons retired from the agency in 1913, Progressive disagreements forced Governor McGovern to nominate moderate Republican state senator Fred M. Wilcox instead of his legislative protégé Thomas J. Mahon. After voters elected Stalwart Republican Emanuel Philipp as Wisconsin governor in 1914, Philipp replaced Progressive commissioner Charles Crownhart with Taft-Republican assemblyman George Hambrecht, and then removed Joseph Beck to appoint former Congressman Thomas Konop. An ally of conservative stock insurance firms and manufacturers, Philipp hoped to stifle aggressive compensation law enforcement, especially by Crownhart. Opposition stopped him from removing Wilcox in 1919, but by then, the governor had transformed the commission into an apolitical body distinct from reformers.[8]

Budgetary cutbacks also forced the Wisconsin Industrial Commission toward political neutrality. The commission's legislative appropriation had risen from $75,000 to $120,000 during Francis McGovern's term, but Philipp trimmed the agency's budget back to $105,750. This retrenchment signaled a nationwide trend after 1915 to streamline new Progressive boards and commissions, and contain their costs. To reduce expenses and achieve efficiency, for example, Illinois consolidated its cluster of labor laws under a new Illinois Department of Labor in 1917, though it also established a separate Department of Mining and Minerals. In Wisconsin, wartime exigencies and strong labor lobbying during World War I halted this fiscal conservatism, but only temporarily.[9]

Despite fiscal retrenchment and leadership changes after 1915, Wisconsin's Industrial Commission retained enough independence to institutionalize the administrative philosophy established by its first three commissioners, Commons, Crownhart, and Beck. Especially important was John Commons. In works on administrative labor law, he and his students portrayed agencies like Wisconsin's Industrial Commission as moral forces to guide business toward socially ethical behavior. With his fellow commissioners' support, Commons put this idea to work in 1911 by launching a state-sponsored safety campaign.

The "Safety First" Campaign in Wisconsin

When they started their work in 1911, Wisconsin's industrial commissioners used persuasion rather than coercion to seek employer compliance with safety laws. It was not wise, Commissioner John Commons said, "to start a pressure upon [manufacturers] beyond what can be worked out cooperatively with them." It was better to let employers gradually adjust to safety measures than impose new regulations on them.[10] An educational campaign, Commons and his fellow commissioners hoped, would facilitate that adjustment. Education could teach employers the emerging new logic of accident prevention and persuade them that state government could help.

To direct its educational campaign, Wisconsin's commission hired renowned International Harvester safety expert C. W. Price in March 1912. A "safety evangelist" and widely experienced "practical man of knowledge" known for dramatic results, Price enjoyed great credibility among manufacturers for his respect and understanding of business needs. From 1912 to 1916, he toured Wisconsin on the commission's behalf, lecturing business groups, visiting large employers, and appearing as featured speaker at company safety banquets. Wherever he went, Price used the Progressive language of social efficiency to condemn work accidents. Speaking to butter makers in 1915, he reported that 12,000 Wisconsin workers had been injured in 1914 and that 163 had died due to accidents, costing employers $2.1 million in compensation payments. If manufacturers had installed safety devices, Price argued, then they might have stopped one-third of those injuries. If workers had been more careful, then they might have avoided the others.[11]

To eliminate accidents, Price urged Wisconsin employers to create plant safety organizations, an idea drawn from big firms like International Har-

vester. He asked companies to form safety committees composed of superintendents and foremen, and then hire professional safety inspectors
to coordinate committee work. Price also advised employers to assemble
worker-composed inspection teams to investigate hazards and make recommendations, a plan that solicited workers' participation in company operations while keeping them under management control. Price finally recommended that foremen carefully tutor new employees, especially non-English
speakers, about job hazards. Like corporate managers, Price saw foremen as
agents of management who could "Americanize" immigrant workers for safe
and efficient labor.[12]

Price especially advocated worker-manager cooperation in factory safety
programs. Like corporate welfare schemes, Price's plan emulated the Victorian family's hierarchical relationships and "reciprocal" obligations. He
urged employers to "get next to their men," and not to expect much worker
cooperation unless they sincerely did their own part to prevent accidents.
As for workers, Price advised employers to assign them to factory inspection
teams. Workers might dismiss managerial warnings about factory risks as
"bunc," he said, but once they saw shop dangers for themselves, they would
become the best safety watchdogs in their plants.[13]

Many Wisconsin manufacturers received Price's message enthusiastically
with good results. At his firm's safety banquet in Kenosha, for instance, the
Simmons Manufacturing Company's owner vowed "to cut out these accidents," and then he organized inspection committees and tacked up safety
posters. His firm subsequently cut days lost due to accidents from 269 in July
and August 1912 to 146 during the same months of 1913. Similarly, Milwaukee's
Pfister & Vogel Leather Company decreased days lost from 371 days to 106,
and the city's Bucyrus Company lessened them from 379 to 172 during the
same period. Moreover, the Illinois Steel Company (Milwaukee) reduced its
accident level by 70 percent in ten years, Fairbanks & Morse (Beloit) lowered
accident rates by 72 percent in a single year, and the Northwestern Railroad
Company curtailed deaths and injuries by 33 and 27 percent, respectively.[14]

Price's effectiveness resulted largely from his ability to address employers'
needs. First, he showed managers how safety plans could save them money,
especially after Wisconsin's 1911 workers' compensation law established general employer responsibility for work accidents at highly uncertain initial
costs. In addition, Price's shop safety plan supported employer efforts to
break independent worker attitudes engendered by craft union traditions,
and secure worker collaboration with managerial goals instead. The Illinois
Steel Company's William Duff Haynie illustrated this point with a story told

in 1909 about a former plant superintendent who returned to visit his old rail mill. A safety conscious worker warned his former boss that his long coat might get caught in machine gears. In Haynie's view, this incident showed how effectively plant safety programs inculcated workers with managerial notions of safety and efficiency.[15]

Indeed, Wisconsin's Industrial Commission itself cultivated worker safety consciousness. The agency distributed posters reminding workers to be cautious and to obey commission safety regulations. One poster depicted a man on a ladder propped up precariously on a pile of bricks. "If this man should fall," the caption said, "it would be hard to convince him that it was not due to his bum luck." Another poster displayed thirty-six safety orders for loggers.[16]

The commission's safety campaign appealed especially to large locally owned firms that had not fully adopted big-industry managerial techniques. Whereas nationally integrated corporations like U.S. Steel and International Harvester instigated safety programs around 1906–1908, local Wisconsin manufacturers perpetuated nineteenth-century views of accidents and labor relations. Only after labor unrest, socialist agitation, and workers' compensation legislation changed the environment did local firms become interested in the new managerial logic of accident prevention. Wisconsin's Industrial Commission exploited this situation by adopting John Commons's and John Andrews's strategy of allowing the "best" employers to set the example for the "worst." That is, the agency dispatched C. W. Price to convey advanced firms' "enlightened" safety techniques to less sophisticated companies.[17] Price's safety campaign thus founded the Wisconsin commission's role as a technical adviser for local industry.

Other Progressive Era state labor departments developed their own safety campaigns, but with local variations. For instance, New York's Industrial Commission endorsed company safety organizations only in a perfunctory 1918 technical pamphlet, a reticence that likely acknowledged trade-union influence. New York labor leaders resisted plant safety programs' managerial paternalism, and aggressively defended craft workers' traditional control over workshop operations. As New York State Federation of Labor (NYSFL) president J. P. Holland remarked in 1916, "The organized wage worker is the only one equipped with power to require remedial revision of working conditions." Unionized workers should join plant safety committees, he said, not just to build up manager-worker trust but also to apply union standards. Other labor leaders bluntly affirmed managerial responsibility for lowering accident rates. At New York's 1917 safety congress, American Federation of

Labor legislative agent Arthur Holder maintained that speedups, long hours, and other means of "driving men beyond their powers of endurance" were accidents' primary causes. He argued that little would be done to reduce accidents until employers ended these exhausting work practices.[18]

Not only did unionized workers discourage New York labor law administrators from embracing corporate welfare practices, but so too did the nature of local industry. Compared to Wisconsin, New York State had triple the number of workers and six times the number of factories, most of them too small to organize safety work. According to New York deputy industrial commissioner James L. Gernon, small manufacturers found safety costs to be "a serious problem" and the hiring of safety experts to be too expensive. And although "cooperation is ideal," Gernon added, it "is not always possible to secure." He and other New York officials recommended safety code enforcement to achieve improvements, not plant safety programs.[19]

Ohio's and California's safety campaigns were more like Wisconsin's. In 1915, Ohio's Industrial Commission hired safety director Victor Noonan to lead the kind of educational program that C. W. Price orchestrated in Wisconsin. Like Price, Noonan went from factory to factory to encourage cooperation and promote plant safety organizations, appealing especially to big foundries and iron and steel firms predisposed to welfare work.[20] Even more like Wisconsin's program, the California Industrial Accident Commission's campaign initially concentrated on "getting acquainted" with manufacturers, not just enforcing rules. The California agency held mass meetings, sponsored shop talks, organized a "Safety Bear Club" among state miners, and by 1916 began urging managers to form shop safety committees. Like its Wisconsin counterpart, the California commission conducted educational work in the spirit of "cooperation and service," dispensing "useful information."[21]

Most state safety first campaigns paid little attention to women's workplace protection. Like big industry, state safety directors including Wisconsin's Price focused on injuries covered by workers' compensation laws, rather than on the wages, hours, and job exclusion questions that worried female protective law advocates. Safety experts were certainly learning about gender-specific job hazards, including industrial toxins, but until World War I, they addressed those problems in terms of female workers' "fatigue" and work hours, not in terms of job injuries. Women's groups lobbied state lawmakers and administrators to improve women's protective measures through the 1910s, but they exerted little influence on the safety campaign.[22]

Beyond plant safety programs, some state labor departments recruited the infant motion picture industry to raise safety consciousness. In Wisconsin, deputy Al Kroes reported business interest in the idea, but Commissioner Joseph Beck initially rejected it. Because the NAM had produced most safety films and because state manufacturers' groups were distributing them, Beck wanted to avoid the appearance of "pushing the National Association of Manufacturers along." Nonetheless, Wisconsin's commission soon embraced the movie idea, securing most films from MMAM and NAM libraries, and others from small distributors.[23]

Safety movies depicted everyday factory dangers, while emphasizing worker responsibility for stopping accidents. *The Workman's Lesson* created by the Illinois Steel Company told the tale of a young man who refused to use a machine safeguard, and had his arm crushed as a result. *The Crime of Carelessness,* an NAM production, involved a worker who thoughtlessly threw a match into a corner, producing a fire that consumed the entire factory. Accompanied by factory safety lectures, such movies were an instant hit. They played to overflow crowds of company executives, supervisors, and workers nearly everywhere. Even Beck had to admit: "This picture show biz is the best thing ever pulled off." Employers were especially appreciative. "A bigger bunch of loudly enthusiastic manufacturers I never saw," Kroes reported. "They came up to me after the shows and praised it [sic] sky-high."[24]

Wisconsin manufacturers' enthusiasm for safety films represented more than the business community's conversion to the safety gospel. Like plant safety committees and industrial welfare programs, the movies were an effective medium for conveying employers' message of cooperation and cultivating feelings of a Victorian corporate family. Employers furnished movie halls, paid the cost of printing tickets and posters, invited workers and their families, and then personally acted as ushers at the films' showings. At a November 1913 Milwaukee event, for instance, the MMAM rented an auditorium, solicited donations from its members, set up a program committee, and hired the International Harvester Company band for entertainment.[25]

Industrial safety movies caught on elsewhere. In Ohio, Industrial Commission safety director Victor Noonan circulated a long list of movies covering industrial hazards, public safety, railroad safety, electrical dangers, and fire prevention. In California, the Industrial Accident Commission joined the state's Redwood Association and White and Sugar Pine Manufacturers' Association to produce its own movie, *Preventable Accidents in the Lumber Industry,* an eight-reel film dramatizing commonplace accidents in logging operations.[26] Like others, this movie stressed worker responsibility.

Back in Wisconsin, the Industrial Commission extended C. W. Price's mission by building a display of practical safety devices, something mandated by state law. To do so, he drew upon the resources of big employers, business groups, and industry-employed engineers who had assembled safety museums of their own. Price secured more than a thousand photographs of safety equipment from U.S. Steel, International Harvester, Remington Typewriter, National Cash Register, and the Link Belt Company, as well as from the NAM. Wisconsin's commission then organized a committee of employers and workers to review the pictures, and that group mounted the best ones on twelve racks according to industry, highlighting machine safeguards in bright red and providing explanations for each. Duplicated in three copies to circulate around the state, this display complemented worker education with lessons in "engineering revision" for employers.[27]

The commission dispatched a deputy with each exhibit to stir up interest. Deputy Al Kroes, a former advertiser, did this zealously. Before arriving in Fond du Lac with his display in 1917, Kroes solicited money from local businessmen to lease a tent at the county fair and place an advertisement in the local newspaper. "Watch Your Step," the ad exhorted. "Safety First, Last, and Always." At Fond du Lac and other sites, the exhibit was a "regular hummer," according to Kroes. "Well, the boss told us to go here and learn a thing or two. What have you got?" one worker asked Kroes. "We saw a notice on the shop bulletin board calling attention to this," said another. Workers' wives also visited the exhibit. "John is sometimes so d-mn careless," one lady told Kroes.[28]

Such expositions were a regular part of state programs during the 1910s, though other states presented them differently. Starting in 1916, New York's Industrial Commission sponsored "safety shows" in conjunction with the agency's annual industrial safety congresses. Usually set up in hotel ballrooms, New York shows gave private exhibitors space to display safety and sanitary appliances—goggles, machine guards, ladders, floor gratings, and overalls "male and female."[29] The Ohio Industrial Commission also organized safety expositions, but it did so by showcasing big firms' work in separate company booths rather than distilling industry's "best practices," as Wisconsin's exhibit did.[30] California's Industrial Accident Commission, meanwhile, established a safety museum of actual equipment in San Francisco that much resembled Wisconsin's photo display.[31]

For its part, Wisconsin's Industrial Commission also promoted "engineering revision" in monthly "shop bulletins." Issued from 1912 to 1917, these publications covered typical injuries and suggested remedies. One 1913 bulletin, for instance, displayed charts and graphs to show that accidents rarely occurred when jointers, power-driven woodworking machines, were equipped with cylindrical safety heads rather than old-fashioned square heads. Whereas unguarded and partially guarded jointers severed workers' fingers, the bulletin reported, jointers equipped with safety heads never caused such injuries, even when accidents happened. Presenting pictures and diagrams to demonstrate safety heads' installation, the bulletin added that such equipment was remarkably cheap, just fifty dollars per machine. At that price, employers could install safety heads on more than three hundred jointers for the cost of compensating seventy-five injured workers. Clearly, safe equipment saved money.[32]

Wisconsin shop bulletins always reminded workers and employers both to be vigilant. "Put your soul into your work," one bulletin exhorted workers, "not your hand or foot." "You need two eyes—look out!" proclaimed another. Bulletins also admonished employers to install safety devices diligently. One publication described an incident where a worker adjusting sprockets on his printing press had his leg fatally caught in unguarded gears seven feet above the floor. Employers might scoff at covering such apparatus, but this event showed that employers had to guard even "ridiculous" hazards. Wisconsin bulletins especially reminded employers to obey state safety requirements and reprinted machine, sanitation, and elevator codes regularly.[33]

Wisconsin shop bulletins contained innovative statistical analysis. Before 1910, both industry and government accounts of accidents were unreliable, but after 1910, workers' compensation reporting requirements made accurate accident measurement possible. Embracing new statistical methods, Wisconsin's Industrial Commission assigned Ben Kuechle to analyze accident reports filed under the state's new workers' compensation law during 1911 and 1912. Kuechle filled out three-by-five cards for nearly eight thousand accidents reported, and then "played solitaire," piling cards into various categories, logging injuries here and elevator accidents there. From Kuechle's tabulations, the commission pinpointed many accident causes, such as the dangerous square jointer heads. Wisconsin shop bulletins then used statistical analysis to recommend mechanical safety devices, not rely on "rule-of-thumb" techniques that skilled workers and foremen had previously used. Bulletins not only promoted better machine design but also endorsed industrial management's statistical methods of reducing accident risks.[34]

Other states' labor departments unevenly disseminated comparable literature. Before 1920, New York's Industrial Commission published almost two dozen shop bulletins explaining its Industrial Code, while California's Industrial Accident Commission printed technical reports in its agency safety magazine. Much like Wisconsin's commission, these two labor departments acted as service agencies conveying technical information to medium-size local firms. In a state dominated by big firms not needing engineering advice, however, Ohio's Industrial Commission published few shop bulletins. Devoted just to law enforcement, meanwhile, Illinois's Factory Inspection Department issued no technical literature at all.[35]

In 1918, Wisconsin's Industrial Commission replaced shop bulletins with the *Wisconsin Safety Review,* a journal published irregularly until 1923. This new publication did more than just polish up the bulletins. Like the *California Safety News* dispersed by California's Industrial Accident Commission starting in 1917, Wisconsin's *Safety Review* directed its analysis of work hazards to foremen and superintendents. Such attention to middle managers reflected a subtle change in the Wisconsin agency's safety philosophy. Earlier, the commission's statistical work had revealed that only one-third of all accidents had mechanical causes, while most others resulted from human failures, such as falls, handling objects, and carelessness. Wisconsin shop bulletins subsequently supplemented its advocacy of safeguards with reminders to inculcate workers with the "habits of safety and caution." The *Safety Review* enlarged on this advice. Just as big industry moved from engineering revision to scientific management and then personnel management, Wisconsin's *Safety Review* shifted focus from installing safety devices to managing workers as the means to prevent accidents.[36]

The *Wisconsin Safety Review* portrayed foremen as key figures in the managerial hierarchy for teaching workers safe habits. For instance, the *Safety Review* advised foremen to keep an eye on novice punch-press operators until they mastered their work, because punch presses were especially dangerous machines. Foreman also had to watch out for reckless and hazardous worker behavior, such as when a laborer washed his hands in naphtha and then thoughtlessly struck a match to light his pipe, igniting an explosion. The *Safety Review* finally contended that foremen had to train workers to respond quickly and effectively to dangers. In the case of boilers, for example, the best guarantee against accidents was a "thoroughly organized crew operating according to a definite daily schedule as to time, method, and purpose." When danger appeared, "there is no time for speculation or reference to rules; [workers'] action must be intuitive and immediate."[37]

The educational tone of Wisconsin's *Safety Review,* as well as the spirit of its shop bulletins, photo display, and plant safety campaign, carried over into factory inspection. "The day has long passed when good work can be done by force," Commissioner Beck told Deputy Dave Evans. Deputies had to persuade, not threaten, employers. To prepare for the persuasive approach, deputies visited modern factories and attended safety roundtables. They then studied accident reports to be "armed with facts" about needed safety measures. Finally, deputies prepared to inform manufacturers about safety exhibits, distribute posters and leaflets, and tell employers about other businesses successfully using safety devices.[38]

Although deputies and factory inspectors in Ohio and California also gave unprecedented attention to persuasion rather than pure law enforcement, Wisconsin deputies exhibited an unusual flair for educational work. Deputy Al Kroes perfected this "soft" approach. Finding most employers ignorant of the commission's work, he took along newspaper clippings about the safety campaign. After reading such reports, business executives generally cooperated. As for foremen, Kroes said that there was nothing like "getting next to your man first." That meant standing in the shop with the foreman and discussing his work with him. Once convinced that Kroes wanted to help, rather than to disturb, foremen yielded willingly.[39]

Wisconsin deputies seemingly changed many employer minds. When deputy Henry Schreiber visited Fond du Lac's Guerny Refrigerator Company, the superintendent asked him, "Well, how many [inspectors] has the state got running around harassing manufacturers? I haven't found one that knows anything." In conversation, however, Schreiber found the superintendent to be ignorant about safety, despite thirty-five years on the job. Admitting at last that his shop could be safer, the supervisor asked Schreiber to identify danger points. Schreiber's experience was typical. Wisconsin deputies initially met resistance, but reported that most employers eventually complied.[40]

Although Wisconsin employers usually reacted positively to the Industrial Commission's barrage of bulletins, safety magazines, and factory inspections, upstate lumber companies remained hostile. In response to a circular condemning the lumber industry's miserable accident record, a few company owners fiercely rejected the agency's call for action. Although acknowledging the Wisconsin commission's "wonderful" assistance, lumber company operators contended that the real cause of accidents in their business was the "class of labor . . . now in the woods"—a lazy and drunken bunch of "happy-go-lucky fellows" who cared little about safety. Woodsmen's addiction to "booze" was pivotal. Operators insisted that nothing would reduce lumber industry accidents more effectively than curbing worker alcoholism.[41]

More fundamentally, lumbermen rejected accident law's new presumption of managerial responsibility. As company president E. S. Hammond saw it, Wisconsin statutes wrongly placed accident liability on lumber firms, while the careless "bums and tramps" hired as workers assumed little of the burden. "If you tax the employer for the accidents that occur to the employee," he asserted, "[then] the employee does not care how many accidents occur as the other fellow has to pay for it." Viewing Wisconsin's compensation and industrial commission laws as stacked against employers, Hammond advised the state's commission to take those measures "back to Germany."[42]

Lumbermen's belligerence contrasted starkly with Milwaukee industrialists' receptivity. The difference reflected Wisconsin business's transition from feisty old-fashioned upstate proprietors who resisted modern regulation to emerging downstate corporate capitalists who accommodated it. On the one hand, most Wisconsin lumbermen were entrepreneurs who wanted to perpetuate nineteenth-century common-law worker-employer relations, unlike the enlightened "Wausau Group" founders of the Employers Mutual Liability Insurance Company and West Coast lumber operators who championed workers' compensation legislation.[43] On the other hand, Milwaukee manufacturers accepted the managerial reforms, welfare policies, and even administrative regulation that typified modern industrial labor relations. During the 1910s, at least, Wisconsin's growing southeastern industrial community gave the Industrial Commission's safety campaign readier support.

The Realities of Success

By the late 1910s, safety law administrators in Wisconsin, California, and other states reported changes in employer attitudes toward workplace safety, mostly crediting themselves for the improvement. Wisconsin industrial commissioner George Hambrecht lauded the "personal work done by the commissioners and inspectors in spreading the gospel of safety," resulting in employers' "hearty cooperation." Will J. French of California's Industrial Accident Commission exuded the same satisfaction when he observed growing awareness of safety in his state: whereas most California firms ignored safety work in 1910, most took it for granted by the early 1920s. He commended his agency for this progress.[44]

Hambrecht's and French's remarks typified Progressive reformers' faith in administration government's capacity to reform the business system. Like John Commons and John Andrews in *Principles of Labor Legislation,* state officials assumed that new state commissions could realign private manufacturing by deploying what Daniel Carpenter today calls their "bureaucratic autonomy" and "reputation." When compensation laws financially induced company safety work, officials reasoned, specialized agencies like Wisconsin and California commissions would teach employers how to implement industry's "best" safety and health practices. The proliferation of such state-level administrative bodies, local officials concluded, did indeed account for better working conditions.[45]

Progressive agencies' actual impact on occupational safety was likely more complicated. Recent scholars contend, for instance, that state-sponsored safety education succeeded to the extent that it converged with private-sector developments such as factory rationalization, big employers' safety first movement, new insurance practices, and machine makers' effort to design new apparatus for the rising "safety market."[46] Moreover, state labor departments' educational programs varied widely in character and effectiveness, depending on local circumstances. Like most of the Deep South, Alabama employers and politicians left big corporations like U.S. Steel subsidiary Tennessee Coal, Iron, and Railroad Company to spearhead Alabama's industrial safety work, not a state labor bureau. In Illinois, political rivalries and antistatist business-labor politics confined that state's Factory Inspection Department to labor law enforcement. Private firms such as U.S. Steel's South Chicago works and industry groups like the NAM led Illinois's safety campaign.[47]

In New York, meanwhile, the state Industrial Commission actively fostered safety work, but in ways that accommodated interest groups. The New York commission did not hire a safety director or directly endorse corporate welfare schemes, but instead published technical manuals, staged an annual safety show, and, importantly, sponsored yearly industrial safety congresses that assembled civic groups, insurance companies, industry associations, labor unions, engineering societies, women's groups, and the American Association for Labor Legislation. Such representativeness especially benefited workers, reformers, and women. Female labor law reformers were well organized in New York, though they still perpetuated the commonplace marriage-and-motherhood view of women's "special role" in society. When World War I boosted demand for female labor, women reformers organized safety congress panels to promote special workplace precautions for working women, such as adequate training, less strenuous work, shorter work hours, better wages, and protection against toxic chemicals.[48]

In Ohio, however, highly concentrated iron, steel, mining, and oil businesses, moderate OSFL leadership, and stability-minded politicians all steered that state's Industrial Commission into partnership with local industrialists. With little attention to women's issues, Ohio's commission encouraged big-industry welfare plans in banquets, shop talks, plant safety organizations, and movies while producing few technical manuals. In safety expositions, the agency left large employers to exhibit new apparatus rather than adopting Wisconsin's and California's museum approach to displaying safety technology.[49]

California's Industrial Accident Commission, by comparison, accommodated influential labor groups and downstate employers with an educational campaign that encompassed both engineering improvements and welfare-oriented managerial schemes. The agency established a safety museum and published bulletins and magazines to inform local employers about safety techniques. Moreover, the commission also sponsored banquets, plant organizations, shop talks, and movies to foster worker-employer cooperation for industrial safety. Notably, neither side of California's safety campaign addressed female work hazards, though women's groups secured creation of the separate Industrial Welfare Commission to regulate women's working conditions in 1913.[50] Still, California's safety education work addressed both worker and employer interests.

Wisconsin's Industrial Commission developed the most flamboyant industrial safety campaign. Unlike New York, and even more than California, Wisconsin's educational program intertwined with big industry's safety first movement when it employed C. W. Price and duplicated corporate safety techniques such as banquets, plant safety plans, posters, and movies. And unlike Ohio, but much like California, Wisconsin's campaign itself disseminated new safety technology by erecting photo exhibits, publishing shop bulletins, and circulating foremen's magazines, though it neglected women's safety issues. Overall, Wisconsin's commission aimed to stand above the politicking between the state's socialist-oriented labor movement and Milwaukee industrial community. Composed of "impartial" administrators, rather than interest-group representatives, Wisconsin's agency positioned itself as an independent technical body that utilized big-industry managerial and engineering practices to serve the state goal of industrial stability. The commission's posture was, however, more "political" than administrators admitted, as safety code proceedings would show.

4

The First Safety Codes

On May 3, 1912, the Wisconsin Industrial Commission approved the General Orders on Safety, the first of ten industrial safety codes that it would issue in the 1910s. The General Orders illustrated a critical feature of state governments' bureaucratic expansion during the Progressive years: a mass of administrative rules regulating business and other features of industrial society. Though unevenly, order making spread as a new mode of work safety and health lawmaking, and some states, like Wisconsin and California, extended it to women's work hours and wages. The new process completed work safety and health law's transition from "Classical" to "Progressive" form, removing safety policy from courts and legislatures and assigning it instead to regulatory agencies. In Wisconsin and a few other states, moreover, order making activated local safe place statutes that raised safety standards above common-law norms and fulfilled the emerging presumption of employer responsibility for work safety. In some states, safety order making's sheer volume gave Progressive administrative government real force, more than historians acknowledge.

Although recent commentators have generally viewed Progressive rule-making agencies as political bodies that either served the public interest, mediated pluralistic interest-group conflicts, or fell prey to "capture" by powerful business organizations, Wisconsin's Industrial Commission and similar agencies were also legal institutions shaped by the rising field of administrative law. Rooted in Classical legal thought and common law, early-twentieth-century administrative law imposed external judicial controls on proliferating administrative agencies, notably by applying the "delegation" and "reasonableness" doctrines to administrative decisions. Later, in the

1940s, conservative legal reformers would place even greater restrictions on the massive New Deal administrative system in the form of state and federal administrative procedure acts. Yet back in the 1910s, administrative law tolerated wide statutory grants of agency power and broad judicial deference, a brief administrative triumph over the old Classical court-dominated legal order.[1] This momentary triumph coincided, significantly, with the Progressive Era's corporate transformation of American industry. The economic impact of safety order making in Wisconsin and other states thus reflected the interaction of growing administrative apparatus and evolving corporate technology and management.

A product of this moment of exceptional administrative influence, Wisconsin's Industrial Commission received one of the most open-ended mandates of all new Progressive Era labor departments. For one thing, the Wisconsin safe place statute provided only a nebulous guide for safety order making, and the Wisconsin Supreme Court at first construed that measure broadly, much in line with the Progressive Era judiciary's general approval of protective legislation, notwithstanding several notorious anti-labor-law decisions by state and federal courts during the early 1900s.[2] For another thing, Wisconsin's safety order–making procedures themselves were open-ended. American courts had not yet specified procedural requirements for quasi-legislative administrative decision making, while statutes provided little instruction for its use. Consequently, Wisconsin labor law officials improvised advisory committees and public hearings as the mainstays for order making, using the principles of "prevailing good practice" and the "representation of interests" as guides for those proceedings.

Based on their reading of the safe place statute, Wisconsin administrators designated "prevailing good practice" as the general standard for safety code making, a "political choice" among different early-twentieth-century manufacturing technologies. Across the country, large scientifically advanced and centrally managed corporations had begun rationalizing production operations, while ordinary factory owners, shop foremen, and craft workers sustained nineteenth-century "shop culture."[3] Divided between Milwaukee-area industrialists and numerous small and medium-sized, mainly upstate, proprietary firms, Wisconsin stood astride this transition in the 1910s, unlike the Ohio and Illinois economies already dominated by heavy industry.[4] By adopting "good practice," Wisconsin's industrial commissioners chose to modernize the state's many proprietary businesses along big industry's technological and managerial lines, though initially, administrators embraced local factory precautions, since corporate safety science remained rudimentary.

Meanwhile, administrators applied the "representation of interests" principle to Wisconsin's safety code advisory committee process. Commentators frequently associate this plan with Wisconsin School labor economists who founded the New Deal–era system of "industrial pluralism" that utilized "tripartite" councils of workers, employers, and the public to conciliate labor relations.[5] In practice, however, Wisconsin's industrial commissioners manipulated committee membership to serve the "good practice" standard. While placing local employers, business associations, and labor organizations (though never women's groups) on advisory panels, Wisconsin's commissioners gravitated over time toward advanced manufacturers and professional "experts." Increasingly, committees embraced corporate safety standards while promoting local employer and worker acceptance of them.

Legal and political differences begot alternative uses of the safety code process elsewhere, even when other states adopted programs resembling Wisconsin's. California's system reconciled business, labor, and insurance interests when it assembled advisory committees. Ohio's program, however, accommodated large manufacturers by initially avoiding the code-making process, while its courts emasculated the state's safe place statute. New York's plan, by contrast, required business-labor representation in safety code proceedings and thus allowed strong labor and reform influence on that state's regulations. And most states, including Illinois and Alabama, still did not vest their labor departments with safety code–making powers at all. The nation's patchwork of safety law institutions continued.

The Political and Legal Environment for Commission Government

When Wisconsin's Industrial Commission and other states' agencies began formulating safety orders in the 1910s, commission government's legitimacy remained questionable. Although Progressive reformers established administrative bodies at an unprecedented pace, and although courts generally upheld administrative power, the level of political and legal support for such authority varied from state to state, and an undercurrent of hostility persisted everywhere.

In Wisconsin, the Industrial Commission immediately encountered political opposition. In 1913, La Crosse assemblyman C. L. Hood offered a bill to repeal sections of the Industrial Commission Act authorizing the agency to make safety orders. Like his conservative western Wisconsin constituents,

Hood objected to the commission's "excessive power" and "useless expense."[6] Simultaneously, a mysterious source paid a reporter to arouse resistance to the agency around the state. Commissioner Joseph Beck speculated that manufacturers had instigated this campaign, but National Association of Manufacturers (NAM) chairman Fred C. Schwedtman blamed old-line stock insurance companies that wanted to sabotage the state's compensation law.[7]

The Hood proposition jolted workers and manufacturers into action. At a February 1913 Milwaukee mass meeting, trade unions passed a resolution directing Hood to "keep his hands off measures enacted on behalf of labor."[8] At the same time, industrialists like Milwaukee's Thomas Neacy and Kenosha's Chester Barnes announced their "unalterable opposition" to Hood's commission law modifications, and warmly endorsed the new agency and its functions. The state assembly subsequently killed Hood's proposal, revealing how Milwaukee-region workers and manufacturers coalesced against old-line insurance firms and upstate employers to safeguard commission authority.[9]

Then, surprisingly, Wisconsin's commission faced a challenge from reformers. Having strongly endorsed commission government in 1911, Wisconsin Legislative Reference Library chief Charles McCarthy revived former Governor Robert La Follette's populistic brand of Progressivism two years later when he complained that commissioners "do not listen to the voice of the people." Industrial commissioner John Commons then suggested legislation allowing governors to suspend commissioners pending recall elections, while McCarthy himself advocated a recall bill. State legislators rejected both proposals, thereby reaffirming urban-industrial Progressives' support of administrative authority. Indeed, prompted by women's groups, lawmakers expanded the Industrial Commission's power to prescribe rules for women's work hours and wages.[10]

Controversy characterized administrative politics in other states as well. Californians and Ohioans robustly supported labor commissions, but Alabamians still resisted state administration, as well as labor laws in general. Illinois reformers confronted especially contentious conditions, because that state's tradition of antiexecutive partisan politics had left local government notoriously decentralized in a maze of office, boards, departments, and special-interest laws. Such was the case with labor legislation. Lagging behind Wisconsin's administrative legislation, Illinois's old-fashioned Factory Inspection Department—itself not created until 1907—struggled to enforce ten separate statutes. As part of his effort to streamline state government in 1917, hence, Illinois governor Frank Lowden spearheaded the consolidation of labor law administration under a new Department of Labor (reducing factory inspection to a "division"), but lawmakers stopped short of granting that body code-making authority.[11]

In New York, meanwhile, turbulent politics produced a strong industrial commission but constrained its authority. During 1913, in the aftermath of the awful Triangle Shirtwaist Factory fire, a Democrat-controlled legislature created an industrial board with safety code powers in the new Department of Labor. Following Democratic governor William Sulzer's subsequent impeachment and Republican Charles Whitman's election, however, a conservative Republican legislature in 1915 reorganized state labor laws under a new industrial commission that imposed restraints on safety code making. Lawmakers not only renewed the 1913 statute's narrow prescription of employer duties but also mandated employer and worker representation on advisory committees, required all safety orders to be examined by an industrial council composed equally of employer and employee members, and broadened employer rights to secure judicial review of safety orders. This 1915 reform reaffirmed administrative authority but subjected it to interest-group influence.[12]

Wisconsin's Industrial Commission, by contrast, enjoyed greater political freedom, and that independence was reinforced by law. By the 1910s, courts in Wisconsin and elsewhere had accepted regulatory commissions' constitutionality and had validated rule-making authority, so long as it involved "fact finding" rather than legislative policy making. Rare test cases involving prohibited occupations for minors, women's work hours, and women's wages did arise, most after 1920, but even then state courts mainly confined themselves to policing order making's conformity to statutory guidelines. Still, the courts remained an indirect force in safety law administration. At the time, federal courts found new ways to limit administrative power and retain some influence over substantive policy making, such as scrutinizing administrative decisions for unreasonableness.[13] Similarly, state courts intervened indirectly into work safety administration when they addressed the meaning of safe place statutes, procedures for safety code making, and the scope of judicial review over safety orders.

Safe place statutes defined the scope of safety order making. Wisconsin's law required businesses to supply "safe" "employment" and a "safe" "place of employment," where "safe" meant "such freedom from danger . . . as the nature of the employment will reasonably permit." Applying to all manufacturing processes, except domestic pursuits and nonmechanized agriculture, Wisconsin's safe place statute protected all "employees" and "frequenters" who entered places of employment, and it vested "employers" who exercised "control or custody" over employment with responsibility for keeping employment safe. Importantly, this statutory scheme established liability in work accident lawsuits involving business visitors ("frequenters") or employers either not electing or excluded from compensation coverage, while it also created criteria for safety order making.[14]

Ohio and California safe place statutes copied Wisconsin's law almost word for word, but New York's legislation framed employer duties more narrowly. Stipulating that all factories "be constructed, equipped, arranged, operated and conducted" as to make them "reasonably and adequately" safe, New York statutes emphasized physical apparatus and "methods and operations" of production, not Wisconsin law's broader notion of "employment." New York law also perpetuated the lower standard of "reasonable and adequate" protection, rather than Wisconsin's higher requirement that employment be as safe as reasonably possible. Crucially, by placing these standards in the rule-making section, New York law did not establish them as independent grounds for work injury litigation, not even to amplify employer duties (as factory statutes like New York's scaffold law did) in common-law cases originating before compensation reform or in maritime cases outside of compensation's jurisdiction. Instead, New York law declared that the state's Industrial Commission would "make such rules and regulations as will effectuate" the policy for reasonable and adequate safety, thereby linking that policy to bureaucratic action.[15]

Back in Wisconsin, the state supreme court quickly recognized safe place legislation in work injury cases against employers not covered by Wisconsin's voluntary compensation program. The court ruled that the legislation applied to management of productive processes, as well as the physical environment, and confirmed that it elevated safety standards. Gone were the common-law due care principle and the 1910 *West v. Bayfield Mill Company* ruling that restricted factory laws to ordinary prudence. "It is obvious that these provisions make some radical changes in the common law," Associate Justice John Barnes affirmed. Wisconsin's safe place law imposed "an absolute duty" upon employers to make workplaces as free from danger as employment reasonably permitted. This analysis signaled the demise of nineteenth-century Classical legal norms and the rise of Progressive social thinking. Wisconsin Supreme Court decisions accepted emerging views about social interconnectedness and rationalized new safety standards as mere links in the chain of industrial production.[16]

Ohio's conservative supreme court, meanwhile, refused to acknowledge that state's safe place statute as independent grounds for work injury litigation.

Based on a constitutional amendment, another Ohio law allowed workers to sue employers in exemption from compensation's "exclusive" coverage when injury resulted from employer violations of any "lawful [safety] requirement." Yet a bitterly divided Ohio court in *American Woodenware Manufacturing Company v. Schorling* (1917) denied that safe place legislation constituted such a requirement, asserting instead that it merely obliged employers to obey the Ohio Industrial Commission's safety orders, much like the New York safety law that was "effectuated" by administrative action. In dissent, Associate Justice R. M. Wanamaker condemned such analysis as "weasel construction" that "suck[ed] out . . . the meat" of Ohio's safe place statute. He chastised his colleagues for not imitating their "progressive humanitarian" Wisconsin counterparts who interpreted safe place legislation as a separate new requirement.[17] For the moment, however, the *Schorling* majority prevented Ohio's safe place law from enlarging employers' liability.

Even in Progressive Wisconsin, judges disagreed about how high the state's safe place statute raised safety standards in work injury suits. Uncertainty persisted in regard to the new statute's provision for "reasonableness." In *Olson v. Whitney Brothers Company* (1915), Wisconsin Justice Aad J. Vinje decided that the safe place law did not place "an impossible or unreasonable burden" on employers but only compelled them to perform work "in the manner in which an ordinarily prudent person might anticipate it may be done." Associate Justice John Barnes complained in dissent that such interpretation "practically gets us back to the common-law rule" of ordinary employer care, and ignored the "entirely new scheme" created by Wisconsin legislation. Still, Vinje's ruling kept Classical common-law views of reasonableness alive, as did decisions in other states. In California, where state statutes allowed workers to sue employers for alleged "gross negligence and willful misconduct," a state appeals court held that California's safe place law required employers only to provide "such safety devices and appliances as are '*reasonably*' adequate to render the place of employment safe," not the highest level of safety. Looking back in 1940, Wisconsin attorney Henry Reuss blamed such judicial backtracking on the imprecision of safe place legislation itself. Not only should the legislation have applied just to safety order making not litigation, he contended, but its "litigation-provoking verbal ambiguities" required a "metaphysician" to distinguish old common-law standards from new safe place requirements.[18] Given the ambiguity, increasingly conservative courts would narrow safe place statutes' meaning, eventually affecting safety code–making power.

On the second issue that they faced—administrative procedure—the courts offered little more enlightenment. Lacking Continental Europe's administrative tradition, American judges drew on the medieval English judicial writ system to control regulatory agencies. Courts policed administrative bodies externally by demanding conformity to legislative mandates, use of "appropriate" procedures, and avoidance of "unreasonable" decisions. Courts left internal decision making to agency discretion, though they gradually demanded trial-like hearings in quasi-judicial proceedings. For quasi-legislative rule making, courts imposed few restraints during the Progressive years. Not until the New Deal era would judicial rulings and administrative procedure acts tighten quasi-legislative proceedings, and not until the 1970s would courts mandate "fair representation."[19]

Industrial commissions in Wisconsin, Ohio, California, and New York, consequently, began safety code formulation in an age of statutory imprecision and judicial neglect. Wisconsin law directed the state's Industrial Commission to "ascertain and fix such reasonable standards" and "prescribe, modify and enforce such reasonable orders" "to carry out" safe place statute requirements, but it did not clearly specify how. Without requiring public hearings, Wisconsin legislation instructed the agency only to publish its internal rules, not enforce safety orders until thirty days after publication, and entertain employers' petitions to review orders. Wisconsin law did allow commissioners to appoint unpaid "advisers" to assist them with their duties, but it did not prescribe advisory committees or define their composition and operation. Wisconsin's supreme court sanctioned this arrangement by holding earlier that commission procedures "are not expected to be as formal and cumbrous as the proceedings of courts," since agencies needed "greater flexibility" "to discharge their duties."[20]

Other states defined commission procedures more explicitly. California law authorized the appointment of unpaid advisers but specifically added that they would "assist the [Industrial Accident Commission] in establishing standards of safety." California law also empowered that agency to issue safety rules, but only "after a hearing upon its own motion or upon complaint."[21] Going further, New York's 1915 statute allowed that state's Industrial Commission to set up "committees composed of employers, employees and experts" to frame safety regulations, and required public hearings and scrutiny by the agency's joint worker-employer industrial council.[22]

On the third area of court involvement with commission operations—judicial review of safety orders—state legislation was more specific. Duplicating other commission laws, Wisconsin's Industrial Commission Act allowed employers to seek review of commission orders in Dane County Circuit Court (site of the state capital), whereupon the court could vacate orders found unreasonable. In other proceedings, especially safety order enforcement actions, the act required courts to treat orders as prima facie lawful. Other states imposed greater judicial controls. Ohio legislation gave that state's supreme court authority to amend safety regulations, and California statutes allowed that state's high court to review and modify orders on specified grounds. California law even empowered courts to determine whether findings of fact supported safety orders, while preserving the state commission's conclusive authority over what the facts were. Most intrusively, New York law allowed jury trials to consider questions of fact in safety code controversies, though courts could only determine orders' reasonableness or validity, not amend them. These variations made little difference. Employers appealed few safety orders to the courts in any state, and then mainly after 1920.[23] Diverging from Classical law, commission statutes everywhere sharply curtailed local courts' power over substantive administrative decisions. In the mid-1920s, Wisconsin firms would circumvent this restriction by attacking orders indirectly in workers' compensation proceedings.

During the 1910s, hence, Wisconsin's Industrial Commission enjoyed remarkable power. More than any other state labor department, the Wisconsin agency received local political support, statutory license, and judicial deference to advance its own interpretation of state safety policy and devise its own procedures. By comparison, interest-group politics and legal regulations more greatly hampered Ohio, California, and New York labor commissions, whereas local politics prevented such agencies' creation in Illinois and Alabama. Wisconsin's commission thus had exceptional administrative autonomy to formulate its first safety regulations.

The General Orders on Safety

When it first considered safety codes in the summer of 1911, Wisconsin's Industrial Commission adopted "prevailing good practice" as the practical measure of what the safe place statute required. Administrators understood "good practice" to elevate safety standards not to scientifically "ideal" heights but to the level of the best safety techniques proven feasible in industry. "Good practice" consequently stipulated that safety orders be "practicable"—that is, technically workable, definite in meaning, and moderately priced. Wisconsin officials pledged that safety regulations would demand "no safeguard . . . which the commission cannot show how to install." They added that "economy and efficiency of operation will follow a strict observance of [safety] orders, for it is a well-known fact that safety and efficiency are inseparable."[24] Such practicality surely met the judicial reasonableness test.

Wisconsin's industrial commissioners relied mostly on "representative" advisory committees to draw up "practicable" regulations. Unlike the "industrial pluralism" policy ascribed to Wisconsin School labor economists and New Deal labor laws, the "representation of interests" on safety code committees exalted governmental priorities over private labor-management negotiation. Inspired by European industrial councils, reformers led by John Commons and his students advocated interest representation as a means of improving state policy making. Employer and worker committee members would help regulators adjust "purely technical" bureaucratic rules to business's and labor's "practical industrial point of view." The process promised workable regulations and industrial peace. Since "neither the union, the employer nor the politician dominates," advisory committees would meet as "cooperative" "continuous conferences" to "work out the rules and regulations [giving] effect" to legislative policy. The idea was popular, and many states adopted it. New York codified the committee system in its safety laws. Wisconsin and California applied it to women's work hours and wages.[25]

The practice of representing interests on advisory panels, however, differed from theory. In Wisconsin, commissioners selected committee members more to cover political bases and gather technical expertise than to equalize business and labor participation. In the 1910s, businessmen occupied half of Wisconsin's advisory committee seats, government officials got a quarter, while labor representatives held one-tenth, and women served rarely, if ever. In fact, committee membership varied. The safety and sanitation committee included seven businessmen, three government officials, and two workers. The building construction committee evenly divided manufacturers, contractors, and craftsmen. Building code, boiler, and elevator panels were dominated by manufacturers, inspectors, architects, and engineers, with no laborers at all. Wisconsin's advisory committee system subsequently restrained labor influence. When Milwaukee unions proposed their own panels, industrial commissioner Joseph Beck directed them to work through the Wisconsin State Federation of Labor (WSFL) or standing advisory committees instead.[26] Wisconsin's committee process favored state administration over direct labor action.

New York, Ohio, and California commissions better balanced business and labor representation on advisory committees. New York's 1915 industrial commission statute not only subjected safety codes to advice by an industrial council consisting of five business and five worker members but also explicitly mandated safety code advisory committees "composed of employers, employees and experts." Without legislative direction, Ohio's Industrial Commission organized a general advisory committee with equal numbers of employers and workers, whereas California's Industrial Accident Commission established parallel advisory panels in San Francisco and Los Angeles with commensurate business, labor, and casualty insurance representation. The California State Federation of Labor (CSFL) lauded the state commission's "broad-minded spirit."[27]

Wisconsin safety code committees, by comparison, inclined more toward technical specialists and big industry. The Advisory Committee on Safety and Sanitation assembled in 1911, for example, not only accommodated the state's foremost interest groups—the WSFL, the Merchants and Manufacturers Association of Milwaukee (MMAM), the Wisconsin Manufacturers Association, mutual insurance companies, and the City of Milwaukee—but also exhibited a strong business slant, with six manufacturers, two union spokesmen, one insurance company executive, two government inspectors, and Industrial Commission special assistant C. W. Price. Two of the manufacturers were upstate agricultural implement and furniture producers, but more important were big Milwaukee firms exemplifying new corporate managerial and welfare practices: Pfister & Vogel Leather Company (leather garments), and Fairbanks-Morse Company (diesel engines). With West Allis milling-machine maker Edward J. Kearney as committee chair, large firms took the lead.[28]

Despite its composition, the safety and sanitation committee did not just adopt big industry's safety rules. Committee members voted to compose regulations themselves, thereby broadening the code-making process beyond the Wisconsin Railroad Commission's piecemeal practice of adjusting industry-made standards. And rather than initiate legislature-like interest-group bargaining, committee members established what historian Samuel Hays calls "symbiotic relations of mutual support."[29] Amid rapid industrial change, panelists collaborated as a team to devise technical solutions for job risks.

Both on the committee and off, Wisconsin's Industrial Commission gathered information informally. Building on regulations developed at the Wisconsin labor bureau's 1910–1911 safety conferences, the commission's deputies submitted suggestions based on inspection work, conducted experiments on new equipment, and visited woodworking and paper plants to investigate hazards and protective measures. Meanwhile, the advisory committee broke into small groups to study logging accidents, toilet room equipment, shop lighting, and sawmill dangers. Still, the commission called repeatedly on big manufacturers. Committee members conferred with the MMAM's Safety Committee and secured safety manuals from large Milwaukee and Beloit companies. Deputies attended MMAM safety roundtables. Commissioner Beck asked the NAM for the design of glazing-machine gear covers. And the advisory committee allowed paper-mill owners to invent a winding reel guard.[30]

When the safety and sanitation committee convened to draft regulations, the "reasonableness" principle dominated its deliberations. To ensure that proposed orders were "practicable," committee members drew on their own experience, conducted trial-and-error workshop experiments, and sought advice from manufacturers. While considering safety heads for jointers, for instance, the panel visited a machine manufacturer's salesroom, and then sent deputy Ira Lockney to ascertain whether all jointer designs could take safety heads. Once reassured, the panel approved the device. Wisconsin industrial commissioners followed up by mailing copies of proposed regulations to local trade associations, industrialists, and labor unions for reactions and suggestions.[31]

Public hearings also appraised the reasonableness of proposed safety orders. Unlike California and New York statutes, Wisconsin (and Ohio) law did not require hearings before safety regulations' formal approval, though Wisconsin (and Ohio) legislation did stipulate that employers could request hearings once orders were published. Nonetheless, public hearings became "regular procedure" everywhere. The purpose, John Andrews observed, was "to gauge the extent and importance of opposition" and to modify orders in response to criticism.[32]

The safety and sanitation committee's first hearing in Milwaukee on January 27, 1912, set the example. With committee chair Kearney presiding and Commissioner Beck in attendance, a business audience (apparently without workingmen) debated the first Wisconsin safety code's reasonableness. Right away, employers criticized the proposed regulations' ambiguity. Order No. 1, for example, provided that all belts, shafts, and chains be "properly protected" where "exposed to contact." The Milwaukee Malting Company's Bruno Fink wanted to know who would determine what was "proper": businessmen or factory inspectors? Other employers wondered what "exposed to contact" meant. If it included shafting ten feet above the floor, was that reasonable? Kearney and Beck responded that common sense governed the rules' meaning. Orders were "self-interpreting." Where there was doubt, the Industrial Commission could decide whether a plant complied with the law.[33]

Businessmen also claimed that proposed regulations demanded excessive changes. One employer argued, for instance, that the order requiring hoods and suction devices on emery wheels was impractical for the "wing grinder" that moved around stationary objects being ground. Contending that installation of suction devices on every emery wheel would be unwieldy and expensive, he requested a more flexible rule requiring exhaust fans for entire rooms. Kearney rejected the idea. Granting exemptions, he said, would emasculate regulations. It was better to adopt a demanding rule and let the Industrial Commission apply it "sensibly."[34]

Yet employers did not know what to expect from "sensible" application. Mr. Clasmann asked the committee to interpret a proposed order requiring a guard on jointer-machine knives. Kearney stated that the rule clearly required "a guard that will protect the hands of the operator," but Clasmann wondered whether such a guard had to be spring-loaded or could be manually operated. A state factory inspector, he explained, had demanded a spring-loaded guard to comply with the state's old factory laws, but Clasmann's machine operator insisted that such an apparatus was more dangerous than a manual device. In the end, Clasmann left the manual guard on his machine. "I was not satisfied, and the inspector was not satisfied," he remarked. Yet Kearney could not advise Clasmann how "sensible" implementation of proposed orders would resolve his quandary.[35]

Some employers dismissed all safety orders as unreasonable. Stating that elevator regulations were not "worth a cent," Mr. Lindsay of Lindsay Brothers (Milwaukee) maintained that required elevator safety brakes never worked, that heavy safety gates were liable to give his operators hernias, and that his compliance with Milwaukee's elevator ordinances cost him two thousand dollars for little improvement. Kearney and Beck reassured Lindsay that the Wisconsin commission's special procedures would guarantee fair and "practicable" regulations. "When the commission gets through with this," Kearney told him, "you will be satisfied with our decision." "We will get in some experts on the elevator proposition and treat the subject as a whole." Beck added that the commission would never impose unreasonable requirements: "If we haven't got some expert who can come and tell you how to do it, and help you out . . ., [then] we haven't any business pushing it onto you."[36]

Kearney and Beck aggressively defended Wisconsin's proposed safety code at the hearing. They deflected criticism, refused exemptions, and exhorted employers to embrace the safe place law's higher new standards. Hence, when manufacturers resisted an order requiring guardrails on elevated walkways since such a precaution had never been used, Kearney defended the requirement. "The objection raised that it has never been done does not appeal to the committee," he argued. "There will be lots of things carried out that have never been done." Without inflicting hardship, he said, his panel intended to improve work conditions.[37]

In the end, Wisconsin officials responded to employer criticism guardedly. They dispatched deputies to study objections, and assembled subcommittees to review contested regulations. After accommodating new information, the industrial commissioners formally issued the General Orders on Safety. Revisions were minor. The commissioners modified Order No. 1 on belts, shafts, and chains by adding three "notes" to clarify the rule for belt-drive covers. They altered the emery-wheel order only to specify exhaust device construction. They left the walkway-guardrail rule intact.[38] Wisconsin's Industrial Commission thus reacted to criticism by refining proposed orders, not by lowering standards. By doing so, the agency affirmed its commitment to transforming work safety science. To satisfy what they regarded as the safe place statute's "prevailing good practice" standard, the commission replaced techniques based on old-fashioned shop culture with regulations utilizing industry's new science-based technology.

Wisconsin labor leaders endorsed the General Orders, though like unionists everywhere, they were skeptical about government protection. WSFL general organizer Frank Weber demanded safety regulation, and insisted that labor groups partake in its administration. "Organized labor," he announced, "stands ready to assist in reducing to a minimum physical injuries caused by industrial accidents." Yet Weber and other Wisconsin union leaders were political moderates willing to settle for practical results. He undoubtedly shared New York State labor federation president James P. Holland's view that "[we] want to protect [our] members . . . with all of the safety appliances that can be put in the factory or mill."[39]

Wisconsin's industrial commissioners placated labor leaders with unprecedented opportunities to participate in administrative proceedings. The commissioners placed labor representatives on the safety and sanitation committee and other advisory panels, and regularly communicated with union officials about pending regulations. Wisconsin officials did not, however, equalize labor's role in safety order making, assigning workers fewer committee seats than big manufacturers and technical specialists got. Like President Woodrow Wilson's labor program, Wisconsin's safety code system opened the policy process to organized labor only partway.[40]

Wisconsin safety order proceedings granted local businesses greater participation but largely used that representation to enlist local firms in cooperating with the commission's new safety science. As Commissioner George Hambrecht remarked in 1917, "We are always glad to get a kicker . . . on our committee, as when he is able to get a broad view of the subject[,] his attitude toward accident prevention changes, and we have a booster instead of a kicker."[41] As for advanced industrialists, however, Wisconsin officials usually deferred to their judgment.

Its rare willingness to waive or "modify" the General Orders demonstrated the Wisconsin commission's resolve to raise local manufacturing standards. By 1918, the agency had received only four such employer requests and approved just one. Simultaneously, Ohio's Industrial Commission answered numerous complaints from "older establishments" by "allowing such modifications or concessions as the conditions and circumstances will warrant." Meanwhile, New York's commission received 205 petitions for "variations" in 1917 and 81 in 1918, half of which it granted.[42] Despite all of the grumbling at public hearings, little initial business opposition to Wisconsin's General Orders arose. The state's Industrial Commission sustained those new standards vigorously.

The Incomplete Transformation

Wisconsin's General Orders symbolize famous aspects of Progressive Era reform: the administrative "transformation of governance" and the "rise of the states" as a bastion of new administrative authority.[43] Yet that transformation was incomplete. Administrative safety law spread unevenly, as most states, including Alabama and Illinois, neglected such a reform, while a few states embracing it chose not to apply administrative powers to mine regulation.[44] And where administrative safety law institutions did spread, they operated differently in a variety of economic contexts.

During the 1910s, only thirteen states aside from Wisconsin established safety code authority, and most either failed to exercise that power or used it only to issue voluntary guidelines or a few scattered orders. Even in "commission" states, administrators applied code-making power diversely. For instance, Pennsylvania's Industrial Board published a broad range of safety codes, whereas Massachusetts officials generally issued nonmandatory "suggestions," allowing inspectors to order workshop changes as they saw fit. As late as 1931, the latter state's labor commissioner affirmed that "Massachusetts is not a code state."[45]

Ohio's Industrial Commission also used its powers sparingly. Although it organized advisory committees to formulate industry-specific rules, the Ohio commission neither held public hearings nor issued formal administrative orders. Instead, the Ohio agency merely published factory inspection "guides," and had trouble enforcing them as a result. "We have experienced considerable difficulty in securing compliance with these orders," the commission admitted in 1917, "because of a feeling prevalent among the employers that these codes are not effective or enforceable until issued as a general order." Coupled with the Ohio Supreme Court's 1917 *Schorling* decision validating Ohio's safe place statute only as part the safety code process, the Ohio agency's practice of decreeing inspection "guides" rather than legally binding "orders" effectively nullified Ohio legislation's higher safety standards. This tepid administrative policy reflected big industry's dominance of Ohio safety regulation, and the Ohio State Federation of Labor's (OSFL) conciliatory attitude then. Rather than impose formal safety orders, Ohio's Industrial Commission deferred to large firms, many with in-house safety programs.[46]

California, by contrast, was a very strong "code state." Starting with factory and mine rules in 1915, California's Industrial Accident Commission promulgated regulations governing engines, laundries, window cleaning, air-pressure tanks, trench construction, electricity, sawmills, quarries, and other hazards. The state's Industrial Welfare Commission simultaneously issued orders regulating women's work hours and wages. California's decentralized manufacturing sector and distinctive politics encouraged such extensive administrative action. Unlike concentrated Ohio industrialists, California manufacturers were mainly small and medium-size proprietary firms needing state guidance on safety improvements. Meanwhile, the state's "intense anti-partisanship" strengthened political interest lobbies like CSFL unions, employer associations, casualty underwriters, and women's groups. California's Industrial Accident Commission adjusted to these realities by bringing manufacturers, workers, and insurance representatives together in safety order proceedings, and then by providing technical assistance to help the state's many small employers modernize their plants.[47]

New York was an even stronger "code state." By June 1919, the state's Industrial Board and successor Industrial Commission had published twenty-two sets of rules for machinery, elevators, steam boilers, mining, fire prevention, sanitation, and other dangers, including regulations to protect female workers in canneries and core rooms. The New York codes' unusual attention to fire, sanitation, and women's issues reflected the state's diverse economy and influential local interest groups, and its particular reaction to the 1911 Triangle Shirtwaist fire. Although New York industry included sizable foundries and large manufacturers, small enterprises with serious fire and sanitation hazards predominated, especially in New York City. Moreover, well-organized trade associations, reform societies, women's groups, and labor unions in Elmira, Buffalo, and New York City persistently brought workplace safety and sanitation questions to public attention. With its statutory requirement to provide worker and employer representation on both its industrial council and its advisory committees, New York's Industrial Commission uniquely accommodated interest groups.[48]

Wisconsin itself was a strong "code state," although not on New York's scale. During its first three years, Wisconsin's Industrial Commission issued safety orders covering machinery, boilers, elevators, building construction, and mines, and then from 1915 to 1920, it added fire-prevention regulations and then appliance codes governing electricity, shop lighting, refrigerators, and compressed-gas apparatus. Starting in 1913, moreover, the commission promulgated orders regulating women's work hours in pea canneries, and later it issued orders on women's hours, night work, and wages. Unlike New York's labor department, however, Wisconsin's agency did not publish separate ventilation, dust, and sanitation codes until the 1920s.[49]

The Wisconsin Industrial Commission's aggressive work safety code development in the 1910s reflected an unusually independent, technocratic response to local economic and political conditions. Wisconsin's interest-group dynamics facilitated this response. More like the politically adroit CSFL than the cooperative OSFL or militant New York unions, Wisconsin's WSFL and Social-Democratic Party labor coalition was an important catalyst for regulatory action, but it did not secure more involvement in administrative proceedings than bureaucratic consultation and a few advisory committee seats. Meanwhile, Wisconsin's big Milwaukee-area manufacturers may have adopted modern corporate welfare plans, but they enjoyed less influence than big eastern firms like Ohio's did, and they consequently welcomed the Industrial Commission's assistance to alleviate work accident costs.[50]

Freed from interest-group domination, bolstered by its 1911 political mandate, and liberated by the flexibility of contemporary administrative law, Wisconsin's Industrial Commission conducted code-making proceedings to serve state goals. Not implementing "industrial pluralism" per se, the agency assembled "representative" advisory committees to fulfill the state purpose of gathering expertise and securing local cooperation. Not succumbing to corporate "capture," the commission's safety code process utilized industry's "best practices" to reduce work accidents and mollify laborers. Distinctively, then, Wisconsin's unusually powerful state administrative apparatus replicated modern industrial safety techniques to attain social stability and economic prosperity.

The Wisconsin Industrial Commission in 1917, featuring Secretary Edwin E. Witte and Commissioners Joseph D. Beck, Fred M. Wilcox, and George P. Hambrecht. (Wisconsin Historical Society Image)

Victims of accidents caused by unguarded machine gears in Wisconsin, 1914. (Industrial Commission of Wisconsin, "Gear Accidents and Their Prevention," *Bulletin,* February 20, 1914)

John R. Commons, economics professor, labor law reformer, coauthor of the Wisconsin Industrial Commission Act, and industrial commissioner, 1911–1913. (Wisconsin Historical Society Image ID Whi 5115)

A trainman gets hurt boarding a locomotive, and a monorail operator suffers an electric shock. Two scenes from the 1917 motion picture *Preventable Accidents in the Lumber Industry,* produced by the California Industrial Accident Commission and local lumber companies. ("Review of Motion Picture," *California Safety News,* April 1918)

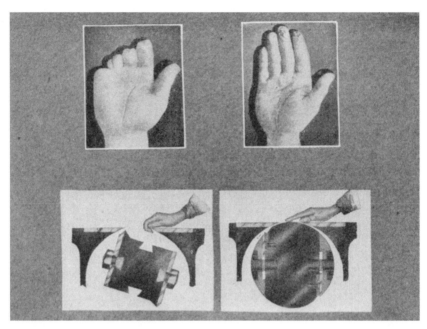

Illustration of Wisconsin General Safety Order No. 200 requiring safety cylinder heads on woodworking machines, 1913. (Industrial Commission of Wisconsin, "Jointer Accidents and Their Prevention," *Bulletin,* February 20, 1913)

Unguarded shafts, belts, and flywheels at the National Enamel and Stamping Company machine shop in Milwaukee, Wisconsin, 1900. (Wisconsin Historical Society Image)

The Puddler Mill Safety Committee, Youngstown Sheet & Tube Company, Ohio. A typical in-house committee of workers formed by big manufacturers during the 1910s to curtail industrial accidents. (Industrial Commission of Ohio, "Papers on Accident Prevention," *Special Bulletin,* March 1915)

Spray painting demonstrated at the Simmons Company, Kenosha, Wisconsin, in the 1920s. (Wisconsin Historical Society Image ID Whi 60854)

A young woman in a New York factory risks entanglement of her hair in an unguarded drill during 1923, potentially one of roughly five thousand child labor accidents in New York annually. (New York State Department of Labor, "Children's Work Accidents," *Special Bulletin No. 116,* January 1923)

Workers sprinkle the floor and atomize the air with water to cut down on silica dust at the Falk Company foundry in Milwaukee, Wisconsin, 1933. ("Good Housekeeping at Falks," *The Foundry,* February 1933)

Lead mine workers next to a steam engine with unguarded drive apparatus, Linden, Wisconsin, 1901. (Wisconsin Historical Society Image ID Whi 60853)

Women on a World War I–era assembly line at Nash Motors, Kenosha, Wisconsin, 1917. (Wisconsin Historical Society Image ID Whi 2044)

5

The Club of the Law

When Wisconsin and other states first enacted workers' compensation statutes during the 1910s, reformers expected that those measures would, like the German system, stimulate factory safety work. According to Wisconsin Industrial Commission special assistant C. W. Price, compensation provided "a tremendous incentive for the promotion of safety as a business proposition." And according to John Common and John Andrews in *Principles of Labor Legislation,* compensation exerted "a universal pecuniary pressure" like taxation that would induce employers to improve work safety conditions and speed recovery of injured laborers.[1]

This view of compensation's accident-prevention stimulus raised reformers' hopes for renovating the inept old factory-law enforcement process. Embracing the social insurance theory that work accidents were not just the fault of negligent employers but the product of an industrial system that treated job injuries as "externalities" to be borne by workers, reformers expected compensation to compel employers as a class to "internalize" accident costs and adopt cost-saving safety measures. Such financial pressures, reformers reasoned, would allow new agencies like Wisconsin's Industrial Commission to eliminate coercive old "command-and-control" factory inspection operations, and then deploy inspectors as state technicians to help employers reduce compensation expenses. As Commons said later, "Let the state furnish the [employer] with the inducements by taxing him proportionate to his employees' loss of wages by accidents, and then employ safety experts, instead of crime detectives and prosecutors, to show him how to make a profit by preventing accidents."[2]

Nowadays, commentators portray compensation reform as creating a relatively uniform new economic climate for safety work across the forty-two states that enacted such legislation in the 1910s, but this is misleading. Although cautioning that various state plans abided wage reductions, waiting periods, modest worker benefits, and benefit caps that weakened employer work safety incentives, modern analysts contend that compensation coaxed employers everywhere to embrace new safety conscious managerial attitudes and adopt new industrial safety practices.[3] Yet Wisconsin and other states made varying arrangements for compensation insurance, employer accountability, and factory inspection, creating quite different legal and financial environments for safety work from state to state. On insurance, for instance, historians generally argue that states endorsed either competitive private insurance or monopolistic state schemes, but it must be added that some states established mixed systems of private and "competitive" state insurance, or strong state insurance regulation.[4] Wisconsin, in particular, founded a unique regulatory plan.

Whatever their nature, local compensation insurance plans raised a problem when they replaced work injury litigation. Compensation may have strengthened employers' responsibility for accidents, but insurance allowed non-self-insured employers to minimize individual liability by pooling risks, shifting claims payments to insurance providers, and submerging accident losses under predictable annual insurance premiums. States consequently devised mechanisms to preserve individual employer accountability. Ohio and (at first) Illinois perpetuated workers' right to sue for injuries caused by safety law violations, while Wisconsin and California developed compensation penalty schemes that increased employer payments for work injuries resulting from such infractions. State programs also instituted compensation insurance "merit-rating" plans to hold employers accountable, particularly the "schedule-rating" system adopted in Wisconsin, and an "experience-based" program introduced in Ohio. The experience plan ultimately prevailed, but controversy raged over which version of merit rating best mobilized compensation insurance to induce safety work.[5]

Along with compensation reform, state labor law officials gradually professionalized factory inspection, although it still varied from place to place, interacting differently with various states' compensation insurance plans. In the "open" atmosphere of competitive private insurance, Illinois's inspectorate remained an old-fashioned corps of law enforcers. In New York State, where the state compensation fund competed with increasingly centralized private insurance providers, a huge inspection bureaucracy backed by organized labor developed. Amid a mixed and regulated compensation insurance system, California's inspection process emphasized cooperation but threatened legal coercion. With monopoly state insurance, Ohio deployed inspectors as technically trained law enforcers. Wisconsin, above all, treated inspection as adjunct to the insurance system. Relying on regulated compensation insurance rates to secure safety code compliance, Wisconsin's tiny staff of "deputies" stressed employer education.

The compensation era, hence, initiated an assortment of compensation insurance and safety law enforcement institutions around the country, not a uniform system. Among the states, Wisconsin established one of the nation's most highly regulated and interconnected insurance and inspection operations. This distinctive insurance-oriented work safety law enforcement system contributed greatly to modernizing Wisconsin workplaces, but it produced mixed results in lowering industrial accident levels.

Financial Incentives and Compensation Insurance Reform

When Wisconsin and other states first implemented workers' compensation statutes, those measures seemed truly to boost business safety consciousness, though a countervailing "moral hazard" of less cautious workers may have arisen too. By 1915, after Wisconsin law had begun assuming that employers with four or more employees had elected coverage, minus those who withdrew, the state's voluntary compensation program reached 13,000 firms, 250,000 workers, and 99.5 percent of the state's industrial accidents, despite exclusions. Safety work accelerated. Administrators reported surging business interest in state-mandated safety devices and growing employer creation of workshop safety plans.[6]

Raging behind these auspicious developments was a battle over compensation insurance. Except among self-insured firms, insurance was the compensation system's principal method of holding employers responsible for working conditions, with insurance costs' correlation to accident rates serving as the financial inducement for safety work. Yet states provided for compensation insurance in varied and contested ways. Wisconsin at first allowed manufacturers to purchase private insurance or self-insure, as did Alabama and Illinois, while New York, California and Ohio initially offered a choice of private, self- or state insurance. Ohio and a few other states eventually adopted monopolistic state plans. Almost everywhere, old-line stock insurance companies resisted compensation, since it fostered mutual firms, rate regulation, and state insurance funds that threatened their position.[7]

State-to-state differences and political strife created a chaotic insurance environment, not a stable one encouraging accident prevention work.

Some states preserved private compensation insurance. In Illinois, reformers, employers, academics, and the moderate Illinois State Federation of Labor (ISFL) united behind elective compensation legislation in 1911 that placated the militant Chicago Federation of Labor (CFL) by sustaining injured workers' right to sue employers and mollified large companies by allowing arbitrated settlements reviewable in court. In 1913, the ISFL and reformers secured another law over employer and insurance firm opposition that created an industrial board to adjudicate claims. Yet manufacturers and insurers blocked the ISFL's 1919 proposal for a monopoly state fund, thereby preserving employers' freedom to shop for lower premiums among competing private carriers.[8]

Alabama also remained an "open state." As elsewhere in the Deep South, Alabama reformers and even labor leaders avoided labor legislation that might discourage regional economic growth, leaving worker protection to paternalistic industrialists. Hence, Alabama enacted a weak workers' compensation law in 1919 in which the courts, rather than a commission, adjudicated claims, and in which the state regulated insurance only to prevent rate discrimination. Alabama employers chose between private stock or mutual insurance, while insurers established private rating bureaus.[9]

New York's turbulent politics, by contrast, expanded state involvement in compensation insurance, but even here competition prevailed. After the New York Court of Appeals struck down the state's original compulsory workers' compensation scheme in 1911, a constitutional amendment authorized a compulsory prolabor measure in 1913 that afforded high worker benefits, a strong workers' compensation commission, and a competitive state insurance fund. Yet after conservatives captured control of New York's legislature in 1914, social progressives affiliated with the American Association for Labor Legislation (AALL) joined business groups to consolidate compensation and other programs under a new industrial commission, and to authorize a "private agreement" system that resolved claims outside of the administrative system.[10]

Insurance law developments complicated the evolving New York program's inducements for safety work. First of all, 1912 and 1914 laws modeled after fire insurance statutes may have sanctioned compensation rate-making associations and required the state insurance commissioner to approve rates, but they still allowed employers to shop for bargains among self-insurance, mutual plans, private casualty companies, and the state fund rather than stress safety precautions. Second, New York's plan did not administratively link compensation insurance rates to industrial safety codes. Employers faced conflicting insurance and safety code requirements, breeding cynicism toward both.[11]

California regulated compensation insurance under controversial circumstances. Reformers originally expected California's voluntary 1911 compensation law to replace fault-based work injury litigation with a public welfare approach to job accidents, but insurers' pricing tactics discouraged employer cooperation. When California then established compulsory compensation and a competitive state insurance fund in 1913, competition raged between stock, mutual, and state insurance, largely because insurers manipulated rates and sometimes left benefits unpaid. In 1915, hence, California legislation eliminated rate competition by ordering the state insurance commissioner to issue or approve uniform classifications and premiums for the state fund, private carriers, and emerging rate-making associations alike. Though the commissioner usually approved rather than initiated rates, he did guard rate equality, as when he denied insurers' request for a 5 percent increase during World War I. Uniformity bolstered the California compensation law's inducements for safety work by eliminating premium shopping.[12]

Ohio also standardized compensation insurance rates. After Ohio lawmakers created an elective compensation plan in 1911 that allowed manufacturers to choose between self-insurance, private insurance, or a state program, a 1912 constitutional amendment allowed the reform-minded 1913 legislative to establish a compulsory, monopolistic state insurance fund. Unlike situations elsewhere, Ohio industrialists, labor leaders, and Progressives all favored a centralized state plan to exclude profit-oriented stock insurance. For their part, stock companies unsuccessfully sought the Ohio plan's repeal, and then sold policies under friendly legal rulings. Finally, Progressive Democratic governor James Cox secured legislation in 1917 banning private compensation insurance altogether. With competition gone, Ohio's state fund uniformly induced safety work.[13]

More like California than Ohio, Wisconsin abolished compensation insurance competition in response to stock firm agitation, and did so through regulation. Wisconsin law initially created a competitive insurance environment like those in Alabama and Illinois, but stock companies attempted to sabotage that program and undermine emerging mutual firms like Employers Mutual Liability Insurance Company of Wausau by manipulating rates and discriminating among customers.[14] In 1915, moreover, stock firms lobbied to oust industrial commissioner Charles Crownhart, a vigorous compensation law administrator and energetic supporter of fledgling mutuals. Conservative governor Emanuel Philipp subsequently failed to renominate him, leaving fellow commissioner Joseph Beck dismayed at how "special interests" never failed "to go the limit, blindly and bull-headedly go the limit." Such instability undercut the Wisconsin compensation system's capacity to induce safety work.[15]

Stock company shenanigans inspired action that paralleled developments in the life and fire insurance fields. Wisconsin reformers chose first to nurture local mutual firms like Employers Mutual, which the state's Industrial Commission supported as a regulatory partner. Then Wisconsin lawmakers instituted compensation insurance rate regulation. In 1913, the legislature banned rebates, prohibited rate discrimination, and instructed insurance companies to file classifications and rates with the Industrial Commission. Perhaps imitating Pennsylvania's plan, Wisconsin lawmakers next in 1917 required insurance firms to join the privately run Rating and Inspection Bureau to formulate classifications and rates that were then approved by the new Compensation Insurance Board. Consistent with Progressive Era Wisconsin's "positive service state," these measures replaced Wisconsin's free compensation insurance market with a tightly regulated system. Casualty insurance became a quasi-public utility in which private insurers charged standard state-sanctioned rates.[16]

Wisconsin's Rating and Inspection Bureau calculated rates in familiar ways with important modifications. The bureau divided industry into roughly six hundred classifications and used the accident experience of each to project basic insurance costs, or "pure premiums," to which it added profits and expenses to determine "rates" charged to employers. Theoretically, insurance rates' sensitivity to accident trends, both within and across classifications, would encourage companies to reduce accident costs. To overcome risk spreading's moderating effects on rates, moreover, Wisconsin law appended "schedule rating" to the system in 1917. First appearing in 1913, "schedule rating" was Wisconsin's and California's initial form of "merit rating," unlike Ohio's "experience rating" and New York's combination of the two. It allowed insurers to lower employers' premiums for guarding machinery and increase premiums for neglecting safeguards. Wisconsin State Federation of Labor (WSFL) general organizer Frank Weber endorsed schedule rating, and Wisconsin administrators proclaimed its success, especially after the Rating and Inspection Board started to base schedule rates on safety code compliance in 1919. Insurance industry reports subsequently confirmed the schedule plan's effectiveness in reducing point-of-operation machine accidents.[17]

Ohio state compensation fund administrators countered, however, that "experience rating" was a better merit plan. Under their scheme, Ohio's Industrial Commission calculated each manufacturer's standard insurance premium, and then subtracted financial credits or added penalties, depending on whether injury and death rates fell below or above predicted levels. More flexible than schedule rating, the experience plan reputedly "gave recognition to an employer for the production of a good experience," while it penalized "the excess losses of the careless employer." Ohio tempered experience rating, however, by lowering the penalty for deaths and by imposing a 24 percent cap on increases during any six-month period. The experience-based merit system also abided moral hazards among insurers and employers who neglected factory safeguarding, contested claims, and underreported injuries. Still, American industry's shift from engineering to personnel accident-prevention methods caused most big employers, insurance carriers, and state compensation funds to embrace Ohio's experience-based approach by 1940.[18]

Wisconsin resisted this trend. In 1921, Progressive Republican governor John J. Blaine vetoed legislation authorizing the experience plan, agreeing with labor leaders that it would encourage manufacturers to avoid accidents by hiring agile younger workers in place of older union men rather than by installing legally mandated safeguards. Unlike Ohio, where big industrial plants invited experience rating, Wisconsin's comparably smaller industrial operations, stronger labor politics, and pervasive suspicion of out-of-state insurance firms sustained schedule rating's direct workplace regulation. Blaine's veto thus preserved insurance practices that fostered safety code compliance.[19]

Along with insurance rating schemes, Wisconsin and other states also developed substitutes for employers' liability previously sustained by workers' common-law right to file work injury lawsuits, a right removed when compensation became the "exclusive" remedy for job accidents. Some states preserved workers' right to sue under special circumstances. In Illinois, CFL leaders demanded perpetuation of that right to make it "so expensive for employers to kill their workmen that every safety appliance known to science will be installed." Illinois lawmakers subsequently retained the right in 1911 compensation legislation for cases involving willful employer violation of safety statutes, though Illinois employers had that provision abolished two years later.[20] Ohio lawmakers, meanwhile, made compensation injured workers' exclusive remedy in 1911 and 1913 compensation acts, but accommodated labor with an exception. When injury resulted from "wilful" employer actions or safety law infractions, Ohio law allowed workers either to accept compensation or sue in court, provided that employers regained common-law defenses.[21]

Wisconsin itself allowed worker lawsuits for "wilful" employer misconduct, but then substituted an administrative plan. In 1913, Wisconsin's legislature voted to penalize employers with a 15 percent increase in compensation payments for accidents caused by safety law violations or "wilful" employer neglect, and to penalize laborers with a 15 percent reduction in benefits when accidents resulted from their own "wilful" failures. In 1917, lawmakers provided further that employers, not insurers, pay compensation increases, guaranteeing direct employer penalties for defying the law. Obscure in its origins, Wisconsin's 15 percent penalty clause focused financial pressure on derelict firms. The scheme was not implemented until 1919, but it instantly caused insurance companies to warn policyholders about penalties, and immediately prompted businesses to request copies of safety orders.[22]

California developed a similar penalty clause. Initially, like Ohio law, California's 1911 workers' compensation statute allowed injured workers either to claim compensation or sue in cases where employers' personal "gross negligence," "wilful misconduct," or safety law violations caused accidents. In 1917, however, California lawmakers substituted a new remedy: in cases of "serious and wilful [employer] misconduct," injured workers recovered 50 percent increased compensation up to twenty-five hundred dollars payable by employers rather than insurers. Imitating Wisconsin, California also reduced worker benefits by 50 percent for "serious and wilful misconduct."[23]

Although other states duplicated particular features of Wisconsin law, Wisconsin's program consolidated compensation and safety law administration exceptionally. Wisconsin's quasi-public method of fixing uniform compensation insurance rates, its schedule-rating plan, and its 15 percent penalty clause all focused compensation insurance on inducing manufacturers to install state-mandated safeguards. Other states' programs were comparatively incomplete. Alabama, Illinois, and even New York maintained competitive insurance systems that vitiated compensation inducements for safety law compliance. Ohio's monopolistic state fund implemented uniform rates, but payment caps, experience rating, and the lack of binding safety codes mitigated their stimulus. California's rate regulation program and compensation penalties paralleled Wisconsin's plan, but they failed to integrate insurance rate making and safety order formulation. Workers' compensation insurance reform thus sustained a patchwork of local institutions with uneven incentives for safety work. Diverse insurance plans affected local factory inspection.

The New Approach to Factory Inspection

Because compensation laws created a new financial environment for safety work, many reformers hoped to recast factory inspection. Reformers hoped to break away from the cumbersome kind of inspection programs that persisted in places like Illinois and prevailed in state coal mining regulation. Despite expansive new laws enacted after 1900, state mine inspectorates in particular remained underfunded, spotty in qualified personnel, lacking in enforcement powers, and weak in diligence. In Wisconsin's 1911 Industrial Commission Act, by contrast, reformers attempted to revolutionize inspection operations in the new post of "deputy." In some ways, deputies remained like the old inspectors: they were to enter workplaces and examine safety and health conditions for law violations. Yet whereas factory laws designated inspectors as all-purpose police officers who could order employers to "guard and protect" machinery, the new commission law defined deputies as persons with "special, technical, scientific, managerial or personal abilities" who would help employers meet legal requirements. Deputies' job, Wisconsin's commission declared in 1914, "is *not* to ferret out points of danger and to tabulate them, but it is chiefly to do *constructive educational work.*"[24]

Given deputies' "educational" mission, and their freedom from the kind of union pressure that organized labor exerted in New York, Wisconsin's industrial commissioners tried to cultivate the new inspectorate as an "apolitical" corps of technicians. To deflect employer criticism that deputies were "labor politicians," the commissioners diligently followed civil service requirements that prospective deputies pass competitive examinations (with exceptions), and they rejected outside business or labor connections that threatened deputies' impartiality.[25] Commissioners also systematized deputies' inspection routine. They directed deputies to follow careful itineraries, fill out "certificates of inspection" for every plant visited, and submit weekly reports recounting hours on the job. They briskly dismissed deputies who neglected these instructions.[26]

Wisconsin commissioners groomed deputies to be educators. They sent deputies to safety lectures, ordered them to visit the best plants in Wisconsin and Chicago, gave them a full set of agency publications, and held regular meetings to go over problems and exchange ideas. The commissioners especially coached holdovers from Wisconsin's Bureau of Labor Statistics to adopt the "friendly approach." Old-timer Dave Evans warned administrators that manufacturers would disregard safety work unless state officials used "the club [of the law]," but Commissioner Joseph Beck advised him to be more tactful.[27] As everywhere, of course, Wisconsin's inspection process combined "coercion" with "education." Yet, on balance, Wisconsin's plan downplayed coercive "command-and-control" enforcement methods and emphasized employers' technical indoctrination.

Other states' labor law officials struck different balances. Ohio law duplicated Wisconsin's "technical" and "scientific" post of "deputy," but Ohio's Industrial Commission treated deputies mainly as enforcers of Ohio's General Code and the agency's unofficial safety guidelines.[28] California law, meantime, classified inspectors specifically as labor law enforcers, and beginning in 1919 authorized them to "tag" and decommission dangerous machinery while the state commission sought court injunctions to shut such equipment down. Still, California's commission directed inspectors to inform employers about the state's safety museum and pass out safety literature. As commission counsel Warren Pillsbury observed, "Educational methods and the good will of employers and employees" secured more safety code compliance than criminal prosecution.[29]

New York factory inspectors also mixed education with coercion, though administrators favored the latter. Inspector J. J. McSherry contended that an effective inspector was a "student of accident causes" who showed manufacturers why accidents occurred and how safety code compliance prevented them. William Gorman maintained that inspectors had to help employers to incorporate safety into their "industrial philosophy." John Hanlon said that inspectors had to convince factory owners that code compliance improved plant efficiency. New York's Factory Inspection Division, however, accentuated legal pressure. Officials directed inspectors to cite code violations and conduct "compliance visits" until manufacturers installed the required devices. New York administrators also prosecuted more employers than other states' officials did, and began a "tagging" process like California's in 1921.[30]

Illinois perpetuated a comparatively antiquated inspection operation. Although Illinois lawmakers created a state inspectorate under the 1893 sweatshop act, they did not establish a factory inspection department until 1907, and did not consolidate it with the state's Workmen's Compensation Commission and Bureau of Labor Statistics until 1917. Despite civil service laws, political patronage dominated the selection and supervision of Illinois inspectors, especially in Cook County. With little training, inspectors performed perfunctory factory visits. In 1930, an AALL investigator still complained that Illinois inspectors did no educational work, and that they shockingly lacked engineering expertise.[31]

As was true with workers' compensation insurance, then, lawmakers and administrators from state to state cultivated distinctive factory inspection regimes. Indeed, each state's inspection system intertwined with its particular insurance arrangement to create a unique legal environment for safety law enforcement. In New York and Illinois, where competitive insurance attenuated compensation's inducements for safety work, labor law officials inclined toward coercive inspection tactics. In California's regulated insurance market, administrators mixed compensation penalty, "tagging," and injunction procedures with cooperative inspection techniques. In the context of monopolistic state compensation insurance, Ohio's specially qualified "deputies" executed a routine but brisk inspection program, despite Ohio safety orders' nonbinding character. And building on tightly regulated compensation insurance rates and a 15 percent compensation penalty, Wisconsin deputies served as specialized technicians who used persuasion and education to secure employer safety code accordance. Indeed, Wisconsin interwove insurance incentives and educational inspection in a uniquely powerful way.

Factory Inspection and Accident Prevention

During the 1910s, Wisconsin and other states launched their various new safety law enforcement programs in local economies at different stages of industrial advancement. In Wisconsin, administrators gradually professionalized the state's factory inspectorate in a slowly modernizing economy that featured traditional upstate lumber and furniture plants as well as growing safety conscious Milwaukee-area machinery, iron and steel, brewing, and meatpacking factories. Retaining eight inspectors from the state's old labor bureau, Wisconsin's Industrial Commission incrementally hired building, boiler, illumination, mining, hygiene, and fire prevention specialists. Only in 1916 did it create a separate safety and sanitation division and, under pressure from women's groups, a separate woman and child labor division with female staff members. In 1917, Wisconsin employed twelve safety and sanitation deputies, one for every 22,000 workers.[32]

Other states' inspectorates varied. In Alabama, where iron and steel and textile industries grew, and mining disasters caused the appointment of mine inspectors, a 1907 factory law merely directed the jails and almshouse inspector to report on sanitary and child labor conditions in cotton mills, a duty repealed in 1911.[33] In Illinois, amid big Chicago firms aligned with the antiregulatory Illinois Manufacturers Association, an outmoded factory inspection department deployed thirty inspectors (one for every 22,000 workers) to enforce ten separate health, safety, woman, and child labor laws without specialization, while the Department of Mining and Minerals picked up mining inspection starting in 1917.[34]

In very different economies, California and Ohio professionalized their staffs. Confronting a surging but still decentralized manufacturing sector lacking many big, advanced firms, California's Industrial Accident Commission added building, mining, elevator, and boiler experts to five regular inspectors, providing just one inspector for every 24,730 factory workers. California's Industrial Welfare Commission, backed by women lobbyists, enforced female protective laws.[35] Meanwhile, in an economy dominated by large corporate manufacturers sympathetic to welfare plans but still featuring decentralized urban trades like garment making, Ohio's Industrial Commission separated inspection into mining, boiler, steam engine, and workshop divisions. The workshop division employed 3 bakeshop inspectors, 1 explosives expert, 25 regular deputies, and 8 "lady visitors," one inspector for every 20,000 workers.[36]

In a sprawling economy exhibiting big industrial plants in Buffalo, Elmira, and elsewhere as well as many small printing and clothing shops in New York City, New York State's inspectorate was large and bureaucratic. New York law established the Bureau of Inspection in the state's new labor department (reorganized as the Industrial Commission in 1915), and authorized the Bureau of Women in Industry by 1918. New York statutes partitioned the inspection bureau into homework, mercantile, hygiene, and factory (including mining) divisions, and classified inspectors into seven civil service grades based on seniority and expertise, with the joint worker-employer industrial council approving all inspectors' appointments as of 1915. Through this bureaucratic maze, the state Industrial Commission assembled a force of 125 factory inspectors, 20 mercantile inspectors, and 10 boiler inspectors, "not more" than 30 of whom could be women. New York thus employed one inspector for every 10,000 factory workers, a sign of trade unionists', women's, and social progressives' influence.[37]

Paltry in number, by comparison, Wisconsin deputies visited only a fraction of state workplaces in the 1910s. Of nearly 10,000 industrial operations (including mines and quarries), deputies inspected only 2,658 sites in fiscal year 1913–1914 and only 2,879 in 1914–1915. Many shops "have never been inspected," Wisconsin's Industrial Commission admitted, and large plant visits were "infrequent." Municipal inspectors and part-time "special inspectors" only partially filled the gap. When they did inspect, though, Wisconsin deputies focused on mechanical safeguarding. Out of 51,800 labor law violations cited during 1913–1914, 31,996 involved belts, chains, clutches, and other transmission gear, while elevators (8,690), sanitation (4,892), and woman and child labor protection (399) preponderated the rest. Total citations dropped to 26,432 in 1915 and declined thereafter, but transmission devices and elevators still led violations.[38]

Labor departments elsewhere set different priorities. Illinois's Factory Inspection Division conducted more than 70,000 inspections during 1918–1919, mostly to enforce woman and child labor laws. Still, as in Wisconsin, Illinois inspectors issued some 4,000 citations for unguarded transmission apparatus.[39] In Ohio, meanwhile, deputies visited nearly half of their state's 16,000 industrial establishments in 1915, but issued only 9,618 "requirements" ordering safety law compliance. Perhaps concentrating on urban workshops rather than big factories, Ohio deputies attended more to sanitation and fire prevention than Wisconsin inspectors did, issuing only half of their citations for risky machinery.[40]

Back in Wisconsin, deputies had good reason to stress mechanical safeguarding. Conditions were hazardous, especially outside of Milwaukee. "It's a wonderful thing that there are not a lot more serious accidents than there are," deputy Ira Lockney said about small-town factories. Moreover, deputies met manufacturers who dismissed them as political hacks, rejected state regulation, and condemned state leaders. Commissioner Beck instructed deputies to avoid politics and concentrate on inspection, but some found employer intransigence to be unbearable. When Al Kroes inspected the Sanitary Bedding and Hammond Company in Marshfield in 1914, manager C. H. Stuck "denounce[d] 'Bob' [La Follette] and all he had done." Griping that everything coming from state government was unnecessary, Stuck remarked that he wanted to "throw [La Follette] in the poor house." "Then I started at him," Kroes admitted. "I am but human." As for Stuck's factory, it was "the poorest plant" that Kroes had seen.[41]

To overcome such resistance, deputies offered technical assistance. Deputy John Humphrey reassured factory owners that he was their "servant," not their "boss." He and other deputies gave employers safety manuals, displayed safety devices, described advanced firms' programs, and outlined safety work's financial benefits. "When a kick comes as being a political job holder," deputy Dave Evans explained, "I try to explain the amount of reduction in accidents [and] hours lost" that safeguarding would bring. If employers claimed that safety law compliance was too costly, he offered to demonstrate safeguards' practicality. "When employers object to guarding," Evans advised his fellow deputies, "name a place where the guard is in use;

[explain] that without a guard it raises their insurance rate; [suggest] that without the guard [employers] would be assessed 15 percent extra compensation if a person was injured for want of a guard; and show them how a guard can be applied as seen elsewhere."[42]

Persuasive tactics had problems, however. Wisconsin deputies did not become technical educators overnight, and one observer found them in 1914 still to be the "negative type" who inspected for code violations, unlike Ohio deputies "whose [technical] services are solicited." Indeed, after some Wisconsin deputies initially told employers just to "read the rules," Commissioner Joseph Beck admonished them to offer more technical assistance. Wisconsin deputies also learned that technical advice was disputable. In 1913, for example, Ira Lockney found a "small hidden gear" at one shop that was "not a bad exposure and not a very great hazard," but he recommended encasing it despite company objections. At another factory, Lockney proposed a disc to cover "slow-moving spokes of a power saw," although the firm protested.[43]

Deputies sometimes had trouble persuading manufacturers that practicable safety devices were available. Consequently, Wisconsin's Industrial Commission gave deputies photographs and a sample of standardized safeguards for display. To develop new appliances, moreover, the agency directed deputies to study every accident to determine "what kind of guard would probably have prevented the accident, . . . how much such guard would cost, . . . [and] whether any guard has been provided to prevent recurrence of such accident, etc."[44]

Wisconsin's commission also coaxed machine makers themselves to invent safeguards with mixed success. Local firms like Fairbanks-Morse (Beloit) sold protected machinery, but out-of-state suppliers resisted Wisconsin standards. A Vermont company told the commission that a cover on the belt and gears of its shearing machine was unnecessary, since "states here in the east do not require it." Engineer R. J. Solensten replied that Wisconsin safety codes still stipulated complete belt and gear guarding, and that machines meeting Wisconsin standards would satisfy laws anywhere else.[45]

Given its emphasis on education, however, Wisconsin's commission was flexible about employer adherence to state rules. Officially, the agency required strict and immediate safety code compliance: it demanded gear guards fifteen feet above the floor "where no one ever goes," insisted on safety goggle use, though molders liked to remove them, and ordered deputies to enforce elevator rules not yet effective, due to dangers presented.[46] Yet Wisconsin's inspection process relaxed standards in practice. Deputies interpreted safety rules in conflicting ways, cited only "unduly hazardous" violations, and occasionally did sloppy work.[47] The commission itself sometimes delayed enforcement of safety orders or suspended them for unusual circumstances, including World War I.[48]

Despite this educational and flexible approach to code enforcement, some "foxey old" employers tried to evade the law. "Many times the person in charge will let violations go," deputy Joseph Vallier reported, "in the hope that they will be passed over in time."[49] Wisconsin's commission addressed such defiance with gradually escalating pressure. If employers failed to remove hazards while deputies remained on company premises, the agency mailed follow-up correspondence and ordered reinspections. Then, if necessary, commission officials met with company officers to arrange a timetable for compliance. Only as a last resort did the commission take legal action, prosecuting only four employers for safety code violations from 1912 to 1917. "It has been the aim of the commission to administer the laws without unnecessary severity," the agency announced.[50]

Some manufacturers exploited this leniency, as did the Rundle-Spence Manufacturing Company, a Milwaukee plumbing-supply firm. After inspections starting in 1913 produced no improvements, Wisconsin's commission scheduled a meeting with company officers, but they did not show up. In 1919, a reinspection showed conditions to be as bad as ever, and the agency initiated prosecution. Rundle-Spence subsequently claimed to have made corrections, but another inspection showed this not to be true. The company then dragged out court hearings and finally paid a one hundred–dollar forfeiture in 1921. Still, the firm did not comply, and the commission commenced prosecution again in 1925. Prosecution was a meek enforcement tool in Wisconsin, a poor backup to compensation's financial inducements.[51]

Other states' labor law programs similarly avoided safety law prosecution. California's Industrial Accident Commission reported "infrequent" legal action and rare use of injunction proceedings, though the mere threat of enjoining unsafe machinery apparently chastened many employers. Even Illinois's old-fashioned Department of Labor only sporadically prosecuted firms for violating state safety legislation, despite hundreds of yearly woman and child labor lawsuits. Ohio's Industrial Commission prosecuted just one firm for safety law violations in 1915, while indicting sixty-nine others for woman and child labor law infractions.[52]

New York's inspection bureaucracy, however, relied heavily on compulsory processes, a result of organized labor's exceptional influence and inspectors' difficulties with urban workshops. By law, New York inspectors conducted thousands of yearly "complaint investigations" triggered by worker grievances, most involving sanitation or woman and child labor infractions, but some entailing fire and accident hazards. New York's commission also prosecuted thousands of employers for labor law violations annually, though how many for safety law noncompliance is unclear.[53]

Wisconsin's educational strategy may have seemed less potent than this, but deputies repeatedly supplied evidence that persuasion produced results. Employers might initially ignore safety requirements, deputy August Kaems reported in 1913, "but when this subject is put before them and explained to them, I do not know of a single case in which they did not get interested at once." Ultimately, the Merchants and Manufacturers Association of Milwaukee commended the Wisconsin agency's work, as did the state laundrymen's association. "We recognize the new order of things," one employer announced. "It requires no argument to convince us that great good has already been accomplished and we welcome the change."[54] Wisconsin administrators assumed that such new attitudes would reduce accident levels, but that expectation proved overly optimistic.

The New Enforcement Regime and Accident Rates

In its 1916–1917 annual report, Ohio's Industrial Commission identified a big administrative problem facing all safety law enforcers. "The [factory inspection] division has been seriously handicapped in [its] accident prevention work," the agency reported, "in that no reliable statistics, regarding accidents occurring in the state," were available. This was true everywhere. Workers' compensation programs improved accident tabulation during the 1910s, but few state agencies provided complete information or calculated frequency (accidents per number of workers or work hours) and severity (workdays lost per accident) rates. Few organizations published reliable nationwide data before 1926, although the U.S. Bureau of Mines began gathering coal mining fatality rates from state data, and the U.S. Bureau of Labor Statistics started to assemble information on steel-industry accidents. Overall, statistics in the 1910s revealed little about safety law's impact on the accident toll. Modern historians consequently discount safety law administration's role in reducing accident levels, attributing improvements mainly to compensation laws and private-sector developments such as workforce stability, factory electrification, and plant safety programs.[55]

Scattered state statistics imply, however, that state safety regulation had an ironic effect on industrial work conditions in the 1910s: it likely contributed to the improvement of factory engineering but failed to reduce accident rates in manufacturing significantly. From state to state, administrators and inspectors reported advances in safeguarding. Wisconsin deputies cited two-thirds fewer law violations in 1917 than they had detected back in 1913, whereas Ohio deputies halved their number of inspection orders between the mid-1910s and early 1920s. The drop suggests that most Wisconsin and Ohio employers obeyed safety laws and installed protective devices.[56]

Nonetheless, accident levels apparently remained high through the 1910s. Though state fatality reports, a good statistical measure, were too fragmentary to gauge accident rates for that decade, statistics still corroborate economic historian Mark Aldrich's view that "there were probably only very modest improvements" in manufacturing accident levels until the 1920s. Without singling out fatalities (about 1 percent of all mishaps), Wisconsin's Industrial Commission reported that accidents haltingly declined between 1915 and 1919, only to rise again in 1920. For its part, the California commission cited a gradual increase of accidents from 62,211 in 1914 to 108,947 in 1919, with growth in temporarily disabling injuries and decline in fatalities. Ohio's agency recorded only 53,368 accidents during 1913–1914, but then tallied 105,667 in 1915–1916, 185,223 during 1916–1917, and 171,691 in 1917–1918. Simultaneously, fatality rates in coal mining also remained high, but that was likely due to the distinctive characteristics of that industry, such as the enhanced danger of explosions brought by mechanization and electrification, the continuance of deep mining, neglect of roof collapses, and an industry-wide decline that encouraged operators and miners alike to boost production at the cost of safety.[57] By comparison, injury levels in manufacturing establishments seemingly should have dropped more readily in response to improved safeguarding.

Why did factory accidents persist despite improving employer safety law compliance? Booming World War I factory employment might explain this contradiction, but a 1921 Wisconsin study revealed a more complicated scenario. Innovatively calculating monthly accident "frequency" rates, Wisconsin's Industrial Commission determined that increased prewar hiring partly explained rising accident levels but that accidents grew faster than wartime employment, injury frequency dropped slower than employment did afterward, and, surprisingly, accidents surged again in late 1919 and 1920 when the labor force shrank. To explain these developments, Wisconsin administrators reasoned that the hiring of "green hands" boosted the wartime accident toll. They also deduced that in slack postwar years, manufacturers worsened accident risks by "speeding up" work processes, while unemployed laborers filed more claims for minor injuries. Most important, Wisconsin officials determined that although machinery-induced accidents (a major cause of fatalities) were dwindling, "minor" injuries produced by simple causes such as "handling objects" and "falls of persons" were increasing.[58]

Other states verified Wisconsin's findings. In 1916, New York's Industrial Commission reported that machinery accounted for a large proportion of factory accidents (26.7 percent) and nearly half of fatalities (42.3 percent) put that "weights and falling objects" (24.4 percent), "hand tools" (10.8 percent), "falls of persons" (9.6 percent), and other nonmechanical causes explained most injuries. Significantly, the New York agency discovered that whereas most machine injuries involved unprotected apparatus, a large minority (38.4 percent) occurred on fully guarded equipment.[59] Meanwhile, California's Industrial Accident Commission discerned that personal falls and falling or flying objects produced more lost work time and compensation payments than machinery did. Simultaneously, Ohio's Industrial Commission found that falling and shifting objects caused more accidents than did machines.[60]

Crude as they were, statistics indicated that various states' safety law enforcement regimes had a complex effect on factory working conditions in the 1910s. While workers' compensation, corporate safety plans and private sector forces stimulated prevention work in large plants, Wisconsin's enforcement operation probably enhanced mechanical safeguarding among out-of-the-mainstream small-town upstate employers, just as California's agency probably influenced predominantly mid-size companies, and Ohio and New York agencies likely succeeded among small urban firms. Statistics happily suggested that such engineering improvements did indeed diminish severe and fatal injuries, but they also revealed that safeguarding alone would not prevent all accidents, and that nonmechanical risks and "minor" injuries were on the rise. These developments would bedevil regulators in Wisconsin and elsewhere after 1920.

6

Politics and Work Safety Education
in the Interwar Economy

After World War I, Wisconsin's Industrial Commission surrendered industrial safety publicity to the business community. The commission suspended the *Wisconsin Safety Review,* issued fewer technical bulletins, abandoned photo displays, quit showing safety movies, and failed to replace publicist C. W. Price with an education director until 1928. Only in speeches, news releases, and support of business activities did the agency still advertise safety. Subsequently, insurance companies intervened with safety magazines, shop talks, and safety movies, while urban manufacturers organized safety schools and conferences. Wisconsin's commission patronized these business efforts. By doing so, it seemingly repositioned itself as a service agency in line with the "associative state," Herbert Hoover's popular notion in the 1920s that urged government power to "reinforce private ordering."[1]

Crucially, the Wisconsin commission's retreat from safety education diminished its ideological influence on labor relations. In the 1910s, the agency's safety first campaign had not only exhorted manufacturers to assume responsibility for working conditions and obey safety codes but also broadcast an ameliorative approach to employment relations that emphasized paternalistic government and business protection of workers. After 1920, however, the commission's withdrawal from educational work reduced government promotion of amelioration, allowing corporate personnel practices and later industrial unionism to define employer-employee relations instead.

The decline of the Wisconsin commission's educational and ideological functions seems to confirm historical theories that shifting economic power crippled Progressive regulatory programs after World War I. "Corporate liberal" scholars argue, for instance, that Progressive regulatory reforms initially favored business, and still did so in the 1920s. "Capture" theorists maintain that corporate interests used their power to seize control of administrative agencies after the war, if not before. Disciples of the "organizational synthesis" assert that regulators aligned with corporate technocrats to govern society by the war's end. "Pluralist" historians contend that an array of interest groups contributed to regulatory plans but concede that compromises left regulatory bodies too weak to counter business influence after 1920.[2] All of these theories suggest that Wisconsin's commission sacrificed its educational mission in deference to growing business dominance and business programs.

In fact, broader political and economic developments reduced Wisconsin's educational campaign. In the 1920s, with labor influence dropping and Progressivism faltering, Wisconsin's political leaders embraced conservative fiscal policies that prevailed everywhere, causing them to temper Industrial Commission leadership and constrict agency budgets. Then in the Depression-ridden 1930s, budgetary retrenchment and political change altered commission priorities. Educational expenditures were cut. Meanwhile, industrial developments transformed labor relations and robbed state-sponsored safety education of its appeal. In the 1920s, when modest job growth, rising mass-production employment, open-shop drives, and antiunion court decisions all militated against labor unionism, big firms and industry associations promoted their own safety and welfare schemes to improve working conditions. When the Depression brought massive unemployment and strident labor militancy in the 1930s, reformers did not revive ameliorative Progressive programs or seek a national plan like the modern Occupational Safety and Health Administration, but instead accommodated workers with collective-bargaining legislation. This policy shift from protective labor measures to corporate welfare and finally to publicly regulated unionism gradually marginalized Wisconsin's educational work and its ameliorative philosophy.

Nationwide, however, state-sponsored safety education continued to vary from place to place. Laggard states like Alabama and Illinois still promoted no educational activity, while New York, California, and Ohio labor departments sustained educational programs, sometimes into the 1930s. The demise of Wisconsin's robust educational campaign was thus exceptionally abrupt, a reflection of how localized safety policy remained.

The Politics of Restraint

During the 1920s, support for state labor law expenditures eroded all around the country. Overpowering labor and reform lobbyists, business interests discouraged government programs that hampered private enterprise, though this did not simply cause retrenchment but shifted priorities to education, highways, and parks.[3] In Wisconsin and other states, the new mood stagnated labor department budgets and reshaped administrative leadership. Progressives and organized labor retained just enough political influence to keep work safety programs intact.

In Wisconsin, support for state administrative spending began to weaken in 1915. Republican governor Emanuel Philipp and conservative lawmakers trimmed Industrial Commission appropriations so much that the agency could not hire enough deputies to enforce safety laws or keep prized employees like Al Kroes on the state payroll. World War I stretched the commission even more thinly. Industrial expansion increased the need for factory inspection, while the legislature assigned the agency new duties. Even so, Philipp declined emergency-funding requests.[4]

In 1919, however, the Industrial Commission got a reprieve. A coalition of reformers, workers, socialists, and German Americans elected a strong Progressive legislature, while postwar inflation and employers' union-busting activities stimulated labor unrest. Subsequently, the Wisconsin State Federation of Labor (WSFL) and socialist legislators invited commission chair George Hambrecht to testify about his agency's budgetary plight, leading to a doubling of appropriations, despite objections from Philipp and his supporters. Though increases went partly to other programs like woman and child labor, safety expenditures rebounded. The Safety Department hired eleven new regular deputies or specialists and expanded operations.[5]

Progressive sentiment lingered in Wisconsin, though the tide was turning. In 1920, another reform coalition elected Robert La Follette supporter John J. Blaine as governor. Blaine immediately caused a furor by nominating WSFL executive board member and La Crosse plumbers union head Reuben G. Knutson to replace outgoing Industrial Commission chair Hambrecht. Claiming that a labor leader could not fill an administrative post impartially, conservative state senators rejected the nomination. Blaine, however, resubmitted Knutson's name and persuaded senators to confirm him on the grounds that his work experience would help the agency. With state bureaucrats anxiously awaiting his arrival—one newspaper speculated that they expected a wild-eyed, bewhiskered plumber dressed in overalls—Knutson reassured skeptics that "I am quite capable of looking at all sides of a question."[6] Still, Blaine learned a lesson. Although labor appointments were common elsewhere—New York's James M. Lynch, Ohio's Thomas J. Duffy, and California's Will J. French all became industrial commissioners due to labor backgrounds—Blaine found that Wisconsin's politics tolerated only "disinterested" administrators on state boards.

That outlook prevailed in 1922. Wisconsin's reform coalition collapsed, voters elected a conservative legislature, and Blaine ran for reelection as the "Economy Governor." Meanwhile, the state's labor movement struggled, as socialists splintered, trade unions lost members, and the WSFL remained only a modest legislative lobby. In this conservative climate, Blaine held the Industrial Commission's appropriation to $285,000 a year, and appointed proven "professionals" to lead the agency—workers' compensation hearing examiner Lucius A. Tarrell and incumbent commissioner Fred Wilcox—with easy senate confirmation.[7] In response, the Industrial Commission economized by discontinuing the *Wisconsin Safety Review*, withholding pay raises, and curtailing travel expenses. Industrial expansion and new responsibilities subsequently strained the agency's capabilities, leaving a backlog of compensation claims and unfilled deputies' positions.[8]

Cautious politics continued after Blaine left office in 1927. Progressive ranks crumbled after the 1925 death of Robert La Follette, and Stalwart Republicans elected Fred Zimmerman as governor in 1926 and then antilabor industrialist Walter Kohler as governor in 1928. Both men permitted moderate budgetary growth and appointed politically safe state commissioners. Consequently, the Industrial Commission's appropriations rose to $300,000 in 1927 and $320,000 in 1929, allowing the agency to hire another deputy and several clerks but not a new education director, which the agency had to arrange under allocations to the state Vocational Education Board. Meanwhile, Zimmerman renominated Knutson as industrial commissioner and then replaced Tarrell with hearing examiner Voyta Wrabetz, whom he named for a full term in 1929. The senate approved these appointments without a fuss.[9] Staid bureaucrats took over the agency's leadership.

Labor departments elsewhere faced similar restraints. Conservative politics sustained the absence of a labor bureau in Alabama and stifled calls to modernize Illinois's labor department, leaving those states still without state-sponsored safety education. In New York, by contrast, labor law supporters fought off administrative downsizing. In 1921, newly elected Republican governor Nathan Miller and a conservative legislature streamlined state labor laws, replacing the five-member Industrial Commission with the new Department of Labor under a single labor commissioner, and then slashing programs. Fifty-two inspectors were dismissed. Prolabor Democrat Al Smith returned as governor in 1923, however, and promptly named former New York Industrial Commission counsel Bernard L. Shientag as labor commissioner to repair the damage. Apologizing to New York's State Federation of Labor (NYSFL) for the previous administration's "false conception of economy," Shientag rehired the inspectors and boosted the inspectorate back to 178. He and his successors James Hamilton and Frances Perkins carefully cultivated organized labor. During their terms, New York's annual industrial safety congresses regularly included the NYSFL and local labor councils as well as industrialists, local chambers of commerce, reform groups, and insurance firms.[10]

Despite contentious politics, New York's labor department sustained industrial safety education through the 1920s. The department's yearly conferences not only drew diverse participants but also featured discussions about accident prevention, plant safety programs, industrial hygiene, and woman and child work hazards. Its special bulletins, moreover, analyzed industrial health, workshop accidents, women's occupational risks, and child labor. Finally, New York's Compensation Insurance Fund helped employers to lower accident rates and insurance costs with statistical studies, safety literature, and plant safety programs, just as private insurers did.[11]

In Ohio, by comparison, safety education underwent a transformation. Initially, Republican governor Harry L. Davis's election in 1920 produced legislation that reorganized Ohio's Industrial Commission and other programs under the new Department of Industrial Relations, resulting in a $28,000 cut from the annual factory inspection budget and the removal of six assistant safety directors. Little educational work occurred due to "lack of funds," factory inspection chief Thomas Kearns reported in 1923. Even "Veto Vic" Donahey's election as Democratic governor in 1922 failed to overcome the Republican-dominated legislature's stinginess. In November 1923, however, the Ohio State Federation of Labor (OSFL) and reformers secured ratification of a constitutional amendment that not only strengthened Ohio's workers' compensation plan but also authorized a new accident and disease prevention board. Significantly, the amendment imposed compulsory assessments on Ohio manufacturers to cover the new board's expenses. The legislature subsequently set up the Division of Safety and Hygiene in Ohio's labor department, funded at $100,000 annually.[12]

Increases in budget and staff subsequently improved Ohio's educational safety work. From 1923 to 1926, the state's factory inspection division sponsored safety lectures, "no-accident weeks," and safety exhibits, along with bulletins, posters, pamphlets, and shop talks in local foundries. Then, starting in 1926, the new safety and hygiene department enlarged a statistical laboratory previously created by Ohio's Compensation Fund and hired engineers to study accidents and devise prevention measures. Targeting workers and small employers, the department created a speakers bureau, published a safety magazine, issued safety posters, and held yearly safety congresses.[13]

The creation of Ohio's Division of Safety and Hygiene exhibited remarkable union involvement in educational activities in this era of supposed labor declension. The OSFL was a major force behind the 1923 amendment that constitutionalized funding for safety and health work. "No legislature would have appropriated $100,000 for industrial safety," OSFL secretary Thomas J. Donnelly observed. "[Now,] this money is coming in year after year, independent of the Legislature." Moreover, Ohio unions worked closely with new safety and hygiene director Thomas Kearns to organize worker-manager safety and health meetings and to promote union educational work.[14] Wisconsin's WSFL, by contrast, left such activity almost entirely to private industry and state administrators.

In California, meanwhile, state officials and labor leaders fought their own battle against administrative downsizing. A conservative Republican state legislature consolidated California's Industrial Accident Commission with other labor bureaus under the new Department of Labor and Industrial Relations in 1921, reducing factory inspection expenditures to $150,000 annually, "a totally inadequate amount," according to commission chair Will French. Then in 1922, voters elected ultraconservative Republican governor Friend Richardson, whose promise of "sweeping retrenchment" yielded cuts in every state agency, despite the California State Federation of Labor's (CSFL) objections. Temporarily, California law protected safety education from budget cutting. A 1913 state statute had created an "accident prevention fund" based on fines collected from safety law prosecutions, while a 1917 amendment had provided further that the accident account receive automatic contributions from the state's Compensation Insurance Fund. This financial arrangement stabilized safety law education and enforcement work during changing political times, but conservatives abolished it in 1927.[15]

Although politics ran against social welfare spending in the 1920s, then, its impact on state-sponsored workplace safety education varied from state to state. New York's and Ohio's educational arrangements withstood conservative politics, whereas California's lasted briefly and Wisconsin's sharply declined, the latter due to direct dependence on legislative budgets. Yet whatever their educational activities, state administrators everywhere faced an economic environment dominated by business.

Business Dominance and Safety Education in the 1920s

After 1920, complex economic developments strengthened business control over the industrial work environment. On the one hand, labor market shifts, a hostile legal climate, employers' open-shop drives, and corporate welfare schemes all weakened trade unionism, though labor activism persisted.[16] On the other hand, a fragmented manufacturing sector sought greater order in the marketplace and workplaces. Manufacturers consolidated and formed business associations. Meanwhile, advanced firms adopted new chemical and engineering technology to improve productivity, and then introduced industrial psychology and personnel management to govern workers. In mass-production consumer-goods industries especially, leading companies expanded employee representation plans and "welfare capitalism" programs to lessen labor turnover, discourage unionism, and preempt government regulation.[17]

Business-sponsored safety activities accompanied these developments, as employers sought to minimize the cost of recently enacted compensation and safety laws. Advanced firms incorporated safety into managerial operations, reengineered factories with safer machinery, and improved the training and supervision of workers. Simultaneously, insurers fostered accident-prevention work among policyholders, industry associations disseminated technical literature, and trade journals published safety information. Even coal mining operators, faced with economic decline and production problems unique to their industry, very consciously duplicated big manufacturers' safety techniques, such as improved engineering, greater attention to safe practices, and better training and motivation of miners. Across the board, these endeavors affirmed statisticians' and engineers' theory that good management prevented accidents.[18]

The extent to which these managerial practices penetrated business, however, depended on local economies. In Alabama, for instance, where antiunion cotton textile companies emerged as the state's top business and iron and steel firms continued paternalistic employee relations, an antistatist business culture allowed firms to neglect safety work.[19] In California, by comparison, the booming economy (except for foundries) remained decentralized, local business federations promoted welfare measures and open-shop plans, CSFL-affiliated unions struggled to influence employment conditions, and only scattered firms adopted protective work.[20] In Illinois, meantime, traditional industries consolidated and new electrical machine manufacturers grew, yet union weakness, open-shop drives, corporate welfare plans, and the Illinois Manufacturers Association's opposition to labor laws all left safety education to big Chicago employers.[21] In New York, by contrast, small workshops continued as an economic mainstay in New York City, while elsewhere large firms organized regional business associations and Schenectady's General Electric plant established antiunion corporate welfare programs, but New York's labor movement remained powerful and sustained workplace safety as a prominent issue.[22] And in Ohio, where old industries consolidated and new ones like autos and electrical machinery accelerated, the state's great corporations quashed strikes, started open-shop campaigns, and expanded antiunion welfare plans, though the OSFL stayed politically active and kept the safety issue alive.[23]

As for Wisconsin, manufacturing changed and grew. Old businesses like lumber and furniture shrank, but numerous small upstate firms persisted. Simultaneously, foundries prospered, knit goods and footwear emerged as state leaders, and new manufacturers of autos, farm equipment, and other machinery associated with internal combustion engines spawned big southeastern Wisconsin plants. As elsewhere, such conditions discouraged unions. Unionized crafts declined, while new antiunion mass-production industries expanded. Moreover, business associations conducted open-shop drives, and big firms like Sheboygan's Kohler Company and West Allis's Allis-Chalmers promoted welfare policies. Unions persevered but conceded safety issues to employers and the state.[24]

Indeed, business-sponsored safety work expanded dramatically in Wisconsin. The state's employers started regional business initiatives to disseminate safety practices, just as firms nationwide created trade associations and other "nonstatist coordinating mechanisms" to regulate the business system. Meanwhile, the "safety market" spread protective technology among Wisconsin firms. Machine makers began automatically to safeguard their equipment, while other firms sold eye protectors, ventilation systems, and safety devices, with one company even trying (unsuccessfully) to bribe Industrial Commission deputies to promote its products.[25]

At the same time, Wisconsin insurance companies established safety programs that replaced Industrial Commission activities. Employers Mutual hired former commission deputies Al Kroes and August Kaems to deliver the same kind of shop talks that Wisconsin's commission had previously sponsored. In 1919, moreover, Employers Mutual introduced safety movies, the first being *The Awakening*, portraying a manufacturer's efforts to promote safety work despite misguided workers' resistance. Like Wisconsin's commission, too, Employers Mutual issued magazines covering industrial insurance and shop safety.[26]

Business-sponsored safety conferences, however, became the principal substitute for the Industrial Commission's educational work. These events began before World War I with panel discussions or roundtables sponsored by the Merchants and Manufacturers Association of Milwaukee (MMAM) and similar groups in Green Bay, Oshkosh, and other communities. Later, from 1918 to 1922, the Milwaukee Association of Commerce (MAC) (the

MMAM's successor) joined the Industrial Commission to hold industrial service conferences for plant managers. Finally, in 1919, the MAC started foremen's schools, the most popular safety conferences. Inviting speakers from the Industrial Commission, big local employers and outside groups once a week for five or six weeks, the schools featured "inspirational" sessions and technical programs, as well as motion pictures and skits. Employers in the Rock River Valley, the Upper Mississippi Valley, the Fox River Valley, the Wisconsin River Valley, and southwestern Wisconsin held similar events. Wherever they met, Wisconsin safety schools disseminated the view of Herbert Heinrich and other contemporary safety experts that modern accidents resulted mostly from "mental and emotional faults," not unsafe machines. Safety schools disseminated such ideas to the plethora of small Wisconsin firms that lacked management structures to propagate safety awareness on their own.[27]

Wisconsin's Industrial Commission logistically supported business safety conferences by arranging speakers, providing publicity, and, in 1928, assigning newly appointed education director Henry Menzel to coordinate meetings. Simultaneously, the agency collaborated with the U.S. Bureau of Mines, when it began dispatching its safety car No. 10 to teach miners and quarrymen first aid. The commission mailed letters to mine operators and quarrymen to announce the program, and then appointed Deputy A. H. Findeisen to give lessons. Findeisen used the occasion to discuss Wisconsin's safety orders and to hand out booklets explaining them. The commission later extended this program to utility workers and firemen.[28]

Meanwhile, Wisconsin's industrial commissioners strengthened their ties to the professional safety community, a development allowed by the Wisconsin agency's unusual political independence. Only four men occupied the Wisconsin commission's three seats from 1921 to 1931, and only Fred Wilcox served as commission chair, whereas Ohio's three-member commission had seven members and four different chairpersons and New York's labor department had four commissioners. Wisconsin administrators' long tenure, plus their unusual stature in the International Association of Industrial Accident Boards and Commissions (IAIABC) and International Association of Government Labor Officials, permitted them to cultivate professional relationships. At a time of increasing interstate cooperation, Wilcox and fellow commissioners kept up correspondence with insurance companies, federal officials, and other states' administrators. Secure in their posts, Wisconsin commissioners immersed themselves in safety professionals' technical world.[29]

Annual IAIABC conferences illuminate the professional safety community's membership and outlook. Beginning in 1914, IAIABC meetings assembled workers' compensation and safety law experts from the United States, Canada, and Mexico, drawing them mainly from government agencies, big corporations, insurance companies, and business groups. Focusing more on compensation than on accident prevention, IAIABC gatherings typified the "technical systems" of "empirical professionals" that were emerging in the "corporate economy" and "organizational society" of the 1920s. IAIABC sessions featured engineers, doctors, and administrators, not "nonexpert" labor organizers, workers, or small business owners. Coming from General Electric, Du Pont, Employers Mutual, Travelers, and other big businesses, IAIABC participants saw technical competence as a means both to enlighten industry and to maximize efficiency and profits.[30] Safety professionals consequently followed Herbert Heinrich's and other industrial analysts' view that industrial accidents resulted more from worker behavior than from physical hazards. Like corporate executives who were shifting attention from engineering control to personnel management, IAIABC professionals stressed the "man problem," that is, managing workers better. They advised employers to put "the right man in the right job," train him properly, and motivate him adequately.[31]

Casualty insurance companies also adopted this thinking. Employers Mutual vice president Fred Braun contended, for instance, that safety was a "psychological" problem. Human failure usually caused accidents, not "senseless machines." Yet Braun added that managers, not workers, bore the main burden for accident prevention. "Maintenance, enforcement, and the providing of guards are management problems," he observed. "Leadership to provide proper incentive toward eliminating man-failure must come from management." In agreement, Travelers Insurance Company executive C. B. Auel reported that most workers experienced few injuries but preceded typical accidents with 330 "practice performances" due to improper work procedures. Management, he argued, could correct this problem. Also in agreement, Liberty Mutual engineer A. S. Regula urged employers to use the "psychology of advertising" to promote worker safety. Bulletin boards, rallies, gruesome accident stories in plant magazines, safety slogans inserted in pay envelopes, and no-accident contests all would improve work habits, he said.[32]

In numerous speeches, Wisconsin's industrial commissioners reiterated the professional safety community's managerial outlook. Safety was not just "philosophy" or "bunk," commissioners constantly reminded listeners, but an integral part of industrial production. As a humanitarian proposition, preventing accidents helped workers more than compensating them after the fact. As a business proposition, safety work lowered workers' compensation insurance costs and improved worker productivity.[33]

Like the experts, Wisconsin's commissioners reasoned that most accidents resulted from identifiable causes that good managers could eliminate. Proper machine safeguarding was one method of doing this. "There is nothing forward looking in the guarding of a machine," Fred Wilcox asserted, "if that guard is to be turned back, or hung on the post, or thrown in the corner." Education offered a second way. Wisconsin statistics corroborated industry analysis that more than 80 percent of accidents came from human error, and the commissioners urged worker training to prevent such mishaps. State officials elsewhere agreed. New York's Leonard Hatch saw nonmachine hazards as the fastest-growing cause of accidents, while California elevator inspector D. J. Harris claimed that "a careful man is one of the best safeguards."[34]

Wisconsin commissioners also embraced the expert view that workers were not simply "careless." As Wisconsin workers' compensation director Harry Nelson argued, "We must go behind the use of that stereotyped word, 'carelessness,'" to find and eliminate "the human failure." Like safety experts, Nelson and other administrators held management responsible for that task. The "efficiency that assures accident prevention," Commissioner Voyta Wrabetz explained, required "the foreman or any other person in a supervisory position [to] maintain proper order and discipline."[35] If not captured by industry, Wisconsin officials bought deeply into the managerial viewpoint.

Workers saw things differently. The labor movement retreated in the 1920s, but unions did not consequently accept managerial safety strategies. In New York and Ohio, especially, labor leaders articulated an alternative vision of workplace safety education that assigned critical functions to unions. As progressive unionists who accommodated some aspects of scientific management, New York and Ohio labor officials admitted that 85 percent of injuries came from human causes and that worker education could prevent accidents. According to NYSFL president John Sullivan, worker "vigilance" and "cooperative carefulness by employees and management" constituted the first line of defense against accidents. According to the OSFL's Special Committee on Industrial Safety, Ohio workers did not properly safeguard themselves, and some had become "lamentably careless."[36]

When it came to educating workers, however, labor leaders offered different ideas from those promoted by industry, experts, and administrators. New York and Ohio union officials not only advised workers to be careful but also affirmed unions' role in promoting safe work. Contending that employers installed safeguards only when pressured by workers, the OSFL proposed the creation of permanent committees in union locals to inform workers about safety hazards. Rejecting employer-dominated factory committees, moreover, both Ohio and New York unionists advocated joint "consultative" bodies that bargained collectively about safety and health. NYSFL president Sullivan asserted that workers sought "that ideal and practical form of cooperation where all workplace problems can be considered and solved in conferences between employers and employees." The OSFL's special committee added that 50 percent of injuries and diseases could be avoided by labor-management cooperation on bodies "appointed by both the employer and employee."[37]

Finding that they rarely achieved "ideal" cooperation, New York and Ohio labor leaders concluded that state action was still necessary to enforce safety measures. Yet labor officials doubted that government would act on its own. "The safety laws are ample enough," Sullivan asserted. "The weakness is in the lack of enforcing agencies." OSFL secretary Donnelly agreed. What good were safety laws, he asked, if organized labor did not get them enforced? "We owe it to ourselves," he declared, "to carry the message of the statutory law, relating to safety, into every neighborhood in every industrial center of the State, . . . let all the Union workers of this State know that Organized Labor has passed these laws, not for themselves, but for all the people, and . . . that they can make the employer furnish safe places or employment and pay in every case of injury, whether they are nonunion or Union."[38]

New York and Ohio labor leaders more energetically engaged work safety education than the image of the 1920s as labor's "lean years" implies, but not Wisconsin unionists. Indeed, Ohio and New York administrators supported a union role in education. Ohio safety and hygiene director Kearns asked the OSFL to organize safety committees in union locals. New York labor commissioner Frances Perkins urged the NYSFL to set up committees to teach workers safe machine use and then to pressure managers to cooperate with them to prevent injuries.[39] By contrast, Wisconsin union leaders refrained from such encouragement. Wisconsin's industrial commissioners, consequently, relied on insurance firms and safety professionals to promote safety education, sustaining their ameliorative and managerial approach to workplace safety.

Wisconsin Safety Education's Final Marginalization

The Great Depression transformed labor law administration all around the country. The era's fiscal contraction forced most state labor departments to cut "nonessential" programs like safety conferences and publications. State programs' decline, however, did not inspire national protective laws. Conversely, the era's economic dislocations, business decline, and labor activism realigned political priorities. On the state level, political leaders concentrated on work relief and industrial relations questions, not Progressive-style protective measures. On the national level, the New Deal focused on economic security and industrial unionism, addressing workplace safety and sanitation only marginally in public contracts legislation.[40]

These dramatic developments affected work safety education differently from place to place. In Alabama, lawmakers finally established a state labor department, empowering it to make safety codes and conduct educational programs. In Illinois, legislators finally gave their labor department safety code–making power without mentioning education. New York's Department of Labor suspended safety conferences in 1932, but Ohio's Division of Safety and Hygiene continued accident and disease statistical work. Remarkably, California's Department of Industrial Relations sustained safety museums, magazines, and conferences, while its Compensation Insurance Fund continued shop safety programs. Yet the Wisconsin commission's educational work dropped off to just a few speeches and safety contests. Except for firms like Sheboygan's Kohler Company, Wisconsin manufacturers abandoned "welfare" work for state-sponsored unemployment and labor relations laws. Meliorative educational programs dwindled.[41]

Meantime, the Progressive Republican younger son of Robert La Follette Sr., Philip La Follette, upset sitting governor Walter Kohler in the 1930 election. With help from the WSFL and University of Wisconsin labor economists, Phil La Follette secured unemployment compensation legislation, increases in workers' compensation benefits, a unified state labor code, and modest growth in Industrial Commission appropriations. Worried about political loyalty, however, he left commission chair Fred Wilcox in office only on an interim basis rather than renominate him. Unexpected events sealed Wilcox's fate. In 1932, Kohler defeated La Follette in the Republican primary, and voters elected Democratic gubernatorial candidate Albert Schmedeman along with a Democratic legislature in the wake of Franklin Roosevelt's presidential victory.[42]

Interrupting La Follette's governorship for two years, Schmedeman's election illustrated state politics' nationwide instability during the 1930s, despite overall state administrative expansion. In Illinois, for instance, Governor Henry Horner warred perpetually with urban political interests and federal relief agencies over patronage and relief spending. In Ohio, conservative Democratic and Republican leaders alike feuded with state lawmakers and federal authorities. Despite Upton Sinclair's controversial gubernatorial candidacy, conservatives controlled California. New York governor Herbert Lehman stood out as a rare champion of new labor laws.[43]

Like conservative partisans in Illinois, Ohio, and California, Schmedeman reduced state expenditures. Wisconsin's Industrial Commission suffered the results. With Schmedeman followers wielding the budgetary axe, Wisconsin's conservative 1933 Democratic legislature decreased the agency's annual appropriation to $295,000, lowering safety outlays to only $95,390 a year, while adding huge chores like unemployment relief and a growing docket of workers' compensation claims. Despite cutbacks, the commission retained most of its deputies, avoiding the kind of retrenchment that, for instance, reduced Ohio's Department of Industrial Relations from 600 to 490 employees in 1932.[44]

Schmedeman next shuffled Industrial Commission leadership. To satisfy a "parade of hungry Democrats," as *The Capital Times* put it, the governor replaced commission chair Wilcox with deputy U.S. internal revenue collector Peter Napiecinski, a nomination that the Democratic state senate quickly approved. The action compromised the commission's vaunted autonomy. As Wilcox observed afterward, the Wisconsin agency had long enjoyed respect for impartial labor law administration, but now it had a "political appointment" at its head. Yet Schmedeman then went ahead and replaced Reuben Knutson with Democrat Harry R. McLogan, a Milwaukee lawyer with a family history of labor activism. Again, the Democratic senate obliged the appointment.[45]

By the time La Follette returned to Wisconsin's governorship in 1935, economic conditions and New Deal legislation had stirred up labor militancy and reoriented labor policy across the country. In Alabama, union organizers invaded steel plants, coal mines, and textile mills, though old antiunion attitudes persisted. In big manufacturing states like Illinois and Ohio, industrial unionism surged. In Wisconsin, AFL-affiliated "federal unions" organized rubber and autoworkers, while the militant Congress of Industrial Organizations (CIO) mobilized iron, steel, and other auto company employees. Consequently, policy makers everywhere shifted from the Progressive strategy of protecting workers to the New Deal idea of regulating unionism. Wisconsin's La Follette himself embraced the new approach by championing Wisconsin's "Little Wagner" Labor Relations Act in 1937.[46]

La Follette also threatened the Industrial Commission's political autonomy. Now heading an independent Progressive party, he did not renominate Commissioner Voyta Wrabetz in 1935 but left him as an interim appointee until naming him to replace Peter Napiecinski in 1937. To finish out Wrabetz's term, La Follette installed an interim loyalist, Progressive Republican Mabel Griswold, further damaging the commission's reputation for impartiality. Yet like other Depression-era state leaders around the country, La Follette supported legislation that provided new sources of revenue for financially strapped state agencies. To help the Industrial Commission, he proposed measures to assess insurance firms and manufacturers for the costs of workers' compensation administration, but business groups shot all such legislation down.[47]

The Wisconsin Industrial Commission's fiscal situation thus remained difficult. Legislative appropriations in the La Follette years elevated the agency's safety and sanitation budget back to $103,516 in 1936–1937 and $119,229 in 1938–1939, but the commission could muster educational funds only for an annual blasters school, a "better light–better sight" campaign, and a few news releases.[48] Then, conservative Republican Julius Heil upended La Follette in Wisconsin's 1938 gubernatorial election, causing new cuts: $9,000 in 1939–1940 and $13,000 for 1940–1941. The commission's Safety Department subsequently eliminated a deputy, terminated blaster schools, and reduced travel. "To continue at our present level," the commission warned in 1940, "means that essential service is impaired and basic functions are curtailed."[49] Safety education became a luxury that Wisconsin's commission could not afford.

The Politics of Safety Education's Decline

The Wisconsin Industrial Commission's educational and ideological functions declined through the 1920s and 1930s, but not simply because the balance of power shifted toward industry, or because agency functions duplicated business work. Instead, budgets and partisan politics restrained the commission. Unlike educational programs in New York, Ohio, and California that received automatic funding, Wisconsin's educational work relied fully on legislative appropriations that gradually shrank. Whereas political independence fortified the commission's stature in the 1920s, partisanship compromised it in the 1930s.

Equally important were changes in labor relations. When union influence receded and employer welfare programs expanded during the 1920s, state-sponsored safety education lost its appeal. Then, when economic dislocations and labor militancy grew in the 1930s, policy makers' interest in the commission's old ameliorative approach to labor relations did not revive but focused on accommodating unionism instead. Unlike Ohio safety and hygiene chief Thomas Kearns and New York labor commissioner Frances Perkins, Wisconsin industrial commissioners failed to adapt. Embedded in the professional safety community, Wisconsin administrators perpetuated managerial views of safety that had blossomed in the 1910s and 1920s.

Nonetheless, the demise of the Wisconsin Industrial Commission's educational function did not completely undermine the agency's autonomy and power during the 1920s and 1930s. When it came to administering workers' compensation and propagating safety and health codes, the Wisconsin commission remained a very formidable institution.

7

The Technocrats Take Command

Although state-sponsored safety education waned during the 1920s and 1930s, safety code development expanded. Wisconsin's Industrial Commission updated electrical safety orders; modified sanitation, lighting, boiler, elevator, and mining codes; and initiated new requirements for noxious fumes, spray painting, ventilation, and quarries. Meanwhile, New York's Department of Labor not only revised orders for machinery, boilers, building construction, window cleaning, and fumes but also formulated new regulations for the needle trades, dry cleaning, and inflammable materials. California's Department of Labor and Industrial Relations simultaneously updated boiler, electrical, construction, and mine safety codes, and then developed new standards for petroleum, mine fires, and shipbuilding. In 1924, even Ohio's Industrial Commission formalized regulations for machinery, elevators, laundries, and metalworking, and then instigated orders for pressure piping, first aid, and quarries. Admittedly, some states' labor departments were lackluster in the 1920s, and not all coal mining states were as energetic as Pennsylvania and Utah in issuing mine safety rules. Nonetheless, many state administrative safety law operations were vigorous after World War I, not "highly fragmented and weak," as historians allege.[1]

Those programs operated, however, in a new economic climate. Though much of American business was still fragmented, core industries merged into big corporate enterprises and trade associations. Standardization of safety technology accompanied consolidation. This began in the mining industry in the 1910s, when mine operators and professional engineering societies sought uniform accident statistics and safety standards, first through the

U.S. Bureau of Mines' establishment as a scientific agency, and thereafter through the development of electrical codes and other rules for adoption as consistent state laws. A key objective was to create uniform interstate standards without eliciting federal intervention. Then, in the wake of World War I efforts to streamline industrial practices, safety engineers affiliated with the American Society for Mechanical Engineers, the U.S. Bureau of Standards, Underwriters' Laboratories, and the American Engineering Standards Committee (AESC) began formulating model "national codes" for industrial and state regulatory use, again without invoking federal power. These model codes demonstrated how technological authority "shifted up" into "technical systems" in the 1920s, linking experts in advanced corporations, government bureaus, insurance companies, universities, and professional societies in networks of technocratic power while excluding local firms, workers, and consumers.[2] Wisconsin's Industrial Commission firmly embraced model codes and the technological hierarchy out of which they emanated. Other states' labor departments adopted national standards as well, but reconciled them more to local interests.

The Wisconsin commission's technocratic orientation did not signal "capture" by big corporations but a response to local politics and law. Politics sustained the Wisconsin commission's autonomy as an exponent of technical expertise. Progressives and organized labor still resolutely supported the agency, despite losing ground. Local industrialists remained too decentralized to dominate the commission, despite amassing power. Moreover, administrative law's continuing openness permitted the Wisconsin agency broad discretion. Lax judicial supervision allowed Wisconsin administrators to perpetuate a wide interpretation of the Industrial Commission's mandate and to privilege technical specialists and research-based science in safety code proceedings, despite rhetoric about the "representation of interests." Wisconsin's safety code system subsequently evoked conflict not simply along class lines of worker versus employer but also along "organizational" lines pitting unsophisticated local groups against modernized business and government bureaucracies.[3] Local employers and trade unions emerged as the biggest critics of Wisconsin safety regulations, but Wisconsin's commission pressured such groups to adopt new techniques in the name of modern science.

Nonetheless, local discontent with Wisconsin's technocratic approach to safety code formulation kindled an early phase of the administrative law reform movement that would mature in the late New Deal years. With workers' compensation as the flash point, Wisconsin employers and lawyers revived the Classical nineteenth-century rule-of-law tradition that favored adversarial legal proceedings and viewed administrative government as a despotic threat to judicial supremacy and basic legal rights. In the late 1920s and early 1930s, this "legalist" revolt generated litigation and legislative action to curtail the Industrial Commission's power. Later in the New Deal years, it merged with the nationwide movement that restrained regulatory agencies with state and federal administrative procedure acts (APAs), including Wisconsin's law of 1943.[4]

This legal reform movement ended administrative law's era of openness with surprisingly mild effects on work safety and health regulation. In Wisconsin, courtroom proceedings increased surveillance over safety orders and narrowed those orders' legal scope, but the state's newly enacted APA exempted workers' compensation and imposed minimal restrictions on safety code making. Despite legal reform, Wisconsin's safety code operation still favored technological advance.

The Nationalization of Work Safety Standards

During the 1920s, Wisconsin's State Electrical Code, Acetylene Plant Orders, and Spray-Coating Orders demonstrated the Industrial Commission's maturity as an exponent of modernization. In these codes, the agency championed corporate practices, new technology, uniform national standards, and networks of technical experts more fully than ever before.

Assigned jointly to the state's Industrial and Railroad Commissions, Wisconsin's State Electrical Code reflected the evolving technology of power in industrial plants. Whereas Wisconsin's original General Orders covered the old belt and shaft system, the Electrical Code involved the new science of wires, insulators, circuits, and motors. Because new apparatus underwent constant redesign, moreover, Wisconsin's two agencies continually updated their initial 1917 Electrical Code by issuing revisions in 1922, 1924, 1930, and 1934; producing a supplement in 1936; and then starting another revision in 1938.[5]

To assemble Electrical Code advisory committees, the two Wisconsin commissions perpetuated the prewar practice of recruiting professionally recognized experts rather than equalize worker and employer representation. The 1919 committee included four state and local government officials, two insurance company employees, six spokespersons for electrical contractors or utilities, a U.S. Bureau of Standards electrician, and a single Wisconsin State Federation of Labor (WSFL) representative. By contrast, California's 1918 electrical code panel accommodated five industry representatives, six officials from state and local government, and six members of local unions.[6]

Even with experts on electrical committees, Wisconsin's commissions used the Bureau of Standards model code as the basis for state rules, just as regulators did elsewhere. Formed in the U.S. Commerce Department in 1902, the bureau surveyed national electricity uses and then published the model code to foster uniform practices around the country. In Wisconsin, a high electrical-accident rate during World War I induced officials to incorporate the bureau code into state regulations, thereby endorsing industry-approved standards. After 1920, Wisconsin's agencies kept state electrical rules up to date with national standards to accommodate new equipment and eliminate antiquated local regulations. Indeed, such was the Industrial Commission's reliance on the bureau code that deputy John Hoeveler secured committee approval of updates just through the mail, ironing out disagreements without meetings or formal committee votes. Such informality was not unusual in America's administrative process—a 1942 study reported similar practices in New York's safety code proceedings.[7]

When it came to public hearings on proposed electrical orders, the two Wisconsin commissions sought to inculcate employers with the need to comply with authoritative standards, again a practice repeated in New York. At Wisconsin's very first Electrical Code hearing in 1917, for instance, administrators overrode business suggestions to implement regulations on a trial basis only. Talking tough, industrial commissioner Joseph Beck told employers that worker safety was too urgent for experimentation, especially since manufacturers would likely neglect electrical orders only tentatively adopted. "We get more [accident reports] from the lack of proper safeguards . . . than perhaps from any other cause," he declared. Following Beck's lead, Wisconsin's Industrial and Railroad Commissions approved a legally binding Electrical Code.[8]

At public hearings for the code's second revision in 1924, Wisconsin officials parried growing complaints about binding regulations. Employer criticism focused on vague phrases like "where feasible" and "practical," which left decisions worth thousands of dollars up for grabs. Faulting that terminology's ambiguity, Mr. Jung of Milwaukee's Jung Electric Company remarked that manufacturers would have to consult an "electrical attorney" before installing new equipment. Yet industrial commissioner Fred Wilcox dismissed these charges. Whatever the vagaries of Wisconsin's electrical regulations, he argued, businessmen should be happy to have them as a guide. Competitors elsewhere had to "lug for themselves" without governmental help.[9]

Nonetheless, manufacturers contended that proposed Electrical Code revisions were unreasonable. Employers objected especially to one proposal requiring double switches on every motor to prevent accidental introduction of electrical current into machines under repair. According to Mr. Douglas of the Milwaukee Engineering Society, it was "rotten engineering" and "a joke" to use one switch to turn off another. "When is a switch not a switch?" he asked. Was this not "undue elaboration of protection"? Wilcox disagreed. Overlooking the Wisconsin safe place statute's mandate for "reasonableness," he defended the double-switch requirement on the grounds that "property rights have to give way to personal rights." "An employee has the right to work under safe circumstances," he remarked, "and neither you nor I, nor anyone else, have any license to say to him that he shall work under unsafe circumstances just because it cost [money]."[10]

Manufacturers complained loudest, however, about the rule applying new electrical regulations to all industrial establishments, new and old. Mr. Stark said that his National Knitting Company (Milwaukee) had spent eight thousand dollars to comply with old rules, and that new requirements would antiquate 75 percent of that work. Mr. Frey of Geuder, Paeschke & Frey (Milwaukee) questioned whether it were fair to force him to spend ten to fifteen thousand dollars to comply with revisions when his firm had experienced no accidents. Responding that it made no sense to safeguard only future installations, Commissioner Wilcox contended that all manufacturers had to provide work conditions "as reasonably safe as the law requires." Still, he promised employers "a reasonable time" to implement new regulations. Indeed, Wisconsin's commission enforced the Electrical Code flexibly, granting waivers and sometimes directing deputies not to apply it strictly.[11]

Nonetheless, employer objections to Electrical Code revisions exposed the Wisconsin commission's technological slant. The state's safe place statute required employers to provide the safest equipment reasonably possible, but "reasonableness" depended on transitory industrial conditions. New electrical technology modified safety standards continually. By following national electrical code updates, the Wisconsin commissions imposed technological changes on small local plants for which frequent reinvestment in new equipment was a hardship. Local employers recognized the agency's bias. Milwaukee's Mr. Jung correctly observed that the electrical code advisory committee was stacked with utility representatives and bureaucrats unsympathetic to small firms' problems.[12]

Wisconsin unions also resisted the Electrical Code's forced modernization. They objected to new orders requiring nonmetallic sheathed cable, a product endorsed by Underwriters' Laboratories, the Bureau of Standards, and the AESC. On the advisory committee, commission representatives, big employers, and utilities supported the new cable, but the WSFL, the Milwaukee Building Inspection Department, and several local firms warned that it threatened union jobs and posed a fire hazard. Opponents sought prohibitory legislation, but Wisconsin's commissions authorized the new cable in the Electrical Code, again siding with technologically advanced employers and utilities.[13]

The Wisconsin advisory panel formulating regulations for acetylene cylinder refilling exhibited the same disposition. As elsewhere, Wisconsin's Industrial Commission created this committee after explosions leveled suburban Milwaukee recharging plants during 1918 and 1919. Frightened residents requested legislation, but Wisconsin's industrial commissioners promised safety rules instead. As usual, the agency assembled a committee of specialists: a Milwaukee manufacturer, a National Safety Council representative, a university chemistry professor, a U.S. Bureau of Explosives official, a business association member, one trade unionist, and the Industrial Commission's boiler inspector. Committee members studied acetylene compression, interviewed equipment manufacturers, conducted public hearings, and then promulgated two orders in 1919 that dictated apparatus design, factory layout, and recharging-plant location. Subsequent to these orders, Milwaukee's Gas Tank Company refused demands to move its plant, even after a second explosion. Yet threatened legal action and a meeting with commission representatives eventually persuaded the firm to relocate.[14] State safety policy superseded local business concerns.

Surprisingly, Wisconsin's acetylene orders then inspired a national controversy. In December 1920, the Alexander Milburn Company, a Maryland manufacturer of oxyacetylene cutting and welding apparatus, petitioned the Wisconsin commission to modify rules governing acetylene tanks and acetylene generators. Underlying these requests was Milburn's allegation that Union Carbide & Carbon Corporation, a Milburn competitor, had used Wisconsin's order-making process to monopolize business through technological means. Milburn claimed that it had received no notice of the commission's October 1919 hearing, that all of the hearing's witnesses worked for Union Carbide or its subsidiaries, and that resultant safety order No. 4600 required a special agitator covered by an exclusive Union Carbide patent.[15]

Stung by the criticism, Wisconsin's commission solicited briefs from Milburn and Union Carbide attorneys, heard oral arguments, and ultimately rejected Milburn's allegations. The agency concluded that advisory committee members recognized the witnesses' Union Carbide connections, and that a Union Carbide competitor attended hearings without raising objections.

Moreover, the commissioners decided that Milburn lost little business due to Wisconsin regulations, since local companies easily converted Milburn equipment to satisfy Wisconsin's rules. Most important, the commission ruled that its maximum-pressure requirements applied to acetylene tanks in hazardous recharging factories, not to welding and cutting operations characteristic of Milburn affiliates. Milburn was mistaken, the agency argued, that Wisconsin's acetylene orders inhibited its welding plants.[16]

Determined to prove its point, Alexander Milburn then sued Union Carbide in federal court for antitrust violations. Milburn attorneys alleged that Union Carbide was a giant holding company that monopolized the manufacture, supply, and use of oxyacetylene equipment, but the U.S. District Court in Baltimore and then the Fourth Circuit of Appeals both determined that Milburn had failed to show that Union Carbide had illegally restrained trade. Still, Milburn's complaint had merit. To ascertain prevailing good safety practice, Wisconsin's Industrial Commission had assembled advisory committee members and solicited expert witnesses from a "technical system" dominated by a large corporation. "Expert" advice gathered in this process abetted corporate efforts in the 1920s to control economic markets through patent monopolies and standardized machinery.[17] By accepting such advice, Wisconsin's commission accommodated big industry's technological strategy.

Wisconsin's Spray-Coating Code did not implicate big industry, but it again encouraged new technology. Growing rapidly in furniture, railroad car, and automobile manufacture, spray painting was potentially harmful because it increased the risk that workers might imbibe lead, arsenic, and other toxic paint compounds. Wisconsin's industrial commissioners investigated the matter in 1916, but did nothing until Socialist Milwaukee assemblyman Walter Polakowski offered a bill on behalf of painters' unions in 1921 to ban the process. The commission opposed this legislation on the grounds that the agency already had jurisdiction over the problem, but the state assembly passed a measure requiring the commission to issue appropriate regulations. That bill died in the state senate.[18]

These events signaled unique intervention by labor into occupational medicine. In 1921, New Jersey and Massachusetts painters' unions, along with the Brotherhood of Painters, Decorators, and Paperhangers of America, joined the New York City Workers' Health Bureau founded by Marxists Grace Burnham and Harriet Silverman to promote workplace health. Affiliated with the Science Advisory Board including Alice Hamilton and Emery Hayhurst, the Health Bureau provided medical information from workers' point of view. The bureau condemned spray painting. Subsequently, painters' unions in Massachusetts, Ohio, New Jersey, New York, and Wisconsin lobbied to restrict the process, if not prohibit it.[19]

Despite this agitation, Wisconsin's Industrial Commission failed to organize a spray-painting committee until labor supporters introduced another bill in the state assembly in 1923. Unlike most Wisconsin advisory panels, this one consisted mainly of local representatives: a furniture maker, two union members, a painting contractor, a chemical engineer, the commission's chief engineer, and a U.S. Public Health Service officer. The committee finally formulated orders in 1924, though disagreement nearly scuttled them. At public hearings, journeymen painters demanded severe restraints on spray painting, and presented Workers' Health Bureau spokesperson Harriet Silverman to testify against the process's health hazards. Meanwhile, painter contractors requested a minimum of regulation. A badly divided advisory committee then submitted majority and minority reports, but the industrial commissioners refused to accept them, threatening to adopt no orders whatsoever unless committee members reached a consensus. Preferring some regulation to none at all, labor representatives consented to modest requirements that permitted spray-painting apparatus while regulating spray-painting pressure levels, protective clothing, respirators, and ventilation. Businessmen applauded the new orders' leniency, but union men bitterly complained that spraying would "positively increase" occupational disease.[20]

This mixed reception demonstrated that spray-painting regulation was not simply a "neutral" question of determining "prevailing good practice" but also a political choice between competing technologies. Wisconsin's commission, as usual, sided with science-based assessments of work hazards. One reason was the complete absence of workers' compensation claims showing that spray painting caused illness. Another reason was the Wisconsin agency's general aversion to regulations impairing new manufacturing processes.[21] To retain political credibility, however, the commission had to accommodate business and workers. Furniture makers, auto manufacturers, and painting contractors endorsed spray painting, but journeymen painters opposed it. To appease both sides, Wisconsin's commissioners deflected criticism with rhetoric about the safety code process's "representativeness." When the Brotherhood of Painters objected to new spray-painting orders, Commissioner Wilcox reassured the union that its interests had been addressed, because an advisory committee "made up of representatives of manufacturers, labor and the public" had "unanimously" agreed on new regulations. Wilcox did not disclose how the committee had reached its "unanimous" decision.[22]

The 1924 edition of Wisconsin's General Orders on Spray Coating hardly ended the controversy. The Industrial Commission updated the code in 1925,

1929, and 1939, but painters' unions charged that the agency loosely interpreted regulations and haphazardly enforced them. Even Wilcox admitted that although large manufacturers complied with spray-painting rules, room and house painters "still give us trouble." Consequently, labor representatives proposed restrictive legislation in 1925 and 1927, while WSFL conventions condemned the commission's handling of spray painting in 1929 and 1933. Still, Industrial Commission safety orders sustained this new technology.[23]

Other states' safety code operations exhibited similar frictions between technological modernizers and local groups. According to one report, New York advisory committees usually offered a "who's who" of experts, not strict representation of the industry, labor, and the public. Hence, New York codes generally suited large modern factories rather than small firms. When, for instance, New York's Industrial Board promulgated a rule requiring "interlock" devices on commercial washing machines based on AESC and National Fire Protection Association standards, small local dry-cleaning establishments objected. The Onondaga County Supreme Court eventually confirmed the small employers' complaint that the rule was unreasonable. Small companies typically resisted expensive safety devices and favored employee training instead.[24]

Labor unions everywhere, by contrast, encouraged state safety codes, though they faulted specific regulations. Wisconsin unionists robustly supported code making, whereas New York State Federation of Labor president John Sullivan described it as "indispensable" for workplace safety. The Ohio State Federation of Labor (OSFL) prodded Ohio's Industrial Commission to keep codes up to date. Indeed, some workers attempted to develop their own safety and health regulations to counter business-dominated model codes. In 1927, a few unions joined the New York Workers' Health Bureau in creating National Trade-Union Safety Standards Committees to formulate worker-made safety standards for use in collective-bargaining agreements. Thereafter, California stereotypers urged state federations to collaborate with Standards Committees. Nonetheless, these panels foundered due to inadequate financial support and state federations' opposition to the Health Bureau's "dual unionism."[25]

Female workers, meanwhile, received little attention in 1920s safety codes, despite widespread enforcement of women's hours and wage laws. Alabama and Illinois still lacked labor commissions that could promulgate women's safety orders, whereas Ohio's Industrial Commission just issued regulations governing women's foundry work, California's Industrial Welfare Commission filed orders setting weight-lifting limits, and New York's Department of Labor sustained a code governing women's work in core rooms. Wisconsin's Industrial Commission issued numerous rules on women's work hours and wages, but none on women's safety.[26]

Through the 1920s, hence, safety code making in Wisconsin and other administratively empowered states was prolific, if technologically biased. Wisconsin's commission in particular embraced model national safety codes, favored specialists on advisory panels, and supported big-industry practices, while it parried local criticism with the rhetoric of representativeness and good practice. This had mixed blessings for workers and small employers.

Safety Regulation and New Deal Labor Relations

Wisconsin's safety code process remained entrenched in technical networks through the 1930s, despite that era's vast changes in labor relations. Across the country, Depression-era conditions unleashed militant unionism and prompted new federal and state laws that recognized unions, fostered "industrial pluralism," and promoted collective bargaining.[27] Yet Wisconsin's safety code proceedings did not accommodate these innovations, not even to transform advisory panels into pluralistic union-management councils. Wisconsin administrators continued to manipulate committee membership both to favor technical specialists and to exploit new technology.

Using traditional procedures, Wisconsin's commission created and revised safety codes in the 1930s. It assembled new advisory panels for explosives as well as tunnel and trench construction, and it revived old committees to update workshop, building construction, electrical, spray coating, ventilation, and mining regulations. Technological change prompted most revisions, though employers and unions sometimes inspired improvements. Industry-oriented groups like the American Standards Association, American Society of Safety Engineers, United States of America Standards Institute, and U.S. Bureau of Labor Statistics joined older organizations to formulate new model national codes, and Wisconsin regulators usually adhered to them. On one occasion, Wisconsin's Electrical Code committee actually weakened a state voltage regulation to comply with the national standard. Simultaneously, California's Accident Prevention Bureau adopted national codes "whenever possible," though local "petty jealousies" sometimes prevented full approval.[28]

When it assembled safety code advisory committees in the 1930s, Wisconsin's commission still sought representatives from organizations like the Wisconsin Manufacturers Association (WMA), the Milwaukee Association of Commerce, and the WSFL, choosing members informally. "Please go over the names," commission secretary Arthur Altmeyer told building engineer W. C. Muehlstein about construction committee nominees, and "recommend to the commission the ones you think should be selected." Technicians continued to dominate Wisconsin advisory panels, not local employers or unions. The 1933 building construction committee contained two building contractors, a structural engineer, an architect, one WSFL representative, a Milwaukee building inspector, and a commission deputy. The 1934 State Electrical Code committee included municipal, state, and federal regulators; utility executives; equipment manufacturers; and a trade unionist.[29] New Deal "industrial pluralism" hardly affected Wisconsin's advisory committees.

Other states' safety code operations better accommodated New Deal labor policy. Illinois lawmakers finally granted that state's Industrial Commission safety order–making power in 1936, notably requiring "equal" employer-employee representation on advisory committees. A 1939 Alabama law created the Department of Industrial Relations with advisory panels "composed of employers, employees and experts." New York's advisory committees did favor experts, though state law still recommended employer, employee, and expert members.[30]

Nonetheless, organized labor still guarded labor departments everywhere. Viewing the Industrial Commission as a friend of workers, Wisconsin labor leaders exerted political muscle at crucial moments both to defend agency appropriations and to deflect legislative attacks on agency powers. The commission needed that support, due to the rising administrative law reform movement.

State Labor Commissions and Administrative Law Reform

After World War I, legal authorities accommodated administrative agencies like Wisconsin's Industrial Commission. State and federal courts shed their Classical legal opposition to administrative power, and even conservatives accepted safety code making and workers' compensation as constitutionally sound. Rather than hearing constitutional test cases, consequently, state courts now decided intermediate questions about administrative decisions' reasonableness and conformity with statutory mandates.[31]

A peculiar incident in Wisconsin, however, revealed that antiadministration views persisted. In 1925, the Paine Lumber Company of Oshkosh circulated a pamphlet entitled *Soviet Government in Wisconsin* to defend traditional business rights against the Industrial Commission's alleged administrative despotism. As survivors of a declining cutthroat business, Paine Company owners linked proprietary nineteenth-century business values to Classical legal ideas. Embracing the "rule-of-law" tradition articulated by conservative theorists like A. V. Dicey, their pamphlet treated administrative bodies as engines of tyranny that needed to be subordinated to regular law courts.

Soviet Government in Wisconsin portrayed the state's workers' compensation program as the epitome of administrative tyranny. Under that program, the pamphlet said, the Industrial Commission made frivolous compensation awards (such as one covering an injury from a snowball fight), while requiring company officers to spend hundreds of dollars in travel costs to contest minuscule claims without meaningful judicial review. According to the Paine Company, moreover, Wisconsin's industrial commissioners turned voters against businessmen with "lurid speech" about industrial accidents, and then solicited workers' political favor with lavish compensation payments. Such chicanery rivaled Soviet totalitarianism. "The Red Army of Russia sustains the Bolsheviki control," the company charged, just as "the commissions of Wisconsin, with their political tentacles, are sustaining a similar despotism."[32]

Although business troubles partly explain the Paine Company's angry pamphlet, an emerging legal crisis promoted it as well. As would become clearer in the 1930s, lawyers feared that administrative expansion would displace the adversarial proceedings and judicial institutions that had shaped their profession. During the post–Red Scare 1920s, especially, lawyers equated the judicial rule-of-law tradition with capitalism and Americanism, and easily agreed with A. V. Dicey's old view that administrative bodies based on European models were alien to Anglo-American legal culture. And contemporary theorists did little to allay lawyers' worries. John Dickinson and Felix Frankfurter defended administrative bodies' flexibility and expertise but struggled to reconcile agency procedures with constitutional norms. Administrative institutions remained controversial.[33]

With lawyers' encouragement, hence, employers around the country began challenging safety rules in court, reflecting a growing movement to shift safety regulation out of the confines of the administrative process and back into adversarial judicial proceedings. New York employers used judicial review procedures to secure rulings that upheld one regulation involving commercial laundries, struck down another requiring interlocking devices on laundry machines, and affirmed a third governing fire escapes. Moreover, in *Schumer v. Caplin* (1925), a liability suit involving a subcontracted window washer, the New York Court of Appeals decided that violating a safety rule did not by itself establish negligence, though violating a statute would. This ruling extended judicial authority over safety rules into tort actions and downgraded those rules' legal authority.[34]

Cases elsewhere also expanded judicial jurisdiction over safety orders while limiting those orders' legal weight. In Ohio, the state supreme court ruled that legislation allowing it to hear direct appeals on safety orders also permitted it to review those orders in workers' compensation penalty proceedings. In California, the state supreme court decided in a compensation penalty-clause ruling of its own that an employer's failure to know safety orders did not constitute the "serious and wilful misconduct" needed to justify additional compensation, even though his failure to know a safety statute would. In Wisconsin, the supreme court began evaluating safety rules exclusively in 15 percent compensation penalty cases, thereby circumventing the statutory requirement that courts treat safety orders as prima facie lawful, except in proceedings brought expressly to review them. And as it broadened its jurisdiction, Wisconsin's supreme court began nullifying safety orders. In one case, the court set aside a scaffold order that seemingly required a guarantee of safety rather than reasonable freedom from danger as the state's safe place statute stipulated. In another, the court annulled a trench excavation order as too vague to notify employers about what safeguards were required. In a third ruling, the Wisconsin court invalidated an elevator operation order, ruling that regulations could cover only "inanimate objects," not methods of management.[35]

Simultaneously, Wisconsin Supreme Court decisions constricted the state safe place statute's meaning. In *Miller v. Paine* (1930), a work injury suit, Justice Walter Owen limited the safe place law only to physical apparatus, not managerial activities like providing warnings about dangerous conditions. In a rehearing, attorneys countered with a letter from the elderly John Commons, who reported that lawmakers had purposefully added "employment" to the phrase "place of employment" in the original act, extending it to "the whole of the personal relations between employer and employe[e] and not merely the mechanical place of employment." Acknowledging this, Owen now held that the safe place statute required "that not only the place of employment shall be safe for employees, but [also] that the employment itself shall be safe."[36] Nonetheless, the Wisconsin court limited its view again in *Baker v. Janesville Traction Company* (1931), another liability lawsuit. Here, Justice Edward Fairchild applied the safe place statute only to "the control of those methods and processes which are used in the employer's business," not to dangers intruding from outside that business. In dissent, Justice Oscar Fritz reiterated the second *Miller* ruling that safe place legislation covered the whole of "employment," but to no avail. Like a contemporaneous U.S. Court of Appeals decision that restricted Ohio's safe place statute, Wisconsin's supreme court confined safe place law to production processes under direct company control.[37]

This narrow interpretation tightened the Wisconsin court's treatment of safety orders. In *Fritschler v. Industrial Commission* (1932), a penalty-clause case, Wisconsin's supreme court held that Safety Order No. 1228 governing electrical wires did not apply to an electrocuted construction worker because the high-voltage wires in question were not under his employer's management. Again ignoring *Miller*, the court no longer deemed employment to be a "zone of danger" but instead limited it to employer-controlled work processes. Like other states' courts, Wisconsin's supreme court circumscribed safe place legislation upon which safety order making relied at the same time that it expanded judicial authority to review safety orders. Its decision illuminated the movement to reconcile administrative institutions to the U.S. court-based litigation process and common-law norms.[38]

Legislation introduced in 1929 to strengthen Wisconsin safety orders' legal status paralleled judicial developments. One senate bill proposed safety orders' incorporation into *Wisconsin Statutes* to give them "the same force and effect as though enacted by the legislature." A second recommended that the commission or state Legislative Reference Bureau codify the General Orders on Safety. A third senate bill suggested that all Wisconsin agencies hold public hearings, publish orders thirty days before effective dates, and provide review of orders. A final bill in the assembly advised gubernatorial endorsement of Industrial Commission orders that were "legislative" in character. In New York, simultaneously, statutes required that the state Industrial Board's safety orders receive approval from the industrial commissioner (a gubernatorial appointee), thereby treating safety rules as legislation needing executive acceptance.[39]

In Wisconsin, Industrial Commission chair Fred Wilcox appeared before the State Bar Association of Wisconsin (SBAW) to oppose such restrictions. He denied that his agency was despotic, insisted that safety order making accommodated all affected parties, and rejected pending reforms on grounds that they would hinder the Industrial Commission's operation. Wilcox especially resisted legislative or gubernatorial approval of safety orders, claiming that lawmakers and governors lacked expertise to evaluate orders properly. Organized labor, newspapers, and Progressives agreed. Praising the Industrial Commission as a "bulwark of the rights of labor," *The Capital Times* denounced reform legislation as "a step backward" and "a challenge to the Wisconsin idea." Gubernatorial approval, the paper argued, would impede the order-making process. By the time that the governor acted, it observed, "the child laborers will probably have grown up or the buildings burned down." Lawmakers subsequently defeated all of the 1929 administrative reform bills, along with others in 1931 and 1933.[40]

Still, Wisconsin's legal community dreaded state commissions. After organizing a committee to study agencies' powers and procedures in 1930, SBAW leaders persistently warned that Wisconsin commissions lacked meaning-ful restraints under the checks-and-balances system. According to SBAW president J. G. Hargrove, those agencies commingled functions of all three branches of government and became "a law unto itself." The legislature abet-ted this danger, he added, by granting commissions sweeping powers without providing adequate judicial review, while state courts upheld such authority "with a rather surprising unanimity of opinion." This fervent SBAW discus-sion coincided with the emerging legalist backlash against administrative government as evidenced by John Dickinson's *Administrative Justice and the Supremacy of Law in the United States* (1927) and Lord Hewitt's *New Des-potism* (1929). Meantime, bar associations in California, Ohio, New York, and other states faulted problems of lawyering before commissions, such as competition from lay practitioners, the location of hearings, and inadequate procedures for resolving disputes.[41]

The onset of the Great Depression transformed the administrative law debate, with complicated effects on state work safety and health programs. Addressing the economic crisis, President Franklin Roosevelt's New Deal and parallel state programs enlarged federal and state administrative apparatus, fostering a broad ideological shift in favor of administrative governance. After controversial anti–New Deal decisions in 1935 and 1936, even the U.S. Supreme Court endorsed administrative measures by 1937, seeming to end Classical jurisprudence. Nonetheless, a coalition of conservative Democrats, Republicans, business leaders, and lawyers led by erstwhile legal scholar Roscoe Pound mobilized to restrict burgeoning New Deal–era administra-tive institutions. Lawyers created the American Bar Association's Special Committee on Administrative Law in 1933, while politicians established the American Liberty League in 1934 to oppose New Deal programs. Though divided, conservatives eventually accepted mild reforms to judicialize agency procedures and subject agencies to closer judicial review. President Roosevelt vetoed such legislation in 1940, but similar proposals engendered the federal Administrative Procedure Act of 1946. A "fierce compromise" that strictly governed administrative rule making, adjudication, and judicial review pro-cedures, the federal APA reconciled administrative governance with tradi-tional judicial authority and adversarial practices.[42]

In the states, administrative law evolved unevenly. Despite administra-tive growth through the 1930s, Alabama did not enact an APA until 1981. In Illinois, the legal community began considering reform only in 1938, and lawmakers then incrementally addressed administrative adjudication (1945), publication of rules (1951), and uniform procedures (1975). In New York, anti–New Deal conservatives in state and local bars sought restraints on administrative bodies, but organized labor and liberal Democrats blocked them, except for a 1938 constitutional amendment requiring publication of agency regulations. Not until 1975 did New York enact an APA (exempting workers' compensation). Meanwhile, reform-minded bar associations secured Ohio's APA in 1943, and California statutes mandating publication of administrative rules (1941), uniform adjudicatory procedures (1947), and (exempting safety codes) regularized rule making (1947).[43]

As for Wisconsin, the state's older administrative law reform movement stalled during the mid-1930s, while Progressive Philip La Follette remained governor and New Dealers prevailed. Still, the SBAW committee on commissions churned out reports criticizing state commissions and the courts' lack of effective oversight. Working with University of Wisconsin law professor Ray Brown, the SBAW sought an independent court of administrative review, but without success. Simultaneously, Wisconsin chief justice Marvin Rosenberry publicly admitted that legal power had shifted from the courts and legislature to administrative bodies, imperiling the rights of ordinary citizens. Yet even he could envision no better solution to modern problems than administrative regulation.[44]

Then, Republican Milwaukee industrialist Julius Heil's stunning gubernatorial victory in 1938 opened the door for administrative reform. The conservative new governor quickly joined the WMA and the SBAW to advocate uniform agency procedures and a new administrative review board, though as elsewhere, liberals including state administrators, Wisconsin law school dean Lloyd Garrison, the WSFL, and Railroad Brotherhoods blocked those proposals. When liberal opposition faded during World War II, however, Wisconsin lawmakers enacted a modified version of a model statute proffered by the American Bar Association and the National Conference of Commissioners on Uniform State Laws with SBAW endorsement.[45]

Like other APAs, Wisconsin's law moderately restrained administrative power. Adopting "little that is really new," according to SBAW committee chair Ralph M. Hoyt, the Wisconsin APA brought various boards and commissions under a single statute, giving them uniform rule making, adjudicatory, and judicial review procedures. Wisconsin's APA did not alter the Industrial Commission's advisory committee system, nor as a concession to organized labor and state administrators did it cover workers' compensation hearings. Yet its most controversial section created new judicial review requirements for safety orders and other administrative rules: it empowered courts to "reverse or modify" commission determinations that infringed upon constitutional rights, exceeded agency authority, rested on improper procedure, or, most important, were "unsupported by substantial evidence" or were "arbitrary and capricious." These provisions enlarged the jurisdiction of adversarial judicial proceedings to contest safety orders, and for the first time gave Wisconsin courts authority to alter such orders. Contrary to the theory that a New Deal "constitutional revolution" ended the Classical era of judicial restraint on administrative expansion, Wisconsin's APA and other APAs perpetuated the courts' intervention in economic policy through their supervision of administrative procedure.[46]

Wisconsin's supreme court confirmed the judiciary's new authority over the Industrial Commission. In *Robert A. Johnston v. Industrial Commission* (1943), the Wisconsin court explicitly rationalized its assessment of safety orders in 15 percent of penalty-clause cases. Whereas state statutes provided that safety orders "in conformity with law" could be reviewed only through specified procedures, the *Johnston* decision reasoned that because compensation penalties depended on employers' violation of "lawful" orders, court proceedings could evaluate safety orders' lawfulness in penalty cases. In this instance, consequently, the court rejected an order requiring "secure footing" on floors as overly broad under Wisconsin's safe place law.[47] Along with Wisconsin's APA, hence, *Johnston* curtailed the Industrial Commission's wide-open discretion and invited Wisconsin courts to reassert old common-law notions of safe employment.

Nonetheless, the conservative shift in administrative law failed to weaken the Wisconsin commission's attachment to technocrats. Court decisions and administrative procedure legislation did not open safety code making to greater local employer or union influence. Legal attention to "fair representation" in the administrative process would not emerge until the 1970s.[48] Wisconsin's commission and similar bodies like New York's new Board of Standards thus conducted safety code proceedings as before. In spite of changing labor relations and administrative law, Wisconsin's Industrial Commission sustained its ameliorative, technocratic approach to work safety.

8

The Limits of Law Enforcement

Like its prolific safety code–making operation, the Wisconsin Industrial Commission's safety law enforcement system remained dynamic after 1920. Under Wisconsin's well-integrated program, workers' compensation insurance continued to provide inducements for employer safety code compliance, while Industrial Commission deputies still posed as technicians helping companies to satisfy safety regulations. The state's 15 percent compensation penalty clause, moreover, became Wisconsin's principal tool for compelling safety rule observance. Eventually, this system would erode, due to insurance industry changes, increasing employer challenges to compensation claims, budgetary restrictions, and the routinization of inspection work, but it stayed formidable through the 1920s.

Historical wisdom, however, denies such a portrait of safety law enforcement for the post–World War I period. Historians and social critics alike have depicted compensation as an industry-oriented program that insufficiently covered injured workers' lost earnings (though benefits rose), allowed employers to offset compensation payments by lowering wages, and served mainly to stabilize business costs. Scholars have also condemned post–World War I factory legislation as too "spare, static, underfunded, and underenforced" to be effective. Even those who see workers' compensation as an ongoing stimulus for business-sponsored safety work contend that state factory laws declined as a force. Although state regulatory programs like Wisconsin's may have engendered safety improvements, they argue, employers and private market forces stimulated far more safety activity after 1920 than government action did.[1]

Nonetheless, Wisconsin's and other states' safety law enforcement programs were not just "spare" operations during this period. Although Alabama still provided little protection aside from its modest workers' compensation plan, Wisconsin, New York, Ohio, California, and Illinois all perpetuated the distinctive compensation insurance regimes, factory inspection apparatus, and woman and child labor programs that they had established in the Progressive years. Though diminished, moreover, labor leaders and Progressives still lobbied to protect and even strengthen state safety law departments, especially in New York, Ohio, and Wisconsin. Although administrative reorganization, budgetary restraint, and business opposition took their toll, state enforcement programs did not simply degenerate into impotent bureaucracies, "captured" agencies, or components of Herbert Hoover's probusiness "associative state."[2] Most sustained substantial operations.

If safety law enforcement seemed "weak" after 1920, then that appearance resulted not from inner institutional decay but from the distinctive way in which each state's program interacted with new economic developments. For one thing, insurers established regional and national rating organizations like the National Council on Workmen's Compensation Insurance (NCWCI) to rationalize compensation costs, much as manufacturers consolidated in giant corporations and trade associations to stabilize economic markets. This emergent rating regime created a private system of insurance industry governance over workplace safety that regularized insurance rates and competed with state rate-making institutions.[3] For another thing, better safeguarding, enhanced productivity, and excess capacity discouraged further industrial investment in safety devices. In manufacturing, electrical power and safer machinery lowered the rate of mechanically caused workplace injuries, so that industrial managers increasingly utilized personnel practices rather than engineering techniques to prevent accidents. The spread of experience-based insurance merit-rating schemes reinforced these trends by removing the necessity of machine safeguarding to reduce compensation expenses. Indeed, to minimize costs, employers and insurers began contesting compensation and penalty claims, further disrupting compensation's safety incentives. With help from unions, workers fought back by defending their claims, but this transformed compensation into an adversarial system that adjudicated benefits more than it stimulated safeguarding.[4]

Because of these developments, private and public workplace safety strategies diverged after 1920. While insurance companies and advanced manufacturers attended to the "man problem" underlying industrial accidents, work safety regulators in Wisconsin and other states still, by law, focused on physical safeguarding. Hence, dwindling state influence on mainstream corporate safety practices resulted not just from institutional "spareness" but from a growing mismatch between labor departments' engineering orientation and big industry's managerial emphasis. Wisconsin's Industrial Commission itself remained a strong institution, but its engineering-based enforcement system fell out of step with industrial realities.

The Regularization of Compensation Insurance Rating

Of all state workers' compensation programs in 1920, Wisconsin's remained one of the best structured to induce employer safety work. Whereas insurance rate competition and risk spreading undercut compensation incentives elsewhere, Wisconsin's uniform rates, schedule-rating plan, and 15 percent penalty clause together focused exceptional financial pressure on manufacturers to install legally required safeguards. Yet even Wisconsin's program succumbed to insurance industry changes that weakened compensation's inducements.

Back in the 1910s, insurance company actuaries and state insurance fund managers had struggled to formulate reliable industrial classifications and rate schedules for the emerging patchwork of state compensation programs. For technical help, insurers organized state-level rating associations, and then in the 1920s established the nationally oriented NCWCI, all of which imitated other "nonstatist coordinating mechanisms" of the time by extending an uneven system of private governance over compensation insurance. Insurers' main concern was formulating "pure premiums" or "manual rates" in an accurate, "scientific" way, not just following competitive pressures. As New York statistician Leonard Hatch warned in 1924, competition might push premium levels so low that injured workers would not receive adequate compensation and insurers would face bankruptcy. To avoid that predicament, he advised state fund managers and private insurers to cooperate with professional rating associations like New York's private Compensation Inspection Rating Board. Indeed, to regularize premiums, 1922 New York legislation ordered all private carriers to follow the board's rates. Like Wisconsin's 1917 law that required private insurers to join that state's quasi-public Rating and Inspection Bureau and to follow rates approved by the state's Compensation Insurance Board, New York's measure replaced competitive insurance with a regulated market.[5]

Although local rating bureaus often received official state sanction, the NCWCI emerged as the authoritative national rating organization. With its capacity to reconcile nationwide accident statistics with the maze of state compensation plans, the NCWCI produced an annual manual that fixed insurance rates for all American industry. Though a "form of semi-governmental supervision," the manual had varied influence. In Illinois's competitive market, insurance carriers ignored NCWCI rates. So did Ohio state insurance fund managers. Yet New York's Inspection Rating Board and Wisconsin's Rating and Inspection Bureau followed NCWCI rates closely, particularly for small firms in peripheral industries. Like the Wisconsin Industrial Commission's dependence on model safety codes, the Wisconsin rating bureau's use of the NCWCI manual demonstrated reliance on national authorities for uniform standards. Indeed, manufacturers complained that Wisconsin's Compensation Insurance Board accepted NCWCI rates too readily. By so deferring, Wisconsin officials accommodated the NCWCI's emphasis on insurance industry stability rather than safety work.[6]

Nonetheless, it was not the NCWCI manual that generated controversy in the compensation insurance field in the 1920s but competing "schedule" and "experience" methods of merit rating. Under schedule rating, insurers lowered premiums for factories that installed machine safeguards and established plant safety programs according to a schedule of standards. Wisconsin's Rating and Inspection Bureau used this system to encourage safety order compliance by "harmonizing" schedule standards with state regulations in 1919. New York's Inspection Rating Board similarly applied this scheme to small employers, although it did not link schedule standards to safety codes—a cause of criticism. Stock insurance companies, big manufacturers, and exclusive state compensation fund managers, however, advocated the experience method of merit rating, which reduced or raised insurance premiums according to whether firms experienced fewer accidents and lower compensation costs than predicted or suffered greater mishaps and expenses. Although schedule rating encouraged well-guarded "standard factories," these groups contended, it failed to avert "unexplainable accidents." Experience rating better prevented such casualties, they argued, because it rewarded employers not just for safeguarding but also for avoiding financial losses. Still, schedule rating had defenders. New York rating board manager Leon Senior argued that the schedule system more accurately covered risks in factories with fixed machinery and locations. He endorsed New York's plan of "blending" the two systems, applying schedule rating to permanent operations while using experience rating for utilities, contract work, and large firms.[7]

The merit-rating debate exposed disagreement within the compensation insurance industry. According to Pennsylvania Compensation Rating and Inspection Bureau general manager Gregory C. Kelly, merit rating originated as a stock company device to match dividends paid to policyholders by mutual firms. After mutuals captured business with the schedule plan, stock firms advanced experience rating as a "competitive weapon" to win business back. Large manufacturers themselves embraced experience rating, he added, since it allowed them to evade liability in states with "loosely supervised" compensation programs. Sure enough, as Kelly suggested, when experience rating started in California, employers tried to reduce their insurance liability by contesting and lowering workers' benefits.[8]

In Wisconsin, industry pressure for the experience plan eventually upended schedule rating. Initially, the schedule plan enjoyed both business support and strong roots in Wisconsin rate-making institutions that linked mutual firms to state regulators. Experience-rating legislation in 1921 consequently failed. In 1924, however, Wisconsin's Compensation Insurance Board ordered an across-the-board rate increase, inciting vehement protests from the Milwaukee Association of Commerce and stock companies affiliated with the National Bureau of Casualty and Surety Underwriters. The rate hike, they complained, aimed just to protect mutual firms. Employers Mutual and Lumbermen's Mutual denied the charge, but industrial commissioner Fred Wilcox practically admitted it. He accused stock companies of opposing the increase to ruin the mutuals, and warned that their action would lead to compulsory state insurance. Nonetheless, stock companies and big employers secured experience-rating legislation in 1925. And the stock company versus mutual firm rivalry persisted.[9]

These legislative developments altered Wisconsin's compensation insurance program. Though steadfast defenders of schedule rating, Wisconsin's Compensation Insurance Board and Industrial Commission grudgingly solicited information about experience rating from the NCWCI and stock firms, and then applied the new rating plan to large local employers. In 1934, industrial commissioner Voyta Wrabetz acknowledged that for big firms, at least, "schedule rating might well be dropped entirely, since experience rating will very clearly reflect the employers' success in the prevention of accidents." New York and California legislation also accommodated experience rating in 1926 and 1935, respectively.[10]

A movement to regularize rates accompanied the rise of experience-based merit rating. Some analysts maintained that rate regularization focused insurers' and employers' attention on accident-prevention work as the only real way to lower compensation costs. Yet most commentators stressed the benefits of stability. "You must not merit rate to such an extent as to practically destroy the principle of insurance," a Canadian workers' compensation commissioner cautioned. "Otherwise we get back to the principle of individual liability." Hence, avoiding "violent fluctuations" in insurance premiums became a regular feature of experience-based plans. When Ohio's state compensation fund instituted experience rating in 1911, for example, it tempered employers' accountability by rating them over five years (not just the previous year), by limiting credits earned or penalties paid, and by removing "catastrophic" losses from rate calculation, paying them out of a special reserve fund instead. Experience rating should provide "a profit motive to the employer for eliminating accident cost," Ohio actuary E. I. Evans explained, but penalties "should not be a burden that will result in the employer discontinuing business."[11]

Although workers' compensation supposedly enhanced "enterprise responsibility," then, commentators worried that insurance undermined compensation's inducements for safety work. According to a company welfare director in New York, compensation insurance allowed employers small and large to disregard safeguarding, since they could "[shift] their responsibility on to the insurance company and rest comfortably so far as their liability is concerned." According to New York insurance fund manager Charles G. Smith, moreover, no-fault insurance weakened incentives by allowing the employer "to relieve himself of all additional responsibility or liability for an accident caused by negligence."[12]

In Wisconsin, consequently, labor leaders lobbied to reform compensation insurance. In 1923, 1925, 1927, and 1931, prolabor state assemblymen introduced measures to establish a monopolistic state compensation insurance fund. In 1927, a prolabor lawmaker advanced another law to assign the Compensation Insurance Board's duties to the friendlier Industrial Commission. In 1930, the Wisconsin State Federation of Labor (WSFL) advocated an exclusive state fund, demanded schedule rating, and denounced insurance firms for minimizing worker benefits. Yet even though the WSFL and its political allies boosted compensation benefits through the 1920s, they could not prevent compensation's erosion as an incentive system.[13] As elsewhere, compensation insurance stabilized rates in Wisconsin, thereby transforming compensation into an indemnity plan.

Wisconsin's Unique Enforcement Tool

Despite insurance regularization, Wisconsin's 15 percent penalty clause sustained compensation's safety incentives. Unlike California's penalty provision that allowed that state's Industrial Accident Commission to raise compensation by 50 percent for "serious and wilful [employer] misconduct," Wisconsin's clause targeted accidents caused by employer safety code violations. In such cases, Wisconsin's Industrial Commission sidestepped the insurance system's "shock-absorbing" effects by requiring employers rather than insurers to pay 15 percent extra compensation. Such direct employer accountability, industrial commissioner Voyta Wrabetz asserted, "is a forcible argument for proper guarding and the institution of ample safety measures and organization."[14]

Wisconsin's Industrial Commission used the penalty clause to affirm managerial responsibility. Although it could reduce worker benefits for their own "wilful" safety law violations, the commission did so only for "deliberate" infractions, not "thoughtless" neglect, a position upheld in court. Regarding employers, the agency treated failure to comply, or even failure to make reasonable efforts to comply, with safety orders as sufficient grounds for the penalty. Employers tried to circumvent this policy by seeking written commission certification of code compliance, and by challenging penalty claims, but the agency declined such requests and occasionally extended penalty-clause actions to general safe place statute violations.[15]

Wisconsin's commission imposed compensation penalties regularly. From 1921 to 1931, the agency penalized employers 5,037 times in 2.3 percent of all compensation cases at an average cost of $73 each. This raised employer liability by $366,683.47 out of a total $37,515,973 paid. Penalties for construction, electrical, and elevator code violations predominated, with several for spray painting. By comparison, the commission penalized workers only 68 times between 1923 and 1931, reducing total benefits by $3,949.23, an average of $58 per violation. Significantly, Wisconsin penalty clause rulings against employers changed over time as factory conditions improved. From 1923 to 1926, the Industrial Commission penalized companies an average of 472 times yearly for factory safety infractions, but after 1926, such violations subsided, and the agency increasingly penalized building contractors—50 times annually—for construction code infringements. By the mid-1930s, employers increasingly contested these actions, turning penalty-clause proceedings into an adversarial process.[16]

Compensation penalty clauses received different attention elsewhere. New York, Illinois, and Alabama totally lacked such legislation. California's clause based on "serious and wilful misconduct" did not clearly apply to safety order infractions. Initially, California's supreme court held that safety rule violations were sufficient to find employers guilty of "wilful misconduct," but subsequently it determined that safety order noncompliance could serve only as evidence of such behavior. California's penalty clause became a remedy for employer negligence, not for code violations.[17]

In Ohio, by contrast, legal maneuvering transformed penalty legislation. Originally, Ohio law allowed workers to sue employers for injuries resulting from violations of "lawful [safety] requirements," but Ohio's conservative World War I–era supreme court restricted this right by holding that neither safe place legislation nor factory laws prescribing general courses of safe conduct applied. In 1923, however, a newly impaneled Ohio Supreme Court updated local safety law by declaring that Ohio's safe place statute, factory acts, and administrative regulations did indeed all qualify as "lawful requirements" upon which injured workers might sue. Later that year, moreover, voters replaced the right to sue with a constitutional amendment empowering Ohio's Industrial Commission to increase compensation by 15 to 50 percent whenever injury, disease, or death resulted from employer failure to comply with any "specific requirement" for worker safety and health "enacted by the General Assembly or in the form of an order" issued by the Ohio commission. Unlike Wisconsin and California laws, the amendment directed Ohio's compensation fund to pay increased compensation awards while ordering employers to reimburse the state. Likely a business-labor compromise, Ohio's amendment eliminated costly lawsuits but fortified employers' accountability.[18]

Ohio State Federation of Labor (OSFL) leaders defended the new penalty law. Although conceding that it eliminated workers' right to sue, OSFL secretary Thomas Donnelly contended that the amendment let injured workers increase claims against delinquent employers without risking compensation benefits. It also allowed workers to promote safety. "Hanging over the heads of the employers in this State," Donnelly asserted, "is a certainty that if the workers know the safety laws of the State, . . . and the employers violate any provisions of the law, that penalization or additional awards are a certainty in every case." Donnelly's statement exhibited Ohio workers' exceptional role in safety law enforcement during the 1920s.[19] In few other states, Wisconsin included, did workers mobilize as directly to implement safety legislation.

Ohio workers' leverage dwindled, however, when judges interpreted the state's penalty clause and administrators implemented it. Ohio's supreme court quickly construed the 1923 constitutional amendment to give the state Industrial Commission conclusive power to determine facts in increased compensation cases, but this decision allowed the agency to disapprove increases on narrow grounds without judicial interference. After 1930, Ohio's supreme court itself slipped back into legal conservatism by denying that the state's safe place statute and building construction code were "specific requirements" whose violation justified increases.[20] Ultimately, Ohio's Industrial Commission and supreme court restricted the state penalty clause.

Compensation penalty laws thus had different uses. California and Ohio deployed them as administrative substitutes for injured workers' right to sue. Wisconsin's penalty clause, by contrast, sustained a form of employer liability within the increasingly regularized compensation insurance system that administrators used to enforce safety codes. How well penalty laws anywhere promoted safe workplaces, however, still rested on the efficacy of factory inspection.

The Routinization of Factory Inspection

In contrast to workers' compensation insurance rate making's institutional centralization during the 1920s and 1930s, factory inspection remained a patchwork of autonomous state programs. Rather than defer to national authorities as safety code and insurance rate makers often did, inspection divisions in Wisconsin and other states responded primarily to local circumstances. And state inspection operations stayed remarkably active, despite budgetary and political constraints.

Wisconsin's factory inspectorate entered the new era strongly. World War I–era budget increases allowed the Wisconsin Industrial Commission's new Safety and Sanitation Department to expand the number of district deputies from eight to eleven, and then add building, boiler, ventilation, electrical, mine, fire, and elevator specialists. Wisconsin's inspectorate consequently doubled from twelve to twenty-four deputies, while its Woman and Child Labor Department jumped from one to five deputies. Thus, whereas Wisconsin had employed one safety inspector for every twenty-two thousand workers in 1917, it engaged one for every eleven thousand by 1922.[21]

Wisconsin's commission continued to professionalize its growing inspection staff. The agency not only hired new deputies who met civil service requirements but also employed some with engineering and law degrees, evoking criticism that they were too academic. The commission also perpetuated the deputies' in-house training and kept up their methodical work routines, dispatching them with large portfolios containing safety codes, safety literature, and blank inspection forms. Deputies continued educational work but now concentrated more on "command-and-control" law enforcement. District deputies, a commission report stated, inspected workplaces "to see that they comply with the regulations," and if not, to advise employers about methods of compliance.[22]

Despite World War I–era staff increases and professionalization, Wisconsin's inspection force did not keep up with industrial expansion due to stagnant legislative appropriations. Although they averaged eight to ten thousand inspections annually, administrators figured, deputies visited each plant no more frequently than once every year, and conducted comprehensive inspections much less often. By the early 1930s, observers characterized Wisconsin inspection work as "spotty." They urged lawmakers to fund sixteen more deputies.[23]

To pick up the slack, Wisconsin's Industrial Commission appointed numerous adjunct inspectors, including municipal inspectors in major cities, and nearly a hundred insurance inspectors for the rest of the state. These "special deputies" examined 80 percent of Wisconsin's elevators and 90 percent of the state's boilers during the 1920s and 1930s. Wisconsin officials found, however, that special deputies conducted only "facial inspections" and cited fewer safety code infractions than state deputies did. Even Employers Mutual executives regarded insurance inspections as a "farce." To improve special deputies' work, Wisconsin administrators threatened to revoke insurance company inspection privileges, began certifying insurance inspectors, and initiated follow-up inspections, all to no avail. State officials eventually conceded that special deputies' performance reflected the kind of facilities that they inspected and the greater frequency of their visits, but they remained dissatisfied.[24]

Concerns about special deputies' proficiency reflected the Wisconsin commission's growing focus on law enforcement. The agency increased prosecution actions for safety code infractions, just as it sustained lawsuits against woman and child labor law violations. Significantly, the prosecutions mainly targeted firms not covered by compensation and the 15 percent penalty clause, typically outfits located in cities of fewer than thirty-nine thousand residents, half in towns of fewer than ten thousand inhabitants.[25] Such small-town factories and workshops fell outside of big-industry's safety movement, and Wisconsin enforcement actions modernized them. After encountering numerous plants and creameries that neither carried boiler insurance nor hired qualified operators, for instance, the Industrial Commission's boiler inspector not only looked for safety violations but also taught operators boiler care. Moreover, dedicating one-half of its prosecutions in the early 1930s to spray-coating violations, particularly in small-town automobile body shops, the Wisconsin agency prodded those local firms to adopt modern spray-painting apparatus.[26]

Wisconsin's commission demonstrated even more attention to code enforcement in the 5,037 penalty-clause actions that it instigated in the 1920s, as compared to just 67 prosecutions. Indeed, Wisconsin deputies devoted one-fifth of their annual inspections (about 2,000) to investigating penalty-rule cases.[27] Despite its thinly stretched inspection force, hence, the Wisconsin agency conducted one of the nation's liveliest enforcement systems during the 1920s and 1930s, not the "spare" operation portrayed by historical wisdom.

Other states' factory inspection programs continued to vary in strength and character. With an operation that truly was spare, Alabama lacked a factory inspectorate until the late 1930s.[28] Illinois, by contrast, maintained a staff of forty-six deputies and five special inspectors in the Factory Inspection Division allied with the state Republican Party. Most Illinois deputies were political protégés of the governor, who appointed them "temporarily" to evade civil service rules and rotated them through Cook County to minimize fraternization with rival politicians. Wearing badges and issuing "tickets," moreover, Illinois deputies were purely labor law enforcers. They focused on woman and child protective measures, and devoted only 6 percent of their inspections to workplace safety and health. And although Illinois's inspection division prosecuted hundreds of woman and child labor law infractions, it rarely indicted safety law violators.[29]

Ohio's factory inspection system, meanwhile, persisted as a weak technical advising service. To labor leaders' and state officials' dismay, Ohio's conservative 1921 legislature shifted the factory inspection staff of twenty-four deputies to the new Department of Industrial Relations, leaving workers' compensation and safety code development in the separate Industrial Commission. The commission subsequently enhanced deputies' authority by formalizing safety regulations in 1924, while the state's 1923 penalty clause augmented deputies' power to investigate code violations. Yet Ohio's Industrial Commission failed to support the inspection division with vigorous penalty-clause actions. The division itself rarely prosecuted code violators, though it did sometimes threaten to enjoin hazardous processes. "Co-operation rather than prosecution" prevailed in Ohio's program.[30]

California's inspectorate resembled Wisconsin's more than Ohio's. With a separate Industrial Welfare Commission to administer woman and child labor laws, California created the Bureau of Industrial Accident Prevention that arranged thirty-two supervisors and inspectors into nine specialized "sections" to enforce safety and health codes, supplementing its staff with hundreds of certified elevator and boiler inspectors from private insurance firms. Much like the Wisconsin commission's safety department, the California bureau relied primarily on workers' compensation incentives to induce employer safety code compliance. It rarely prosecuted safety and health law violators or invoked the state's injunction and tagging procedures. Unlike Wisconsin, however, California officials seldom utilized the state's 50 percent penalty clause.[31] While Wisconsin's program grew more coercive, California's remained cooperative.

Comparatively, New York's factory inspectorate remained the largest and most coercive of all. By 1924, after Governor Al Smith and Democratic lawmakers reversed conservative administrative cuts, New York's inspection division rebounded to 188 factory, mining, mercantile, and boiler inspectors in a far-flung operation that enforced safety, health, woman, and child labor laws in sixty-six thousand factories, workshops, and mines, three-quarters of which employed fifty or fewer workers. Indeed, much New York enforcement work involved tiny immigrant-run New York City workshops. And factory inspection continued to function separately from workers' compensation and industrial code proceedings, a bureaucratic distinction reinforced by inspectors' unionization. Inspectors lacked support of a compensation penalty clause, and failed even to get accident reports from the compensation division. Distrust and competition raged between New York labor department divisions.[32]

Meanwhile, inspection division supervisors imposed strict bureaucratic controls. Worried about sloppiness or corruption, supervisors regulated inspectors with "gum-shoeing" techniques (undercover investigations of inspectors' performance) and paperwork. This produced "lock-stepping" discipline but disrupted inspectors' duties. Rigid scheduling precluded New York inspectors from meeting with company officers, while paperwork diverted them from scrutinizing workplaces. The New York factory inspector, one investigator observed, "spends more time in making check-marks [on forms] than he does in making inspections." His "primary purpose," she added, "is to discover violations of the law and report them to headquarters, rather than to educate the employer."[33]

Not surprisingly, New York's inspection division emphasized pressure tactics to enforce labor laws. "We secure a greater measure of compliance with orders because of the state's power to compel respect for its authority," division head James Gagnon reported, "than could be done by any means of education yet conceived." Yearly, consequently, Gagnon's division conducted hundreds of worker-instigated "complaint" inspections, occasionally threatened to "tag" and decommission dangerous machinery, and issued thousands of "counsel letters" ordering safety code compliance. In New York City, moreover, the division held thousands of courtroomlike "calendar hearings" to rebuke employers for chronic safety and health law infractions. And New York inspection officials prosecuted labor law violators more aggressively (thousands of times annually) than did other labor departments, though judges sometimes suspended penalties, particularly in woman and child labor cases. In one 1930 case, a Dutchess County judge investigated a Poughkeepsie sawmill himself and then refused to enforce a sanitation order.[34] Still, New York's inspection division sustained coercive practices.

Altogether, then, various state labor departments continued to exert considerable presence in workshops and factories through the 1920s, though in different ways. And neither Depression-era budgetary retrenchment nor new priorities obliterated that presence in the 1930s. Despite vacant deputy posts, missed inspections, and skipped accident investigations, Wisconsin's Industrial Commission kept its inspectorate intact and returned it to full strength after 1940. Simultaneously, the agency increased safety law prosecutions from only five in 1930 to twenty-two in 1940. Elsewhere, reform politics and militant unionism stimulated expansion of inspection programs. Laggard Alabama lawmakers finally established a weak labor department in 1935, and then created the modern Department of Industrial Relations staffed with factory inspectors in 1939.[35]

Yet ongoing structural change in American industry still attenuated state safety law enforcement. Nationwide, insurance rate systemization, experience rating, and increasing contests over worker claims all relaxed workers' compensation's financial inducements for safety law compliance, while new engineering and managerial practices moved big industry away from mechanical safeguarding. Safety law's impact was consequently complicated, as industrial accident statistics demonstrate.

Accident Trends and Safety Law's Mismatch with Industrial Change

At New York State's ninth annual industrial safety congress in 1925, labor department statistician Leonard Hatch considered whether New York's workplace accident rate was increasing. "This is undoubtedly the most significant question which could be asked and answered in such a Congress as this," Hatch declared, but then he added that "it is the most difficult one to answer with certainty." His remarks were telling: accident statistics remained inadequate everywhere. On the state level, administrators rarely determined worker exposure levels or accident frequency rates. On the federal level, officials seldom calculated national accident trends or made state-to-state comparisons. Only gradually did federal authorities assemble statistics for specific industries, and not until 1929 did the U.S. Bureau of Labor Statistics (USBLS) publish a comprehensive national report.[36]

Despite their deficiencies, state and federal statistics both suggest two long-term national trends that implicated state safety law enforcement. First of all, private business, various state labor departments, and the USBLS all reported declining rates both of fatalities and of injuries, and a big drop in workdays lost from the 1910s to the late 1920s and 1930s. The USBLS qualified this news by revealing that accidents producing "permanent partial" disabilities continued to rise, though deaths and "total" injuries decreased. More important, the USBLS also disclosed trouble in extractive industries, with the frequency of fatal coal mining accidents and the rate of nonfatal metal mine and quarry injuries both escalating. Indeed, coal mining's old problems with deep mining, mechanization, explosions, and roof falls continued after World War I, as operators and miners alike neglected safety to survive the industry's steep decline.[37]

In manufacturing, however, state statistics indicated a second important trend: a dwindling proportion of machinery accidents. Looking only at fatalities from 1913 to 1925, New York's Leonard Hatch reported a steep decrease in machine-related casualties but a sharp increase from other causes, such as motor vehicles, falls, and handling objects. Hatch admitted that growth of nonmechanical employment partly explained this trend, but he mostly credited the safety movement for reducing machine hazards. Likewise, Wisconsin's Industrial Commission discerned that machinery's part in all accidents dropped from 21.1 percent in 1918 to 14.6 percent in 1928. Wisconsin's agency attributed this decline to effective safety code enforcement.[38]

Given that good news, labor law officials cautioned manufacturers against complacency. Although "it has been quite customary to regard the machine as being now so well guarded as to be almost nonhazardous," the USBLS warned iron and steelmakers in 1929, machine accidents remained a problem. Indeed, New York's labor department reported in 1926 that machinery caused a preponderance of child labor injuries. In the 1930s, Ohio's Safety and Hygiene Division consistently ranked machinery among its "Big Six" accident causes. Wisconsin Industrial Commission secretary Arthur Altmeyer warned in 1932 that modernized machinery presented "new unguarded danger points."[39]

Nonetheless, anecdotal and statistical evidence shows that machine safeguarding really had improved after 1920. New York's Department of Labor cited "amazing results" in equipping factories with safer machinery, while Wisconsin's Industrial Commission observed consistent employer cooperation in installing safety devices. By the late 1920s, consequently, New York complaint inspections dwindled, Ohio deputies issued fewer inspection citations, Wisconsin deputies cited a declining number of safety code infractions, Wisconsin's Industrial Commission imposed fewer compensation penalties, and labor law officials everywhere instituted lesser prosecution actions. State officials commended their own "constant" safety law enforcement efforts for machine safety improvements, though recent scholars attribute that success more to private-sector developments, such as corporate safety programs, professional societies' accident-prevention campaigns, insurers' experience-rating schemes, labor market shifts toward safer jobs, and the electrification of industrial plants.[40]

Actually, state safety programs' impact on accident levels is difficult to measure, although a 1929 USBLS report revealed complex state-to-state differences. For fatal injuries, whose local statistics were generally comparable, the report indicated that Wisconsin experienced a relatively low accident rate during the 1920s in foundries, machine tools, and enamel work but fell behind Illinois, New York, and Ohio in the leather, paper, and meatpacking industries, and even behind Alabama overall. Regarding permanent and temporary injuries, where local reporting requirements made only New York's and Illinois's data comparable to Wisconsin's, the USBLS study disclosed that Wisconsin enjoyed lower accident levels in leather, lumber, machine tools, and meatpacking but higher rates in foundries, paper, and enamelware. As for all accidents in the iron and steel industry, the report showed that California's, Illinois's, New York's, and Ohio's records generally improved but that Wisconsin's worsened. The 1929 USBLS report thus failed to single out a superior state safety law program.[41]

In fact, industrial development complicated all states' work safety efforts. By the mid-1920s, the safety movement had spread to a "progressive minority" of big corporations across the country, just as technological rationalization, systematic management, and welfare programs had. Excess capacity, however, pervaded big industry, causing even advanced firms to curtail new equipment investments and to maximize labor use through speedups and hard-driving practices. Simultaneous productivity gains allowed manufacturers to tolerate higher accident levels. In coal mining, especially, modern machinery and management so reduced employment costs that operators saw compensation expenses drop, regardless of safety conditions or accident rates. As Casualty and Surety Underwriters general manager Albert Whitney reported in 1928, "production per man-hour has increased so much more rapidly [than accident frequency] that the hazard in terms of production has decreased." Coal mine operators now measured accidents "in terms of goods produced," not human loss.[42] Added to machine safeguarding, declining machine injury levels, and insurance rate regularization, excess capacity and improved productivity across industry rendered accidents as a manageable business cost. Employers subsequently addressed work safety through human-based personnel schemes rather than engineering-based physical protection.

Labor law officials echoed industry's new personnel-oriented thinking. New York's Leonard Hatch urged employers in 1925 not to exaggerate work safety's "mechanical element," since worker morale, safe conduct, and other "human elements" of industrial production were causes of accidents. That same year, California inspector D. J. Harris told companies that cautious men were always needed to prevent accidents in addition to safeguards. In the 1930s, Ohio's Safety and Hygiene Division reported high numbers of non-machine-related workplace injuries and advocated greater managerial attention to good housekeeping and worker supervision. Noting in 1939 that 87 percent of accidents were not of machine origin, Wisconsin industrial commissioner Voyta Wrabetz contended that safe worker practices and good housekeeping would prevent most industrial injuries.[43]

In day-to-day administration, however, state officials faced a widening gap between industry and government practices. Whereas safety-minded corporate employers increasingly focused on managing workers, institutional inertia and law confined state administrators to enforcing legally mandated physical safeguards. This emerging mismatch transformed the regulatory function of state agencies like Wisconsin's Industrial Commission. Back in the 1910s, various states' labor departments had joined the business-led safety movement to spread machine safeguarding throughout industry. By the mid-1920s, however, with mainstream factories better safeguarded, and with big employers stressing personnel practices, state safety law agencies shifted their attention to peripheral businesses that failed to modernize plants, maintain safety appliances, or safeguard new equipment. Thus, New York's factory inspection division focused on modernizing unsophisticated urban workshops, while Wisconsin's commission concentrated on renovating upstate factories, small-town creameries, and local auto body shops.

Nonetheless, although state safety law enforcement seemed to languish at the end of the 1920s, it gained new vigor in the late 1930s. Both the number of industrial injuries and the volume of safety law compliance activities increased when business revived after the Depression and then began mobilizing for World War II. Subsequently, Wisconsin's Industrial Commission instigated an unprecedented number of prosecution actions to enforce the state's General Orders on Safety and heating and ventilation code.[44] Moreover, an explosion of silicosis cases exposed the growth of industrial disease, a new battleground for state regulation of industrial work conditions.

The Troubled Campaign against Occupational Disease

In 1938, Wisconsin workers' compensation director Harry A. Nelson praised his state's industrial health legislation. Whereas nineteenth-century common law had usually allowed ailing workers "no recovery whatsoever," he boasted, modern Wisconsin legislation compensated workers for all job-related illnesses. And when combined with safety and health code enforcement, he added, Wisconsin's compensation legislation encouraged business to prevent disease. "In Wisconsin," he remarked, "much occupational disease, rampant ten years ago, has already been greatly reduced, and within the next ten years will be substantially stamped out."[1]

Nelson's statement will surely surprise modern observers. For numerous reasons, modern analysts portray industrial illness as a historical arena of extraordinary governmental neglect. A common law–based "medico-legal discourse," they argue, delimited employers' liability for occupational disease in work injury lawsuits and workers' compensation proceedings. Protective laws for women, they add, excluded female workers from unhealthful jobs rather than lessen workplace health risks. And in the absence of effective federal regulation, big industry manipulated law, science, and politics to evade responsibility for occupational health. Yes, there was some state-level disease-prevention legislation, recent scholars admit. Yet in line with theories about corporate influence on Progressive regulatory reform, they conclude that a "corporate/industrial/scientific system" dominated industrial disease policy though the 1920s.[2]

Still, as Harry Nelson suggested, states took more action against industrial illness than scholars recognize. Although most states continued to treat the problem under the nineteenth-century common-law litigation system, Wisconsin and a few others like Illinois, New York, Ohio, and California gradually, if inconsistently, enacted workplace sanitation, disease prevention, and disease compensation laws. Recent scholarship rightfully questions how widely these measures spread among states, and how effectively they performed, but Wisconsin's and other states' programs truly began to extend modern regulation and social insurance over work-related illness.

Economic and scientific developments conditioned state industrial health reforms. Unlike the environment for industrial safety, the rising corporate economy and "organizational society" produced neither disease-prevention "technical networks," a "health first" movement, an ethic of "managerial responsibility," nor an ameliorative "corporate liberal" business policy toward illness. Instead, employers tried to confine occupational disease to the common-law liability system, where rules debarred their responsibility for most ailments and unclear etiology precluded proof that workplaces caused illness. Reform initially emerged, then, from a splintered industrial hygiene movement that spanned insurance firms, reform associations, labor unions, women's groups, and researchers, as well as agencies like the U.S. Department of Labor, the U.S. Public Health Service (USPHS), and the U.S. Bureau of Mines (USBM). By 1920, reformers lobbied government for action, but big industry predominated with practices that served its own needs: business-oriented research, industry-run health associations, medical screening of employees, evasion of liability claims, and opposition to state regulation. Social attitudes, meanwhile, encouraged neglect of female, immigrant, and black workers' health.[3]

Without a unified national movement, disease law evolved in fragmented local politics. Business and political antipathy to protective labor law kept Alabama from enacting disease compensation or preventive measures until the 1930s, just as politics in Deep South, southwestern, plains, and northern New England states impeded such legislation. Disease law reform was stronger in eastern, midwestern, and western industrial states, but disparate legislation resulted. Interest-group struggle in Illinois, New York, and Ohio generated diverse compensation and factory sanitation laws, but Wisconsin and California politics extended commission systems over occupational health.[4]

Wisconsin's program was notable. Like other states, Wisconsin began enacting factory sanitation and women's protective measures in the late 1890s, but its 1911 industrial commission law made it one of the few states to commit a safe place statute and "representative" advisory committees to industrial health code formulation, and its 1919 disease compensation act made it one of only five states before 1930 to extend "blanket" compensation coverage over occupational illnesses. Just as toward accidents, consequently, Wisconsin's system applied social insurance and a technocratic regulatory policy to industrial disease, especially in the silicosis crisis of the 1930s. By contrast, when Illinois, New York, and Ohio also created occupational health institutions, they enacted narrower disease compensation laws and erected uneven disease abatement programs. Most states, meantime, still left work-related illness to the litigation system.

By 1940, then, American industrial health policy was complex. Occupational health law remained in a patchwork of state laws and institutions, and neither Wisconsin's nor any other state's program "substantially stamped out" disease as Harry Nelson cheerfully predicted. Yet contrary to the historical image, disease law did gradually advance, and programs like Wisconsin's and New York's established pockets of strong state action.

Early Industrial Health Regulation

Until 1890, worker-related illness's main legal forum was the judiciary, just as it was for accidents. In work injury lawsuits, courts applied common-law liability principles to disease by requiring workers to assume the risk for "naturally" occurring job-related ailments, and by holding employers accountable only when their negligence afflicted workers. In Wisconsin, hence, courts usually left workers to bear illness's costs, consistent with the developmental nineteenth-century judicial policy that minimized business risks. The social milieu sustained this view of disease liability, though ideas about accidents were changing. Trade agreements, craft-worker attitudes, business indifference, bacteriological medical theory, and localistic politics all discouraged legal expansion of employers' responsibility for disease.[5]

Yet late-nineteenth-century developments did inspire some regulation. Although employers and doctors still ignored the matter, trade unionists, tenement reformers, settlement-house workers, and women's groups secured sanitation and ventilation laws in Ohio, California, New York, Illinois, and seven other states, including Illinois's 1893 sweatshop act and New York's 1895 bakeshop law. Simultaneous Wisconsin health laws merely banned factory overcrowding and required sanitation reports, but factory inspectors took matters into their own hands. After finding fumes, gases, "great amounts of dust," "perfectly sickening" odors, and "nauseous" privies in factories, and after seeing "indisputedly slight" efforts to address such problems, Wisconsin inspectors summarily ordered employers to install hoods, exhaust pipes, ventilation fans, suction devices, and sewage systems in their shops.[6]

Then, in 1899, the emerging Wisconsin State Federation of Labor (WSFL) helped reformers to pass various measures that regulated sweatshops and bakeries, required factory ventilation, and mandated hoods, suction devices, and blowers for dusty operations. These statutes crudely began replacing judicial common-law liability standards. Wisconsin's sanitation act vaguely demanded action against "odor, filth, vermin, decaying matter or other conditions detrimental to health," while its emery-wheel law required installation of ventilation and suction apparatus without fixing standards. Employers' common-law exemptions from disease liability, however, remained intact.[7]

Reforms after 1910 further supplanted the judicially based liability system, but only partially. On the one hand, early workers' compensation laws like Wisconsin's allowed benefits only for "accidental" injuries or deaths. These measures consequently failed to establish "enterprise responsibility" for protracted workplace illnesses or to create incentives for their prevention. Simultaneously, Illinois law perpetuated worker lawsuits for specified illnesses caused by employer violation of health statutes, while that state's compensation act still covered only "accidental injuries," a position affirmed in court. Massachusetts legislation allowed compensation for "personal injuries," but courts applied that language only to diseases not deemed "natural."[8]

On the other hand, Progressive reform expanded state regulation over industrial health. In Illinois, investigations headed by hygienists Alice Hamilton and Emery Hayhurst exposed extraordinary neglect of industrial illness and prompted the 1911 Occupational Diseases Act that regulated "hazardous processes" involving lead, chemicals, and minerals. Thousands of annual sanitation inspections followed. In New York, inquiries by medical inspector C. T. Graham-Rogers and the post–Triangle fire Factory Investigating Commission produced a new state industrial hygiene division that pioneered research into thermometer making and other chemical processes, and inspired new industrial health codes governing sanitation, ventilation, dust removal, women in canneries, and the use of lead and other toxic substances. In Wisconsin and California, meanwhile, state industrial commissions received broad authority to prescribe standards and rules meeting safe place requirements that work be as free from danger to workers' health as the nature of the employment reasonably permitted.[9] At first, however, this wide discretion generated less disease-prevention activity than political agitation in Illinois and New York did, because Wisconsin and California compensation laws initially induced action only against accidents.

Accordingly, Wisconsin's Industrial Commission relegated occupational disease to a sanitation subcommittee in its first safety code advisory panel in 1912. Blending its "representation-of-interests" and "prevailing-good-practice" principles, the commission filled the subcommittee with local experts: Milwaukee's chief sanitation inspector, three company engineers, a union metal polisher, and commission special assistant C. W. Price, though no women. Relying solely on local "good practice," the subcommittee surveyed factories and then fixed rules for toilet facilities, cleanliness, and "noxious" substances. Giving dust removal and "fresh air" special attention, the subcommittee also defined airspace limitations, suction requirements, and exhaust-apparatus design. These rules became the General Orders on Sanitation in January 1913.[10]

Then, paralleling contemporaneous investigations into lead, phosphorous, and toxins by states, hygienists, and the American Association for Labor Legislation, the Wisconsin agency expanded its disease-prevention work. From 1912 to 1916, the commission sponsored inquiries into lead poisoning, even though Wisconsin's compensation law did not cover this ailment. The commission also participated in Merchants and Manufacturers Association of Milwaukee roundtables, drawing on these events to devise a 1913 shop bulletin on infections and a 1915 first-aid handbook. Under separate protective legislation sustaining contemporary concerns about female workers' vulnerability to fatigue, moreover, Wisconsin's Industrial Commission prescribed orders regulating women's work hours, particularly in the pea canning, fruit, and vegetable industries.[11] Wisconsin's commission did not extend regulations to other women's health issues, as New York's labor department did.

Milwaukee unions found these health regulatory efforts to be wanting. Wisconsin's Industrial Commission did consult unions and employers about dust removal, leading it to prescribe better hoods and suction devices, and then direct deputies to monitor exhaust systems.[12] Nonetheless, labor leaders rightfully complained that the commission's disease work lagged far behind its accident-prevention efforts. The agency issued no occupational-disease reports, neglected industrial health in its "safety first" campaign, abandoned a proposed USPHS disease research station, declined to convene its exhaust-systems subcommittee at union request, and did not appoint a health specialist until 1919. Not surprisingly, a 1917 meeting discovered gaps in the agency's sanitation code.[13] On disease, Wisconsin's vaunted commission initially produced meager results.

During the 1915–1925 period, however, medical advances and politics brought legal change. Until then, courts usually applied compensation laws only to "accidental" ailments like gas poisoning and infections, not "natural" or "expected" industrial diseases. After 1920, modern medical tools such as X rays, blood analysis, and statistics clarified the causes of occupational illness, portraying it not as just the "natural" or "inevitable" result of various trades but as the product of metals, chemicals, and dusts that employers introduced into work processes. Hygienists now argued that employers had to take responsibility for work illnesses and pay compensation.[14]

A few states incorporated this thinking into compensation reforms, nudging occupational illness away from the judicial liability system toward social insurance. Five jurisdictions adopted comprehensive disease compensation coverage for varying reasons. Whereas Connecticut accommodated business by enacting a comprehensive law that removed emotional disease cases from potentially costly court proceedings, California's and Wisconsin's mix of strong administrative leadership, weak business opposition, and stout labor lobbying produced "blanket" legislation. After California's supreme court annulled a compensation award for inhaling wood-alcohol fumes, administrators persuaded lawmakers first, in 1915, to drop the rule that compensable "personal injuries" had to be "accidentally sustained," and then, in 1917, to specify that "personal injuries" included disease. Similarly, after denying benefits for lead poisoning and tuberculosis, Wisconsin administrators got legislation in 1919 that extended compensation to "all other injuries, including occupational disease." California and Wisconsin laws thus rejected narrow employer proposals to schedule compensable ailments, though they retained "time of injury" clauses recognizing only instantaneous illnesses.[15]

In New York, Illinois, and Ohio, however, well-organized unions sought comprehensive disease compensation, but powerful business groups resisted such laws as a ruinous form of industry-funded health insurance. Schedule laws resulted as a compromise. New York's 1920 law extended compensation to twenty-three industrial illnesses but limited eligibility to workers afflicted during the previous year. An Illinois statute in 1923 applied compensation to diseases arising from hazardous jobs listed in the state Occupational Diseases Act's Section Two, while retaining workers' right to sue employers for "wilful violation" of health precautions required in Section One. Ohio legislation extended compensation to ailments associated with particular manufacturing processes, though restricting it to workers diseased in the prior year.[16]

New compensation legislation in the 1915–1925 period, hence, broadened employer responsibility for occupational disease, despite modern perceptions that the law still imposed "no clear duty." Blanket laws like California's and Wisconsin's covered all disabling ailments. Schedule statutes in New York, Illinois, and Ohio encompassed most known industrial illnesses—93 percent, according to Ohio authorities—though they omitted radium poisoning, silicosis, and anthracosis (black lung). The exclusion of these latter ailments certainly exposed a huge statutory gap for chronic diseases that reformers, trade unionists, and a few researchers were struggling to establish as major problems in the radium dial, foundry, and mining industries. Nonetheless, at least eleven states had created some new incentives for disease prevention. Old judicially based liability rules still applied everywhere else.[17]

When compensation laws expanded, industrial developments worsened health risks. High-power drilling and sandblasting equipment in foundries, quarries, and mines increased worker exposure to dust, while manufacture of batteries, electrical equipment, radium-dial clocks, and petroleum products subjected workers to lead, radium, and other harmful substances. World War I weapons production awakened hygienists to such dangers, and big employers created research laboratories and health programs to deal with them. Yet employers still resisted government regulation.[18]

In Wisconsin, manufacturers and insurers condemned the state's new disease-compensation law. Its broad language, they complained, would encourage "unjust" disease claims, inflicting burdensome health costs on employers. Commissioner Fred Wilcox reassured employers that Wisconsin's commission would avoid unfair disease awards and that disease compensation would not ultimately cost that much. Experience proved him right. From 1920 to 1929, Wisconsin disease compensation claims accounted for only 3,166 of 200,791 filed, and covered only 1 percent of benefits paid. Similarly, California, Illinois, New York, and Ohio reported that disease coverage raised compensation costs only 0.3 to 1.2 percent in the 1920s.[19] Such minimums perhaps reflected employer evasion and schedule-law exclusions, including radium poisoning, silicosis, and anthracosis.

Nonetheless, Wisconsin administrators expanded disease-prevention efforts. With the Industrial Commission's augmented postwar budget, officials hired hygienist C. W. Keniston, modified forms to accommodate disease reporting, and appointed a new advisory panel to update the state's sanitation code. Reflecting Wisconsin's "representation-of-interests" principle, the committee accommodated the WSFL, the Milwaukee Federated Trades Council (MFTC), the Milwaukee Association of Commerce, the Wisconsin Manufacturers Association (WMA), the state medical society, and a few local doctors and manufacturers. As before, this panel informally investigated Milwaukee factories, and then in 1921 recommended sanitation-code revisions based mainly on local practices. Such localism and medical backwardness typified state labor officials' outlook through the 1920s, as demonstrated by contemporaneous International Association of Industrial Accident Board and Commissions (IAIABC) meetings that rarely addressed occupational disease or women's health.[20]

Yet Wisconsin's commission embraced modern science in 1922 heating and ventilation regulations, after receiving complaints about dusts and fumes in foundries, grain storage elevators, garages, laundries, and slaughterhouses. Unlike the fragmented industrial hygiene field, ventilation had "shifted up" into a national "technical network" of specialists. Administrators consequently staffed the ventilation committee with architects and engineers from national professional societies and business organizations, along with local doctors, employers, and bureaucrats, but no workers at all. This group recommended state-of-the-art exhaust, ventilation, and dust-removal methods over local employers' protests about the costs. Yielding more here than for electrical rules, however, Wisconsin officials modified ventilation orders to meet local objections before finally implementing them in 1925.[21]

Organized labor and women's groups stimulated other commission action toward industrial illness. Union complaints triggered a commission investigation of sanitation-code violations in railroad terminal washrooms from 1921 to 1923. Journeymen painter protests, moreover, caused the Wisconsin agency to develop its 1924 spray-painting orders, although the commission rejected demands to prohibit the practice. Problems in Wisconsin canneries, finally, induced the commission to issue regulations governing female workers' hours and mealtimes, though the agency neglected women's health hazards under safety laws.[22]

Elsewhere, disease-prevention work varied. In California, compensation reports indicated rising levels of triton, benzine, lead, and mercury poisoning, but the state's Industrial Accident Commission failed to adopt a health code, approve spray-painting orders, or enforce 1889 and 1921 factory sanitation acts. Labor leaders protested in vain. In 1924, moreover, a California appeals court weakened compensation incentives for disease prevention by enforcing a six-month statute of limitations on worker claims. The state's Industrial Welfare Commission, by contrast, regulated women's hours of work energetically, particularly in canneries.[23]

In Ohio, organized labor campaigned forcefully for disease prevention, with mixed results. After lawmakers scheduled compensable illnesses in 1921, the Ohio State Federation of Labor (OSFL) lobbied unsuccessfully for comprehensive disease legislation, but did prod Ohio's Industrial Commission to adopt blower and exhaust, polishing and grinding, and pottery industry orders, and then to establish an industrial hygiene bureau. Yet as elsewhere, Ohio courts weakened regulation. Deciding in 1928 that "potter's consumption" was neither an accident under compensation statutes nor a disease enumerated there, nor an actionable common-law injury, an Ohio appeals court obliterated employer incentives to prevent that disease.[24]

In Illinois, authorities vigorously enforced health laws within narrow lines. Illinois's Occupational Diseases Act compelled more medical reporting and sanitation inspection than commission states achieved, although historians question this law's effectiveness. Under the act's Section Two, Illinois inspectors checked factories for white-lead exposure; surveyed battery plants, brass foundries, and chromium-plating shops for metal poisoning; and tutored small firms about health precautions. Under Section One, moreover, worker lawsuits grew as an incentive for preventive work. Late-1920s appellate decisions extended Section One to tuberculosis and silicosis, expanding employers' liability for disease in judicial actions.[25]

In New York, by contrast, interest-group politics generated the country's strongest industrial health operation of the 1920s, despite compensation legislation that covered only enumerated and "accidental" diseases, a position affirmed in court. New York's industrial hygiene division acquired exceptional autonomy with its own advisory committee and scientific staff. The state labor department, moreover, affiliated with the New York City Reconstruction Hospital worker health clinic, addressed work-related disease and women's health at annual safety congresses, and published women's and children's health reports. Having issued sanitation, ventilation, and dust-and-fumes orders in the 1910s, meanwhile, the department's code division promulgated dry-cleaning, dry-dyeing, humidity, and temperature regulations in the 1920s. Its inspection division simultaneously devoted one-fifth of factory visits to sanitation-code compliance. And its hygiene division evaluated exhaust devices, investigated chemical risks, and approved ventilation plans.[26]

By 1929, hence, state action against industrial illness was growing, though state programs together failed to establish an effective national system. Eleven states had disease compensation laws, twenty-two empowered state labor departments to issue work health orders, several published industrial disease statistics, and a few launched investigations. True, judicially based liability law still covered occupational illness in most states. True also, disease research shifted away from government agencies and reform groups toward insurance companies, business-funded university programs, and industry-dominated health conferences. Nonetheless, government attention to industrial health expanded. Consequently, IAIABC members offered their first program about chemical risks in 1928. Medical research and newspaper reports about New Jersey's "radium girls" exposed industry's growing disease toll. New York labor commissioner Frances Perkins acknowledged growing sickness, and then endorsed comprehensive compensation and better occupational health codes. Rising disease compensation claims induced Wisconsin's Industrial

Commission to form a dust- and gas-code committee and request funding for a disease expert.[27] Then, an industrial health disaster struck—silicosis.

The Silicosis Crisis

In the late 1920s and early 1930s, America's workplace safety and health system suffered an epidemic of silicosis, a lung disorder acquired from long-term exposure to silica dust in foundries, quarries, and mines. Depending on the jurisdiction, the outbreak unloosed work injury lawsuits, workers' compensation claims, worker layoffs, and impoverishment—all during the Depression. This catastrophe incited a furious public policy debate. In most states, trade unionists and reformers demanded silicosis's inclusion under common-law liability or compensation systems, but businesses resisted such coverage as financially ruinous, particularly because silicosis's occupational origins remained scientifically murky.[28] Yet in Wisconsin, where compensation already covered work illnesses comprehensively, debate focused uniquely on preserving the state's existing program.

Silicosis came to state regulators' attention gradually. Like other administrators who lacked medical and statistical knowledge of protracted ailments, Wisconsin officials originally focused just on "accidental" work-related diseases. From 1928 to 1934, however, silicosis cases trickled into Wisconsin's compensation system and then flooded it. Latent cases turned into full-fledged silicosis, just as Depression-ridden employers discharged thousands of workers and lawyers jumped into compensation disputes as a new field of practice. With doctors and personal-injury attorneys helping them, workers deluged Wisconsin's Industrial Commission with compensation claims, while afflicted laborers elsewhere inundated courts with work injury lawsuits. Silicosis claims in Wisconsin rose from 23 in 1929 to 321 in 1934, and compensation costs mushroomed among foundries and granite companies. Insurance companies tottered. Rising medical costs, broadened eligibility requirements, and increasing benefits had already upset insurers' carefully calibrated rating schemes. Now, as the Depression shrank payrolls upon which insurance premiums were based, silicosis compensation threatened insurance firms with huge losses. Nationwide, insurers subsequently ordered foundries and quarries either to fire potentially silicotic workers or to forfeit insurance coverage.[29]

Insurer fears were real. After a 1934 Illinois judicial ruling broadened compensation coverage, lawyers canvassed worker neighborhoods and then filed hundreds of silicosis claims. In Wisconsin, Employers Mutual rejected 106 of

279 silicosis claims as frivolous between 1931 and 1935, suggesting that non-disabled workers were seeking compensation as unemployment relief. Some succeeded. Wisconsin compensation director Harry Nelson later admitted that "lack of knowledge and early hysteria" caused employers, insurers, and even state officials to allow "overliberality of settlement" whereby "a much larger amount" was paid "than will ever be paid in the future."[30]

When they processed silicosis claims, moreover, Wisconsin officials encountered a legal problem—the state compensation law's "time of injury" clause that granted benefits only for "accidental" harms occurring during employment, not protracted diseases that matured later. In *Wisconsin Granite Company v. Industrial Commission* (1932), Wisconsin's supreme court observed that lawmakers had retained the original "time of injury" provision when they brought occupational illness under compensation legislation in 1919, but complained that courts now had to "grope in deep twilight" to ascertain when disease compensation entitlement began. Sustaining earlier rulings that related "time of injury" to worker "disability," the supreme court sanctioned disease compensation only if employment existed when disability occurred.[31]

Coincidentally, during Philip La Follette's governorship in 1931, lawmakers amended the "time of injury" clause to include "the date when the disability from occupational disease first occurs." Yet *North End Foundry Co. v. Industrial Commission* (1935) narrowed this amendment's application. In that case, North End dismissed numerous workers, after medical examinations revealed many with nondisabling early silicosis. Some men filed compensation claims. Wisconsin law clearly held employers responsible for disabling illnesses, but here, Wisconsin's supreme court applied "disability" only to harm preventing an employee from "performing his work in the usual and customary way." Compensation, the court held, replaced only wages lost due to occupational disability, not losses from "medical disability."[32]

Meanwhile, granite and metalworking firms campaigned furiously to weaken Wisconsin's compensation law. With foundry workers' support, these companies persuaded the conservative Democratic state assembly elected in 1932 to enact legislation chopping silicosis benefits in half. The bill needed Democratic governor Albert Schmedeman's signature, however, and faced opposition from newspapers, mutual insurers, and state administrators. Heeding the critics, Schmedeman vetoed the measure. "Simple humanity" convinced him that "the state cannot be a party to sending these men to their certain death." To address silicosis, he argued, employers had to remove afflicted men from dusty jobs and clean dusty workplaces. Seemingly courageous, Schmedeman's veto actually demonstrated shrewd recognition that most local farmers and employers still supported Wisconsin's traditional compensation policy. Public sympathy for the Depression's dependent poor fortified his decision.[33]

This political drama coaxed Wisconsin's conservative 1933 legislature to improve silicosis coverage. Addressing employers' practice of firing silicotic workers before disability occurred, lawmakers redefined "time of injury" again to mean "the last day for the last employer whose employment caused injury." This law protected workers who developed full-fledged silicosis after being dismissed, but not those remaining asymptomatic.[34] It thus continued to defend workers' wage-earning power while absolving employers of non-disabling illness.

Elsewhere, the silicosis crisis deepened ferment over compensation reform. Most western mountain and prairie states and most Deep South states resisted disease compensation. Alabama still neglected such legislation when it established a modern labor department empowered to address disease in the late 1930s. Only Alabama's supreme court accommodated liability reform, bending common-law rules in 1938 to treat silicosis as a "continuous tort" resulting from persistent employer negligence. California and three other states, by contrast, joined Wisconsin to preserve "blanket" disease compensation statutes, though California courts still limited coverage to illnesses keeping laborers from work. Most northern industrial states, meanwhile, merely scheduled compensable diseases (without silicosis) until the mid-1930s, despite labor campaigns for "blanket" legislation. After 1935, however, a host of factors—escalating lawsuits, Depression-era relief demands, militant unionism, employer cost-stabilization efforts, and the Gauley Bridge (West Virginia) disaster that killed two thousand mostly rural black workers—compelled northern states to broaden compensation coverage of silicosis.[35]

In New York, after afflicted workers flooded local courts with lawsuits, foundries and insurers secured a bill removing silicosis from litigation and adding it to the state's compensation disease schedule to avoid huge damage awards. Governor Herman Lehman vetoed that measure, however, and supported a union-backed "blanket" compensation law adopted in 1935. Rancor persisted. Lehman, like Wisconsin's Schmedeman, vetoed another business bill to reduce disease benefit levels, prompting employers to dismiss workers showing any sign of lung disorders. With New York State Federation of Labor president George Meany fulminating about the "thousands" laid off, industrial commissioner Elmer F. Andrews assembled employers, workers, insurance executives, and state officials to formulate new legislation in 1936.

A compromise, that law indemnified silicosis, prohibited preemployment exams, required new dust regulations, and allocated funds for research and disease law enforcement but limited benefits and banned claims on pre-1935 exposures. The compromise limited employer costs. New York firms would pay little to silicotic workers.[36]

In Illinois, by contrast, legal developments triggered reform. By 1934, judicial interpretation of the state Occupational Diseases Act's Sections One and Two had unloosed numerous silicosis lawsuits and compensation claims. In 1935 and 1936, however, probusiness judicial rulings abrogated Section One worker lawsuits. With militant unionism growing, Illinois lawmakers formulated corrective legislation that allowed employers to elect disease compensation while still letting workers sue nonelecting firms for ailments caused by negligence or health law violations. Excluding "ordinary diseases of life," the compensation section (as amended in 1937) included only illnesses incidental to employment with "a direct *causal* connection" to work conditions, thereby avoiding general health insurance.[37]

Ohio legislation reflected similar compromises. Under pressure from labor, lawmakers added silicosis to Ohio's compensation disease schedule in 1937, limiting benefits to workers disabled by five years of dust exposure with Ohio employers. Only in 1939 did Ohio lawmakers apply compensation to "all other occupational diseases," though they restricted eligibility to "total" disability.[38]

Viewed nationally, then, the Depression-era silicosis crisis expanded state compensation disease coverage, but left gaps and limitations. The number of "blanket" compensation states doubled to ten, the number of schedule-law states rose to fourteen, and a few states like Illinois allowed some litigation. Twenty-four states still provided no disease compensation, including silicosis-plagued Vermont, Oklahoma, and Kansas. In law and administration, moreover, virtually all states continued to ignore miners' asthma (anthracosis).[39] Even covered states like Wisconsin limited liability to occupational disability.

And disease coverage affected compensation costs minimally. Despite employer and insurer warnings about financial ruin, state and federal agencies reported disease payments to be "negligible." In Wisconsin, only 2.5 percent of 1937 compensation payments went for industrial illness. In Ohio, disease encompassed less than 1 percent of compensation benefits paid, about 2 cents per $100 payroll. In New York, employers disbursed only $99,594 to silicotic workers from 1936 to 1939.[40] Such "negligible" cost was a dubious stimulant for prevention work.

The New Science of Occupational Disease Regulation

In theory, compensation coverage should have inspired disease-prevention efforts. As New York's Elmer Andrews observed in 1935, "It is now of direct cash benefit to employers to see to it that silica and other harmful dusts are not released in their plants." Nowadays, scholars doubt such assertions, arguing that compensation laws were inadequate, state enforcement defective, corporate domination of research overwhelming, and conditions for employer evasion favorable.[41] Yet Wisconsin's program seemed to work according to plan.

Multiplying compensation claims spurred Wisconsin's Industrial Commission to action in 1928, *before* the silicosis crisis struck, *before* scientific authority on silicosis "shifted up" to national institutions, and *before* big industry formed the Air Hygiene Foundation (AHF) to control silicosis research. Characteristically, the agency assembled an advisory committee on dusts and gases consisting of local employers, union men, physicians, and toxicologists. This group studied medical literature and consulted experts. Then, after foundry operators challenged its recommendations, the committee met with a special WMA panel (aided by Metropolitan Life Insurance Company medical staff) to modify its proposals.[42]

Finally appearing in 1932, Wisconsin's first dust code introduced new science and policy. While strengthening extant exhaust and respirator rules, its Order No. 2002 embraced industry-oriented science when it adopted USPHS standards for maximum worker exposure to harmful dusts, fumes, vapors, and gases. The aim was to prevent workers' inhalation of such substances "to an extent" that injury resulted. This policy transformed the Industrial Commission's "prevailing-good-practice" principle from demanding industry's "best" preventive measures to prescribing scientific exposure limits. Employers had to manage hazardous material, not eliminate it.[43]

Despite the dust code's business-oriented science, persisting uncertainty about silicosis's industrial etiology forestalled employer acceptance. Investigations followed. Wisconsin's Industrial Commission sponsored meetings in Chicago and at New York's Trudeau School of Tuberculosis. Employers Mutual hired USBM chemist E. G. Meiter to do research in a Milwaukee laboratory. And Wisconsin's legislature formed a blue-ribbon panel to hold hearings around the state. Though the legislative panel recommended compensation benefit reductions—a suggestion never adopted—all three investigations proposed better housekeeping, regular physical exams, and reassignment of afflicted men to nondusty jobs.[44]

Insurance companies affirmed this strategy. Travelers, Metropolitan Life, and Liberty Mutual conducted their own studies and recommended their own silicosis-abatement tactics. Metropolitan Life's pamphlet *Silicosis*, for instance, surveyed silica-dust hazards across industry. and then suggested dust traps, exhaust systems, periodic silica-air content measurements, and routine medical exams for workers.[45]

Wisconsin's Industrial Commission fostered similar practices. Hearings revealed dust abatement in Milwaukee foundries, but neglect among stone-cutting and metalworking operations. Despite Depression-era budgetary restraints, hence, the commission dispatched deputies to measure local firms' dust levels for compliance with Order No. 2002, an effort helped by the state health board's new Industrial Hygiene Unit created with federal funds in 1937. To dissuade employers from firing nondisabled silicotic workers, moreover, the agency threatened to invoke 1935 legislation allowing compensation up to 70 percent of lost wages. The commission also manipulated compensation insurance rates to discourage dusty conditions, a strategy bolstered by a 1938 court decision barring the insurance commissioner from meddling in that system.[46]

The hygiene unit marked a new federal role in industrial health policy. To improve local workplace health activity, New Deal legislation provided funds for the USPHS to establish industrial hygiene units in nearly thirty state health departments, including Wisconsin's and California's. Labor leaders condemned the plan as feeble, based as it was on the USPHS investigation and educational approach to occupational health rather than labor department regulatory practices. Yet Wisconsin's hygiene unit boosted regulation by facilitating the financially strapped Industrial Commission's factory-dust surveys.[47]

Ultimately, Wisconsin officials declared their silicosis problem to be "solved." Early reports showed that big Milwaukee-area firms had implemented extensive dust-control practices, such as vacuuming plants, atomizing dusty air, wetting down floors, enclosing sandblast rooms, and establishing company medical laboratories. Later studies revealed similar work in foundry, mining, and quarry companies, though a 1938 WMA spot check found several substandard firms. Simultaneously, Wisconsin silicosis compensation claims plummeted from 321 in 1934 to 24 in 1939. "Accrued" cases arising from earlier work conditions explained the 1934 bulge, officials contended, but compensation incentives and dust-code enforcement lowered claim levels thereafter.[48]

Other states acted against silicosis, but not until the mid-1930s, *after* clamor for compensation expansion arose, *after* the Gauley Bridge incident, and *after* industrialists created the AHF. California's Industrial Accident Commission issued its first dust-code in 1936. Illinois lawmakers empowered their reorganized industrial commission to issue orders governing dusts, gases, and fumes. New York lawmakers directed their own Industrial Board to adopt rules abating silica and other dusts, leading an advisory committee stacked with industry and medical experts (and one worker) to formulate rock-drilling and other new dust regulations.[49]

The nationwide impact of such programs was questionable. In Ohio, statistics indicated that claims for fatal and totally disabling silicosis continued to rise after that state adopted silicosis compensation, especially in foundry and ceramics trades.[50] States with severe silicosis problems like Vermont, Oklahoma, and Kansas still provided little disease regulation. By comparison, Wisconsin's silicosis-prevention program was aggressive, a posture accentuated by the state's proposed medical examination plan.

Wisconsin's Medical Examination Program

According to experts, dust removal alone would not eliminate silicosis. To stop it completely, employers had also to perform periodic medical examinations to identify workers suffering early stages of the illness for reassignment to nondusty jobs. Unionists and labor law officials viewed this practice suspiciously, since employers widely abused it. Nonetheless, in 1937, Wisconsin lawmakers enacted legislation authorizing the examination of on-the-job workers upon employer or employee application.[51] In 1939, moreover, Wisconsin's Industrial Commission joined the State Board of Health to inaugurate a voluntary state-regulated screening program. Commission officials saw it as an application of the state's safe place law, but controversially, it shifted the agency into personnel relations.

Medical examinations were hardly new. In the 1910s, big employers instituted the practice to lower compensation risks, reformers urged it to promote working-class health, and insurance and pension plans used it to evaluate applicants. Even Wisconsin's Industrial Commission contemplated a statewide plan until World War I scuttled it. Yet organized labor warned that employers would abuse medical screening to weed out union "troublemakers" and poor health risks. Indeed, Ohio's Industrial Commission reported that industrialists routinely examined employees to identify "fit" workers and refute compensation claims. Wisconsin commissioner Joseph Beck witnessed firms that used screening to build up "Olympian societies" of workers to minimize compensation expenses.[52]

Controversy over medical exams continued into the 1920s. Large employers incorporated medical screening into personnel management and welfare schemes to contain labor costs and control ailments like lead poisoning. Meanwhile, worker opposition deepened. From American Federation of Labor head Samuel Gompers on down, labor leaders saw medical examinations as a managerial tool to eliminate union men, evade responsibility for worker illnesses, and invade workers' privacy. Consequently, Wisconsin, New York, and Ohio unions opposed the procedure.[53]

Wisconsin's 1925 experience-rating bill exposed medical screening's divisiveness. The bill proposed experience rating as the method of figuring employer merit credit in compensation insurance premiums, but workers protested that employers would misuse medical examinations to get favorable rates rather than improve employment conditions to do so. Wisconsin's Industrial Commission sided with workers. With Governor John Blaine's approval, commission officials and labor leaders accepted the bill only after an amendment denied experience-based merit credit to firms using "oppressive" examination programs.[54]

The 1930s silicosis crisis reoriented administrators' thinking. The crisis acquainted them with protracted illnesses gotten from gradual exposure to harmful substances. And it moved them toward the new industry-oriented science of exposure limits, not organized labor's policy of eliminating noxious substances.[55] Agency officials consequently determined that exams could identify afflicted workers before excessive dust exposure disabled them. Administrators still rejected medical screening's abuse, however. When companies dismissed workers whose exams revealed incipient silicosis, Wisconsin officials exhorted those firms to reemploy these men in dust-free jobs, and then secured the 1935 law requiring manufacturers to compensate workers prematurely discharged, a measure rarely invoked.[56]

Then, the agency pursued a state-supervised medical examination plan. Deciding that physical exams were vital to stopping occupational illness, and that state intervention was necessary to eliminate employer abuses and labor-management turmoil, Wisconsin's Industrial Commission surveyed self-insured employers, and then joined the State Board of Health to assemble an advisory committee of manufacturers, workers, and insurers to discuss the matter in 1937. To formulate standards, that panel organized a subcommittee consisting of the hygiene unit chief Dr. Paul Brehm and four doctors chosen by business and labor.[57] The subcommittee's labor-management balance accommodated New Deal–era industrial politics, unlike advisory committees' technocratic orientation.

Under Industrial Commission and health board auspices, the subcommittee proposed voluntary examination procedures in 1939. The panel encouraged employers to perform regular pre- and postemployment physical evaluations, hiring physicians and paying examination fees themselves. It particularly asked doctors to screen workers only for defects affecting job performance or public health. And it urged employers and employees to submit disputed findings to the Industrial Commission for arbitration. Seemingly, the plan reconciled everyone's interests: employers' desire for fit workers, the state's mission to relieve worker illnesses, workers' hope for health protection without risk of arbitrary dismissal, and doctors' wish for professional independence. Wisconsin's industrial commissioners endorsed the program as a "fair" method of fostering labor-management collaboration with state disease-prevention goals. They especially supported arbitration as a noncontroversial way to resolve medical disputes.[58]

Still, the examination program extended state power onto controversial ground. Moving beyond the Industrial Commission's traditional focus on engineering control, the plan implicated employers' right to hire and fire, and workers' claim to job security. Administrators presumed that Wisconsin's safe place statute—stipulating that employers "do every other thing reasonably necessary" to protect worker health—authorized the commission to promote such a voluntary program, perhaps with an administrative order.[59] Yet doctors, workers, and employers balked.

The medical community warmed to the examination program slowly. Like the American Medical Association, Wisconsin's State Medical Society did not address occupational disease or create an industrial health committee until the mid-1930s. Consequently, few doctors immediately backed the state plan, and medical society secretary J. George Crownhart opposed it as a threat to doctors' autonomy. Only after administrators lobbied the organization did it endorse the state screening program.[60]

Organized labor was cooler. Labor leaders knew that medical exams could improve worker health, but they still feared employer abuse of the practice and dreaded state encroachment on hiring and firing questions just when New Deal labor laws were relegating that matter to collective bargaining. Thus, while Commissioner Voyta Wrabetz and Employers Mutual's Ben Kuechle solicited labor groups' support throughout 1937 and 1938, Milwaukee's International Molders' Union Local No. 125 fought the state program. Insisting that employers arbitrarily used physicals to "let certain men out," Local No. 125 followed national leaders' instructions to fine workers submitting to examinations. Wrabetz promised the molders that Wisconsin's plan would avoid other states' "oppressive" practices, especially in arbitration, but he and Kuechle never overcame "the absolute lack of co-operation on the part of the unions."[61]

Even employers greeted the examination plan skeptically. The Industrial Commission got WMA backing and asked self-insured companies to participate, but various firms objected and most just ignored it. Viewing medical exams as a method of controlling labor costs, manufacturers complained that Wisconsin's plan required too many physicals, produced too much paperwork, and, most tellingly, interfered with their "fundamental right" to hire whomever they wanted. Disliking administrative governance anyway, employers found Wisconsin's exam program to be too intrusive.[62]

The Wisconsin commission's main support for the screening plan came from Employers Mutual Insurance Company, the agency's longtime partner in safety and health regulation. Like most insurers, Employers Mutual already championed medical screening. Indeed, company vice president Ben Kuechle spearheaded the state program. Yet the joint Industrial Commission–Employers Mutual effort to promote a state examination plan diverged from contemporary industrial relations. When agency officials considered an administrative order in 1941 to mandate exams, consequently, industrial politics stopped them. Wisconsin's exam program languished.[63]

The Measure of Progress

Despite the screening plan's fate, compensation director Harry Nelson's 1938 boasts about Wisconsin's industrial disease program generally held up. Wisconsin did extend blanket compensation coverage to disabling industrial illnesses, did preserve full benefits for disabled silicotic workers, did promulgate job-related health codes, did stimulate industry action against silicosis, and did sponsor a voluntary medical exam program. Whether these ventures explain Wisconsin's plunging silicosis rate is unclear, but they still document extensive governmental activity.

Admittedly, Wisconsin sustained a developmental and technocratic policy toward workplace illness, just as it did for accidents. Given its concern with compensation costs, Wisconsin's commission focused on diseases that produced financial business losses. And given its "prevailing-good-practice" standard, the agency adopted big-industry health practices such as exposure limits and medical screening, not pure preventive measures urged by unionists and reformers. As with safety, Wisconsin's system supported the "best" health practices consistent with economic progress. Yet Wisconsin's program was still an exceptional administrative bulwark against industrial disease. The state's "blanket" disease compensation law held employers accountable for all disabling industrial illnesses. The Wisconsin Industrial Commission's early dust and fume regulations did not just mollify rising industrial combinations like the AHF but fostered state-of-the-art cleanup work.

Few other states' programs duplicated Wisconsin's industrial health apparatus, even though some, like New York's, were impressive. As of 1939, only twenty-four states had enacted disease compensation laws, fourteen of which merely scheduled occupational illnesses. Most of these laws retained important restrictions. Ohio's eligibility requirements allowed only half of silicotic workers to get benefits. New York's law denied many totally disabled workers their claims. And in states without any disease compensation legislation, common-law liability still excluded most industrial illnesses. Gaps in state laws left ailments like anthracosis and silicosis to spread.[64]

In 1940, Illinois labor department official John M. Falasz condemned this situation. Observing how "unrelated" state efforts to address occupational illness had produced no "semblance of uniformity," Falasz declared national industrial health policy to be in disarray. He thus urged adoption of the Murray Bill pending in Congress to fund state labor department hygiene units and to create a U.S. Bureau of Labor Standards division to draft uniform national disease-prevention regulations.[65] The Murray Bill died, but it exposed American industrial disease law's chaotic condition. Notable programs like Wisconsin's and New York's made progress toward industrial health regulation, but they remained isolated in the nation's decentralized, uneven federal system. A uniform national occupational disease policy would not emerge until the passage of the 1969 Coal Mine Health and Safety Act and the 1970 Occupational Safety and Health Act.

Epilogue
The Road to OSHA

In 1975, amid widespread complaints about state governments' failure to curb industrial accidents and disease, the Wisconsin Industrial Commission's institutional successor—the Department of Industry, Labor, and Human Relations (DILHR)—relinquished industrial safety and health regulation to the newly created Occupational Safety and Health Administration (OSHA). That event marked a watershed in American regulatory history: the demise of Wisconsin's and other states' work safety and health plans and the rise of an unprecedented national program instead. It was a stunning change, but one that politics, law, and administration soon compromised.

Back in the Progressive Era, state-level reform had instituted a patchwork of local institutions to regulate work safety and health, not a uniform regulatory method everywhere. In Alabama, for instance, a coalition of industrialists, politicians, and reformers crushed union agitation and resisted protective labor legislation to maintain competitively low labor costs and sustain local class structure. As a result, Alabama enacted only a weak compensation program to indemnify accidents and did not establish a modern labor department until the union upsurge of the 1930s. In Illinois, meanwhile, partisan politics, divisions within the labor movement, and manufacturer opposition all discouraged modern safety and health institutions. Instead, business versus labor lobbying in Illinois only belatedly prompted the adoption of conventional factory laws and a traditional inspection bureau (quickly politicized by party patronage), but did produce separate compensation legislation and impressive occupational health statutes. In New York's vast economy, simultaneously, a pluralistic array of well-organized manufacturers, trade unions, social reformers, and women's groups begot a huge hybrid bureaucracy that featured separate code making, factory inspection, and compensation functions, including programs that promoted occupational health and compensated most industrial illnesses. In New York, especially, organized labor exerted real force on safety and health regulation, because state law required worker inclusion in code-making proceedings and because unions heavily influenced the state's inspection operation.

In Wisconsin, California, and Ohio, by contrast, voters elected crusading Progressive governors who promoted commission plans to regulate work safety and health. These programs all included safe place statutes, code-making powers, workers' compensation incentives, and technically oriented factory inspection routines, but they operated differently. Whereas pluralistic politics caused California's paternalistic commission to accommodate business, labor, and insurance groups in safety code proceedings and educational programs, big industry versus big labor rivalry induced Ohio's agency alternately to pacify manufacturers with moderate safety regulation and to mollify labor groups with an innovative industrial hygiene plan. In Wisconsin, the state's tradition of strong executive leadership and its unique political challenge from Milwaukee industrialists and socialists encouraged the state's Industrial Commission to adopt a "left corporate liberal" technocratic approach to safety and health regulation. With exceptional independence afforded by local politics and administrative law, that agency incorporated big industry standards into safety codes; enlisted business, expert, and labor cooperation on safety code advisory committees; and deployed workers' compensation insurance to secure employer compliance with state regulations. In all of these ways, Wisconsin's commission advanced state goals of reducing industrial hazards and securing industrial peace.

Economic and political developments weakened these state programs over the years, but support for a centralized national system did not congeal until the 1960s and 1970s. Before then, agitation for federal action had been building, especially in the mining industry. Back in the New Deal era, worker unrest inspired a minor provision for job safety in the 1936 Walsh-Healey Public Contracts Act, and then during the World War II era, lobbying by the United Mine Workers (UMW) and a series of mine disasters prompted the 1941 Federal Coal Mine Safety Act, a 1946 federal mine safety code, and the 1952 Coal Mine Safety Act, the last of which extended federal regulation over mines of fifteen workers or more. About the same time, physicians affiliated with the UMW Welfare and Retirement Fund increased pressure for federal attention to black lung disease. Ferment over mine safety and health problems escalated within UMW ranks, especially among militant young miners, culminating in wildcat strikes and state legislative hearings after the Farmington, West Virginia, mine explosion in 1968. Subsequently, Congress enacted the sweeping Federal Coal Mine Health and Safety Act of 1969 that imposed federal regulation on mines of all sizes, set up a new federal mining law enforcement agency, prescribed rigorous dust regulations, levied stiff fines on violators, and notably established a federal compensation program for black lung victims.[1]

Against this backdrop, interest in federal regulation of manufacturing safety and health grew. During the mid-1960s, unions prodded the U.S. Department of Labor (USDL) first to evaluate piecemeal federal safety laws that had evolved from the New Deal era onward, and then after some bureaucratic squabbling to recommend vast enlargements in legislation to President Lyndon Johnson. Presidential politics derailed Johnson's plans, but the Farmington mine disaster and strikes, along with continuing agitation by AFL-CIO operatives and consumer advocate Ralph Nader, all kept the safety issue alive. With labor and Nader representatives pushing hard for action, congressional Democrats enacted a strong national measure with concessions for conservatives. Newly elected president Richard Nixon signed that bill into law as the Occupational Safety and Health Act (OSHAct) in 1970.[2]

Like environmental laws, consumer protection legislation, and other "new social regulation," the OSHAct reflected the era's distinctive political outlook. For one thing, the OSHAct rested on antibusiness protest politics, not Progressive Era compromise. With business on the defensive, labor unionists, middle-class environmental reformers, and consumer advocates pushed Congress to ensure "corporate responsibility" for workers' "quality of life" rather than continue the Progressive policy of mollifying workers with moderate protections. In addition, antibusiness rhetoric popularized the reigning scholarly critique of the American regulatory state that condemned both state and federal regulatory bodies for their "capture" by corporate interests. Such sentiment encouraged Congress to attempt new regulatory institutions that were powerful enough to avoid industry domination and to change industry behavior.[3]

Seemingly, the OSHAct fulfilled this new regulatory strategy with a plan that was even more powerful than Wisconsin's old "left corporate liberal" approach. The law nationalized work safety and health regulation, thereby dislodging employment conditions from the province of fragmented state legislation. It then imposed a new "general duty" on all private employers engaged in interstate commerce nationwide to furnish employment "free from recognized hazards" that might cause "death or serious physical harm." This duty not only substituted a uniform national standard for old variegated state requirements but also mandated safety and health levels above the common-law "ordinary care" criterion and even above Wisconsin's stipulation that work be "as safe as the nature of the employment reasonably permits." Reformers regarded the general duty clause as a "radical" new guarantee for worker safety and health, though judicial interpretation quickly confined it to "recognizable" and "preventable" hazards.[4]

To realize this apparently "radical" vision, Congress created OSHA, a powerful federal agency entitled to prescribe and enforce national safety and health "standards." Situating it in the prolabor USDL rather than establishing it as an independent commission, lawmakers constructed OSHA with adversarial procedures to prevent its capture by industry. In standard making, the OSHAct authorized OSHA to assemble advisory committees including "an equal number" of employer and worker members, unlike Wisconsin's business- and expert-dominated panels. In judicial review and standard making, the OSHAct gave employers, employees, and all other "affected persons" broad rights to participate, privileges reinforced by judicial decisions affirming affected groups' standing in administrative proceedings.[5]

Congress also tried to actualize the OSHAct's putative "radicalism" with tough "command-and-control" enforcement methods. In the act, Congress directed OSHA "compliance officers" to make unannounced workplace inspections, allowing workers to request such visits and accompany OSHA officials on their rounds. Congress then ordered OSHA compliance officers to issue summary citations and fines up to one thousand dollars for each "serious" safety or health violation, and ten thousand dollars for each "willful and repeated" infraction. Like traffic tickets, these "first instance sanctions" imposed fines *before* formal hearings, though employers could appeal them to the new Occupational Safety and Health Review Commission. The threat of such fines was "strikingly aggressive" compared to most state programs' reliance on workers' compensation incentives, including Wisconsin's old compensation insurance merit rates and penalty-clause actions. Even coercive states like New York and Illinois imposed fines only *after* prosecuting employers.[6]

OSHA's potential as a "radical" agency, however, crumbled in the conservative political reorientation of the 1970s. Initially, when the OSHAct went into effect during 1972–1974, expansive welfare state politics crested, as laborers, consumers, environmentalists, and feminists maximized pressure for OSHA action. Then, after the 1974–1975 recession, the activist coalition collapsed, and corporate conservatives mobilized against regulation. Industry "counterattacked" OSHA's strong leadership during Jimmy Carter's presidency, and then supported President Ronald Reagan's conservative agenda to shrink OSHA and deregulate business. After 1980, stiffening business opposition, declining congressional support, and flagging labor activism undermined OSHA's strong regulatory capacity.[7]

Bureaucratic proceedings paralleled these developments. After early OSHA standards adopted business-sponsored guidelines, technically trained labor unionists, environmentalists, feminists, and hygienists finally surmounted industry's monopoly over occupational health science and secured tough OSHA chemical standards during the mid-1970s, a regulatory breakthrough, particularly for women. Nonetheless, business lobbyists revived, while unions and their allies fragmented. Industry stymied noise rules and other regulations in the 1980s, bringing OSHA standard making virtually to a halt.[8]

Courts reinforced the conservative trend. In early asbestos and vinyl chloride decisions, federal judges agreed that OSHAct "feasibility" language required OSHA to consider economic factors when making rules but denied that compliance costs were excessive in these particular cases. In the subsequent benzene decision, however, the U.S. Supreme Court affirmed that OSHA had to find a "significant risk" to worker health before imposing costly regulations. While rejecting cost-benefit analysis in a cotton-dust case, moreover, the Court confirmed that the OSHAct bound OSHA to economic rationality, a limit similar to the old judicial rule of "reasonableness."[9]

Meanwhile, the states played a much overlooked role in complicating the OSHAct's implementation. Despite widespread criticism of state regulation, some businesses, several unions, state administrators, congressional conservatives, and the Nixon administration secured an OSHAct allowance for states to assume "the fullest responsibility" for administrating the new law. In addition to letting them continue existing programs through December 1972, the OSHAct entitled states to create local plans for the "development and enforcement" of safety and health requirements that were "at least as effective" as national standards, subject to USDL supervision. The act also allowed USDL to hire state personnel to assist OSHA's work. The OSHAct thus did not entirely consolidate federal authority but preserved some decentralization in a federal-state partnership.[10]

The state-plan system had many problems. While the OSHAct did not demand state acceptance of the broad general duty clause, it did require sweeping administrative reforms to ensure state compliance with ever-changing national safety and health standards and coercive OSHA enforcement techniques. Struggling during Richard Nixon's and Gerald Ford's presidencies to establish local programs that met OSHAct criteria, USDL officials accommo-dated state plans in dubious ways. They continued old state programs beyond December 1972 (a policy overturned in court), then accepted "developmental" rather than fully "operational" state plans (without statutory authorization), and finally approved state programs that were questionable. Organized la-bor and Ralph Nader's Health Research Group denounced these actions for avoiding federal guidelines. Later in court, union leaders demanded that state plans comply "absolutely" with OSHAct staffing and funding requirements. The courts agreed, but Reagan-era politics stymied such improvements.[11]

Even local authorities resisted OSHA state plans. Jealous of their shrinking power, state officials condemned the system as a "bureaucratic monstrosity." Instead of a federal-state "partnership," they complained, the new system imposed a "paternalistic relationship" wherein OSHA "dictated" "nitpick-ing" requirements to the states. Facing economic decline, moreover, state lawmakers balked at OSHA state plans. By 1975, recession and deindustrial-ization tightened state budgets, especially in mid-Atlantic and midwestern "Rust Belt" states. Economic hardships pushed industrial states to drop local OSHA programs.[12] Wisconsin was one of them.

Like many states, Wisconsin maintained its old safety and health system until federal courts annulled OSHA's extension of such programs beyond 1972. Wisconsin officials then kept state inspectors at work under an OSHA contract, and finally secured USDL approval for a developmental OSHA program in March 1974. As in other Rust Belt states, however, Wisconsin's financially strapped legislature killed the plan in May 1975. Local safety and health regulation in Wisconsin abruptly stopped.[13]

Behind Wisconsin lawmakers' fiscal calculations were local political at-titudes that had turned harshly against state regulation. *The Capital Times* conferred "straight Fs" upon local regulators for failing to stop rising injury levels, while School for Workers professor Richard Grinnold dismissed the state's old program for not presenting a "credible threat" to recalcitrant em-ployers, due to timid prosecution and paltry fines. Moreover, state AFL-CIO president John Schmitt criticized Wisconsin's educational methods of pro-moting safety for lacking real force.[14] Evidently, Wisconsin's "left corporate liberal" approach to safety and health regulation had collapsed.

Political realignment reinforced popular thinking. Whereas organized labor had earlier backed Wisconsin's Industrial Commission, for instance, the state AFL-CIO now embraced OSHA, despite some workers' defense of local control. Like the national AFL-CIO, state labor leaders feared that local administration would impair enforcement of national standards, allow employers to escape to weaker states, and complicate national AFL-CIO efforts to monitor OSHAct implementation. Similarly, although big employers had originally supported Wisconsin's old program, many now wanted a national plan with uniform standards. Besides, industrialists expected OSHA inspections to be less intrusive than local oversight. One Wisconsin official projected the elimination of 63 out of 80 Wisconsin safety and health inspectors under OSHA administration, just as observers forecast that OSHA takeovers would reduce New York's inspectorate from 450 to 171, New Jersey's from 280 to 52, and Illinois's from 230 to 90. Likewise, Wisconsin politicians surrendered safety and health to federal control. In a recession, Wisconsin lawmakers happily cut $840,000 earmarked biennially for state safety regulation, paralleling decisions in New Jersey, Illinois, New York, and Pennsylvania to withdraw OSHA state plans for fiscal reasons.[15]

Meantime, lonely Wisconsin administrators defended old regulatory methods. DILHR chairman Philip Lehrman contended that Wisconsin's "consultation" method would secure more employer safety law compliance than OSHA's "big stick" would, especially since state regulators understood local industrial needs. Lehrman's successor John Zinas added that a "neophyte" agency like OSHA could hardly be more effective than experienced Wisconsin state institutions.[16] Remarkably, Wisconsin officials rarely cited workers' compensation inducements to justify their state's safety and health program, except in reference to the state's 15 percent penalty clause. By this time, insurance costs had so stabilized and accident claims had become so contested that compensation ceased to stimulate much preventive work.[17] The change undermined Wisconsin's left-liberal social-insurance approach to regulation, leaving just education, cooperation, and minuscule fines to compel safety law implementation.

Wisconsin's brief effort to establish an OSHA state plan exposed the local administration's ongoing weaknesses. According to an OSHA review, Wisconsin's proposed state plan easily satisfied OSHA safety and health standards but suffered extensive defects in enforcement. The OSHA evaluation indicated that Wisconsin factory inspections failed to utilize first instance sanctions, cumbrously divided responsibilities between separate bureaus, neglected to follow uniform procedures, disregarded union representatives, and ignored workplace hygiene.[18] These failures reinforced Wisconsin lawmakers' decision to embrace OSHA's "stronger" national apparatus.

Other Rust Belt states also abandoned local programs. Like four other states, for instance, Ohio never proposed an OSHAct state plan, though it let OSHA hire Ohio inspectors to supplement federal staff. Organized labor supported this increase in federally supervised Ohio inspectors, but state administrators and USDL officials jointly terminated the plan as ineffective, despite charges that Republican governor James Rhodes scrapped it just to placate local manufacturers. The action removed 75 state inspectors and hygienists, confining Ohio safety and health regulation to overworked OSHA officials.[19]

Unlike Ohio, Illinois did submit a developmental OSHA program, but politics and administrative ineptness scuttled it. Although Illinois governor Daniel Walker appointed crusading labor journalist Rachel Scott to implement the plan, the Illinois AFL-CIO, local labor activists, safety experts, and even business leaders soon condemned it for its loopholes, inadequate funding, unfit inspectors, and incompetent leadership. Grim budgetary problems then combined with mounting criticism to persuade Governor Walker to forsake Illinois's proposed program. Illinois safety and health regulation ceased in June 1975.[20]

Like Wisconsin and Illinois, New York initially advanced a developmental OSHA plan, only to forsake it. Seeing direct OSHA regulation as "inflexible bullying," New York's Associated Industries favored the state's OSHA plan, as did the local union representing factory inspectors. As elsewhere, however, the state AFL-CIO rejected local regulation, while New York's legislature resisted it for fiscal reasons. As lawmakers delayed, OSHA head John Stender threatened to revoke USDL approval of New York's tentative program. Governor Hugh Carey then terminated the state's OSHA plan to save $3.5 million in yearly expenses. His decision substituted OSHA's supposedly stronger regulatory apparatus for New York's eighty-year-old bureaucracy.[21]

In contrast to Rust Belt states, West Coast and Rocky Mountain states embraced OSHA state plans, including California. In June 1971, an explosion at the Sylmar Tunnel in the San Fernando Valley killed seventeen workers, sparking conflict between Democratic state lawmakers and Republican governor Ronald Reagan. Having championed bureaucratic cuts, Reagan found himself on the defensive when legislative hearings suggested that understaffing and incompetence in California's industrial safety division had contributed to the Sylmar disaster. Reagan parried Democratic legislative reforms by launching his own investigation, changing safety division leadership, and increasing the division's budget. But then, after California's labor department submitted a developmental OSHA plan to USDL, another accident at the Arroyo Seco Bridge in Pasadena killed six more laborers. In January 1973, hence, Democrats reintroduced their work safety and health reform bill, a measure that doubled as state-plan enabling legislation after USDL approved

California's program in May 1973. Unlike the eastern situation, localism, anti-Washington sentiment, and a state budgetary surplus facilitated passage of a modified version of this bill as the CAL/OSHAct in September 1973.[22]

The CAL/OSHAct created the country's strongest state OSHA plan. Although CAL/OSHA did not adopt the OSHAct's sweeping general duty language, it did establish strict standard-making procedures and did replace California's old "voluntary compliance" system with OSHA-like surprise inspections and first instance sanctions. Early OSHA evaluations advised Californians to strengthen enforcement procedures further, and local politics impelled CAL/OSHA to regulate more vigorously than OSHA itself. With labor support and business opposition in 1981, consequently, CAL/OSHA issued controversial fumigant rules that exceeded OSHA requirements, an action upheld in court.[23]

California's program showed that state operations could fit into the OSHA system, but that was hardly typical. As of July 1975, fifteen states mainly in the Middle Atlantic, Midwest, Lower South, and Great Plains refused local OSHA programs, including old industrial states like New Jersey, New York, Illinois, and Wisconsin. Simultaneously, thirty-five states in the Northeast, Upper Midwest, Upper South, Rocky Mountains, Southwest, and Pacific Coast had state plans in various stages of implementation. By January 1984, however, only twenty-four states and jurisdictions achieved final USDL approval for local programs.[24] Ultimately, states established local OSHA plans only sporadically.

The OSHA system, hence, transformed American occupational safety and health law, but in a problematic way. OSHA dismantled the Progressive patchwork of diverse state work safety and health programs, including Wisconsin's now decrepit "left corporate liberal" institutions, and substituted a national system that instituted high uniform work safety and health standards, a national policy for occupational disease, and tough uniform enforcement methods, extending such oversight especially into jurisdictions like Alabama that had provided little regulation before. Specifically, the OSHAct raised employers' legal duties for work safety and health above the nineteenth-century judge-made due care standard and above Progressive Era safe place statutes, requiring employers nationwide to provide jobs that were "free from recognized hazards," including previously neglected chemical risks. By creating OSHA itself, moreover, the OSHAct shifted safety regulation "up" into a quintessentially bureaucratic federal agency. Federal technocrats and nationally organized business, labor and public-interest lobbies dominated OSHA standard-making deliberations. OSHA inspectors subjected local employers to "by the book" enforcement, legalistically imposing first instance sanctions for violating national standards.[25]

In practice, however, OSHA proved to be surprisingly ineffective. On the local level, most industrial states failed to create OSHA-like programs, and many, like Indiana, Virginia, and New Mexico, established disputable ones. Moreover, controversy raged over what state plans had to do in order to meet OSHAct criteria. Federal courts endorsed organized labor's view that the OSHAct required states to expand safety and health personnel, but politics sabotaged this and other OSHAct mandates for state programs. By 1980, observers regarded few state plans aside from California's to be truly operational.[26]

Simultaneously, national OSHA regulation foundered. By 1980, administrative politics and adversarial proceedings hindered OSHA's formulation and revision of standards, while chronic understaffing crippled OSHA enforcement efforts. Through the 1970s, OSHA inspectors reached only 1 percent of workplaces, covered only 5 percent of workers, stressed mechanical over health risks, and issued citations and fines sparingly. *Marshall v. Barlow's Inc.* (1978) weakened the inspection process further by allowing employers to debar OSHA inspectors lacking warrants. Reagan-era cuts reduced OSHA work even more by closing one-third of OSHA field offices and discharging 20 percent of inspectors, a problem that similarly afflicted the federal Mine Safety and Health Administration.[27]

Topping that off, the OSHA system apparently failed to reduce accident and disease levels. Without data on particular industries or states, aggregate statistics revealed a temporary dip in the national accident rate during the mid-1970s, but analysts attributed that drop to the 1974–1975 business downturn, not to OSHA's inauguration. In the long term, analysts found that serious accidents continued to rise, though minor injuries abated and OSHA regulation seemed to minimize certain kinds of mishaps. On the whole, statistical evidence through the 1970s impugned OSHA's capacity to improve safety and health conditions.[28]

In the end, OSHA dramatically altered the aims, administrative structure, and political functions of American work safety and health law. Yet rather than erecting an efficient national program, OSHA evolved as an awkward amalgamation of imperfect national institutions and uneven state plans. The OSHAct's uniform occupational safety and health policy was idealistic and laudable, but the program's administrative complexity and conservative political environment undermined its "radical" potential just as it got started.

Appendix

Table 1. Accident Cases Reaching Wisconsin's Supreme Court before 1911

Industries	Before 1881	1881– 1890	1891– 1900	1901– 1910	Total
Railroads	17	24	54	32	127
Manufacturing			28	60	88
Paper and lumber mills	.	8	34	30	72
Quarries			6	21	27
Construction	1	5	10	7	23
Shipping and express	1	4	6	2	13
Lumber and logging		1	4	6	11
Mining		2	5	2	9
Utilities			3	6	9
Other			3	1	4

Safety Hazards	Before 1881	1881– 1890	1891– 1900	1901– 1910	Total
Railroads	17	24	54	32	127
Crushed limbs, fingers		2	18	40	60
Falls	1	8	25	16	50
Falling objects	1	6	13	24	44
Striking objects		2	16	19	37
Cuts		2	14	14	30
Explosions			7	10	17
Gas and electricity			3	5	8
Other			3	7	10
Total	19	44	153	167	383

Source: *Wisconsin Reports,* Vols. 11–144 (1860–1910).

Table 2. Accident Case Rulings Reversed in Wisconsin Supreme Court, 1881–1910

	1881–1890	1891–1900	1901–1910	Total
Total Reversals	18	84	73	175
Favoring worker	5	13	21	39
Favoring employer	13	71	52	136

Issues on Reversal	1881–1890	1891–1900	1901–1910	Total
Employer Defenses				
Fellow servant	4	3	1	8
Assumption of risk	3	12	15	30
Contributory negligence	2	8	6	16
Trial Court Error				
Flawed verdict	3	24	10	37
Improper instructions		20	12	32
Triable fact	3	11	21	35
No proximate cause		10	2	12
Flawed evidence	1	2	1	4
Improper dismissal	2	5	5	12
No negligence	3	5	5	13

Source: *Wisconsin Reports*, Vols. 52–144, 1881–1910

Table 3. Representation on Advisory Committees of the Wisconsin Industrial Commission

Interest Represented	Pre-1920	1921–1940	Total
Business	38 (51%)	45 (38%)	83 (43%)
Out-of-state corporations	1	2	
National business groups	1		
Wisconsin companies	29	22	
State business groups	6	17	
State or local utilities	2	3	
Professions	9 (12%)	25 (21%)	34 (18%)
National organizations	1	7	
State organizations	1	4	
Educational institutions	3	2	
Physicians		6	
Insurance firms or groups	4	6	
Organized Labor	8 (11%)	18 (16%)	26 (13%)
Government	20 (26%)	30 (25%)	50 (26%)
Federal agencies	2	3	
State or local	4	11	
Industrial commission	14	16	
Total	75 (100%)	118 (100%)	193 (100%)

Source: Industrial Commission of Wisconsin, *General Orders,* 1912–1940.

Table 4. Profile of Employers against Whom the Wisconsin Industrial Commission
Threatened or Started Prosecution, 1921–1940 (by Census Classification of City or Town)

	First[1]	Second[2]	Third[3]	Fourth[4]	Unknown	Total
1921–1925	8	1	9	18	7	43
1926–1930	6	2	3	11	2	24
1931–1935	1	2	5	17	1	26
1936–1940	3	9	17	44	1	74
Total	18	14	34	90	11	167

Sources: Industrial Commission of Wisconsin, *Minutes*, 1921–1940; Industrial Commission of
Wisconsin, *Biennial Reports* (1920–1922), 55, (1922–1924), 54, (1924–1926), 55, (1926–1928), 59,
(1928–1930), 52, (1930–1932), 56–57.
1. Population over 150,000.
2. Population from 39,000 to 150,000.
3. Population from 10,000 to 39,000.
4. Population under 10,000.

Table 5. Safety Order Violations Prosecuted by the Wisconsin Industrial Commission,
1921–1940 (by selected safety codes)

	1921–1925	1926–1930	1931–1935	1936–1940	Total
Boiler	3	2	2	1	8
Elevator	7	3		1	11
General orders	23	15	7	48	93
Heating/ventilation	2		3	13	18
Spray coating	1	1	13	9	24
Unknown/other	7	3	1	2	13
Total	43	24	26	74	167

Sources: Industrial Commission of Wisconsin, *Minutes,* 1921–1940; Industrial Commission of
Wisconsin, *Biennial Reports* (1920–1922), 55; (1922–1924), 54; (1924–1926), 55; (1926–1928), 59;
(1928–1930), 52; (1930–1932), 56–57.

Notes

Abbreviations

AALL	American Association for Labor Legislation
CBA	California Bar Association
CIAC	California Industrial Accident Commission
CSFL	California State Federation of Labor
IAIABC	International Association of Industrial Accident Boards and Commissions
ICO	Industrial Commission of Ohio
ICW	Numerical "C" Files, Industrial Commission of Wisconsin, preceded by Box/File Folder number (for example, 12/C45 ICW)
ICW General Archives	Industrial Commission of Wisconsin, Government Documents Division, State Historical Society of Wisconsin
IDL	Illinois Department of Labor
MFTC	Milwaukee Federated Trades Council
MMAM	Merchants and Manufacturers Association of Milwaukee
NYSDL	New York State Department of Labor
NYSFL	New York State Federation of Labor
NYSIC	New York State Industrial Commission
ODIR	Ohio Department of Industrial Relations
OSFL	Ohio State Federation of Labor
OSHR	*Occupational Safety and Health Reporter*
SBAW	State Bar Association of Wisconsin
USBLS	U.S. Bureau of Labor Statistics
USDL	U.S. Department of Labor

USDLS U.S. Division of Labor Standards
Wausau
  Archives Wausau Insurance Companies Archives
WBLIS Wisconsin Bureau of Labor and Industrial Statistics
WLRL Wisconsin Legislative Reference Library
WMA Wisconsin Manufacturers Association
WSFL Wisconsin State Federation of Labor

All court cases are fully cited when first mentioned, and abbreviated thereafter, in each chapter.

Introduction

1. Because 1911 Wisconsin legislation, modern federal statutes, and other laws have covered "employment" as well as "places of employment," this book utilizes the terminology "*work* safety and health" rather than "*workplace* safety and health," except where law or policy specifically applied just to the "place" of work.

2. *Milwaukee Free Press,* June 15, 1911, 12; *The Milwaukee Journal,* October 27, 1913, 7.

3. For early commentary, consult Altmeyer, *Industrial Commission;* J. Andrews, *Administrative Labor Legislation,* 73–74; Brandeis, "Administration of Labor Laws," 652–53; Howe, *Wisconsin,* 109–36; and McCarthy, *The Wisconsin Idea,* 162–63. More recent characterizations include Buenker, *The History of Wisconsin,* 300, 527–28, 549–50, 580–98; Harter, *John R. Commons,* 112–13; Lubove, *Struggle for Social Security,* 33–34; and Nesbit, *Wisconsin: A History,* 413, 429–30.

4. Altmeyer, *Industrial Commission,* 106–7; Andrews and Andrews, "Scientific Standards"; I. Andrews, "New Spirit"; Nesbit, *Wisconsin: A History,* 430; Buenker, *The History of Wisconsin,* 550.

5. Harter, *John R. Commons,* 112–13; Brandeis, "Administration of Labor Laws," 645; Frances Perkins, Address to the Wisconsin State Legislature, June 19, 1933, 24/ C412 ICW. On industrial pluralism, see Ernst, "Common Laborers?" 66–68, 80–82; Fink, "'Intellectuals' versus 'Workers,'" 409; Furner, "Knowing Capitalism," 273; and Tomlins, "New Deal," 19–23.

6. MacLaury, "Job Safety Law," 19; Commons, *Myself,* 143.

7. Kolko, *Triumph of Conservatism;* Weinstein, *Corporate Ideal;* M. Sklar, *Corporate Reconstruction.*

8. Wiebe, *Search for Order,* 222–23, 295–96; Galambos, "Emerging Organizational Synthesis"; Hays, "The New Organizational Society," 244–63; Hawley, *Great War,* 9–11.

9. McCraw, "Regulation in America"; Keller, *Regulating a New Society,* 5, 197–98; Link and McCormick, *Progressivism,* 62–66.

10. Keller, *Regulating a New Society,* 197–202; Rosner and Markowitz, *Dying for Work,* xiv–xvi; Graebner, *Coal-Mining Safety,* 72–100; Fishback and Kantor, *Prelude to the Welfare State,* 1–50.

11. Aldrich, *Safety First,* 122–48; Witt, *Accidental Republic,* 103–22.

12. Ashford, *Crisis in the Workplace,* 47–49; MacLaury, "Job Safety Law," 18–19; C. Noble, *Liberalism at Work,* 54–57; D. Berman, *Death on the Job,* 50–62, 68–71.

13. Korman, *Industrialization,* 119–30; Nelson, *Managers and Workers,* 138–39.

14. Stone, "Post-war Paradigm," 1511–80; Tomlins, *State and Unions,* 230–51.

15. Novak, "Legal Origins," 252, 260; Graebner, "Federalism in the Progressive Era"; Weir, Orloff, and Skocpol, *Politics of Social Policy,* 17–27; McCraw, "Regulation in America," 182; Hays, "Political Choice in Regulation," 127–28; Carpenter, *Forging of Bureaucratic Autonomy,* 14–33; Block, *Revising State Theory,* 16–19.

16. See, for instance, Friedman and Ladinsky, "Social Change"; and Witt, *Accidental Republic,* 180–86.

17. Howe, *Wisconsin,* 136; Brandeis, "Administration of Labor Laws," 626–44. See also Link and McCormick, *Progressivism,* 58–66; Skowronek, *New American State;* Horwitz, *Transformation, 1870–1960,* 3–6, 213–16; Wiecek, *Lost World,* 3–16.

18. Rodgers, *Atlantic Crossings,* 245–50; Witt, *Patriots and Cosmopolitans,* 209–13; Kagan, *Adversarial Legalism,* 10–13, 129; Burke, *Lawyers, Lawsuits, and Legal Rights,* 18, 39; Nonet, *Administrative Justice,* 17–18, 29–31.

19. M. Sklar, *Corporate Reconstruction,* 1–33; Nelson, *Managers and Workers,* 55–78; D. Noble, *America by Design,* pt. 1; Hays, "The New Organizational Society," 246; Hawley, *Great War,* 53–55, 81–82, 92; Aldrich, *Safety First,* 90–156; Montgomery, *House of Labor,* 203–81.

20. Buenker, *The History of Wisconsin,* 593–94. On "corporate liberalism" and "corporatism," see Weinstein, *Corporate Ideal,* ix–xiv; M. Sklar, *Corporate Reconstruction,* 34–40; and Schmitter, "Century of Corporatism," 20–41. For an alternative interpretation, see Hawley, "Discovery and Study."

21. Keller, *Regulating a New Society,* 197–202; Rosner and Markowitz, *Dying for Work,* xiv–xvi; Hawley, *Great War,* 100–104.

22. Teaford, *Rise of the States,* 5–6.

Chapter 1: From Common Law to Factory Laws

This chapter is an updated version of my article "From Common Law to Factory Laws: The Transformation of Workplace Safety Law in Wisconsin before Progressivism," *American Journal of Legal History* 37 (1993): 177–213, and appears here in revised form with permission of that journal.

1. Hurst, *Conditions of Freedom,* 5–8, 73, 96.

2. *Murray v. South Carolina Railroad Company* 1 McMul 385 (1841); *Farwell v. Boston and Worcester Rail Road Corporation* 45 Mass. 49 (1842); Schweber, *Creation of American Common Law,* 4–6, 36–37, 111–17, 123–28, 141–44, 189–92, 240–58.

3. Early works offering the "subsidy" thesis include Levy, *Law of the Commonwealth,* 166–82; Friedman and Ladinsky, "Social Change," 50–58; Friedman, *History of American Law,* 300–302; and Horwitz, *Transformation, 1780–1860,* 209–10. See also Wiecek, *Lost World,* 46–48. More recent scholars treat mid-nineteenth-century

industrial torts as a product of the nation's "liberal republican" ideology and emerging free-labor market economy, with severe results for workers. Consult Konesky, "Their Own Interests"; McEvoy, "Freedom of Contract," 205–6; Rabin, "Historical Development," 959–61; G. Schwartz, "Tort Law and the Economy"; G. Schwartz, "Character of Tort Law," 692–715; White, *Tort Law in America*, 12–19, 50–51; Witt, *Accidental Republic*, 13–15, 44–45, 118; and Schweber, *Creation of American Common Law*, 2–7. Other recent scholars see nineteenth-century work-injury law built on the austere preindustrial master-servant regime. See Orren, *Belated Feudalism*, 77–90, 102–17; Simpson, *Leading Cases*, 100–134; Tomlins, "Law and Power," 71–91; and Tomlins, *Law, Labor, and Ideology*, 333–84. Specifically rejecting the subsidy thesis, others argue that legal continuity held enterprisers accountable for many accidental injuries. Note Kaczorowski, "Common-Law Background," 1127–29; and Karsten, *Heart versus Head*, 114–26. On the balancing of values and interests, see Freyer, *Producers versus Capitalists*, 167–95; and Ely, *Railroads and American Law*, 212–15.

4. *Farwell*, 62. Commentary on other founding cases suggests similar reciprocal employer duties. See Simpson, *Leading Cases*, 108; and G. Schwartz, "Character of Tort Law," 694. On the "[moral] law of reciprocity" in accident cases, examine Bailey, *Guardians of the Moral Order*, 160–69.

5. Horwitz, *Transformation, 1870–1960*, 3–4, 9–31; Wiecek, *Lost World*, 3–15, 64–96; Schweber, *Creation of American Common Law*, 2.

6. On employer liability laws, compare Friedman and Ladinsky, "Social Change," 63–65, to Bergstrom, *Courting Danger*, 74–78. On factory laws, see Rosner and Markowitz, *Dying for Work*, xvi; Nelson, *Managers and Workers*, 122–39; Aldrich, *Safety First*, 100–101; Witt, *Accidental Republic*, 19.

7. U.S. Census, 1880, vol. 1, *Population*, 853; vol. 2, *Report of Manufacturers*, 5, 11, 189–91. This report estimated factory labor forces as follows: Alabama, 10,019; California, 43,693; Illinois, 144,727; New York, 531,533; and Ohio, 183,609.

8. *Chamberlain v. The Milwaukee & Mississippi Railroad Company*, 11 Wis. 238, 249–56 (1860); *Moseley v. Chamberlain*, 18 Wis. 700, 705 (1864, decided 1861); *Smith v. The Chicago, Milwaukee & St. Paul Railway Company*, 42 Wis. 520, 526 (1877). See also Tomlins, *Law, Labor, and Ideology*, 368–83; Karsten, *Heart versus Head*, 120–21; and Witt, "Transformation of Work," 1478.

9. Cooley, *Treatise on Torts*, 634–35; "Comment—the Creation of a Common-Law Rule," 597–615; Wiecek, *Lost World*, 7–11, 46–47; Schweber, *Creation of American Common Law*, 2.

10. *Wedgwood v. The Chicago & Northwestern Railway Company*, 41 Wis. 478, 478–84 (1877); *Clarke v. Holmes*, 7 H. & N. 937, 943 (1862); Bronstein, *Caught in the Machinery*, 29. *Wedgwood* emphasized the exercise of adequate employer *care*, not specifically "safe and suitable machinery," an absolute standard of safety. For different treatment, see Friedman and Ladinsky, "Social Change," 62; and Friedman, *History of American Law*, 483. See *Smith*, 526. The New York case was *Laning v. New York Central Railway Co.*, 49 N.Y. 521 (1872).

11. Rabin, "Historical Development," 932; Friedman, *History of American Law*, 299–302; Witt, *Accidental Republic*, 43–51; Bergstrom, *Courting Danger*, 59–65; Schweber, *Creation of American Common Law*, 34–36; Tripp, "Instance of Labor," 532–33. For court cases, see *Read v. Morse*, 34 Wis. 315, 319 (1874); *Carlson v. The Phoenix Bridge Co.*, 132 N.Y. 273, 277 (1892); *Burns v. Sennett and Miller*, 99 Cal. 363 (1893); *Holland v. Tennessee Coal, Iron, & Railroad Co.*, 91 Ala. 444 (1890); *Leonard v. Kinnare*, 174 Ill. 532 (1898); *Cincinnati, Hamilton & Dayton Railroad Co. v. Frye*, 80 O.S. 289 (1909).

12. *Smith*, 525–26; *Dorsey v. The Phillips & Colby Construction Company*, 42 Wis. 583, 597 (1877).

13. Witt, "Toward a New History," 722–25; Witt, *Accidental Republic*, 31–35.

14. Horwitz, *Transformation, 1780–1860*, 208–9, 256; Hurst, *Conditions of Freedom*, 3–32; Zainaldin, *Law in Antebellum Society*, 52–62; Hovenkamp, *Enterprise and American Law*, 1–6; Wharton, *Law of Negligence*, 187–88.

15. Supporting the subsidy theory are Horwitz, *Transformation, 1780–1860*, 209–10, 253–56; and Friedman, *History of American Law*, 467–87. Against this view are G. Schwartz, "Tort Law and the Economy"; Kaczorowski, "Common-Law Background"; Karsten, "Explaining the Fight," 45–92; and Freyer, *Producers versus Capitalists*, 167–95. Out of this latter group, G. Schwartz concedes that workplace accident law was "singularly complex, ungenerous and troublesome" ("Tort Law and the Economy," 1774–75). See also Karsten, *Heart versus Head*, 80–83; Tomlins, *Labor, Law, and Ideology*, 383–84; Wiecek, *Lost World*, 46–48; and Aldrich, *Death Rode the Rails*, 24–25.

16. Friedman and Ladinsky, "Social Change," 62; Simon, "For the Government," 115–16; *Farwell*, 62.

17. J. Byles in *Clarke*, 948, quoted in *Hough v. Railway Co.*, 100 U.S. 213, 222–23 (1879); *Wedgwood*, 482–83; *Laning v. New York*, 521, 533–34. See also Schweber, *Creation of American Common Law*, 125, 141–43.

18. U.S. Census, 1880, vol. 2, *Manufacturers*, 189; vol. 1, *Population*, 853; U.S. Census, 1910, vol. 9, *Manufacturers*, 32, 100, 292, 878, 984, 1356; vol. 4, *Occupational Statistics*, 531–33; Buenker, *The History of Wisconsin*, 80–116.

19. Nelson, *Managers and Workers*, chaps. 1–6; Brody, *Workers in Industrial America*, 3–21; Korman, *Industrialization*, 15–40; Aldrich, *Safety First*, 77–91.

20. U.S. Census, 1910, vol. 9, *Manufacturers*, 29–30, 100–101, 292–93, 878–82, 984–85, 1356–57, 1360; Woodward, *Origins of the New South*, 291–308; Rogers et al., *Alabama*, 277–87; Bean, *California: An Interpretive History*, 271–84; Knepper, *Ohio and Its People*, 286–304; Cayton, *Ohio*, 180–81; Howard, *Illinois*, 391–92; Ellis et al., *New York State*, 461–532.

21. Bergstrom, *Courting Danger*, 11–30. See Table 1 in Appendix.

22. See Table 1 in Appendix.

23. Friedman and Ladinsky, "Social Change," 62; Simon, "For the Government," 115–16, 120–24; Bergstrom, *Courting Danger*, 140–43.

24. On due care, see *Eingartner v. The Illinois Steel Company*, 94 Wis. 70, 79 (1896); *Kreider v. The Wisconsin River Paper & Pulp Company*, 110 Wis. 645, 651 (1901);

Yazdzewski v. Barker, 131 Wis. 494, 498 (1907); *Glenesky v. Kimberly & Clark Company*, 140 Wis. 52, 57 (1909). On warnings, see *Nadau v. White River Lumber Co.*, 76 Wis. 120, 130 (1890); *Kliegel v. Aitken*, 94 Wis. 432, 435 (1896); *Anderson v. Hayes*, 101 Wis. 519, 522 (1899); and *Rankel v. Buckstaff-Edwards Co.*, 138 Wis. 442, 450 (1909). On safe and suitable machinery, consult *Whitman v. The Wisconsin & Minnesota Railroad Co.*, 58 Wis. 408, 413 (1883). On servicing, see *Hobbs v. Stauer and others*, 62 Wis. 108, 110 (1885); *Sherman v. The Menominee River Lumber Co.*, 72 Wis. 122, 127 (1886); and *Montanye v. Northern Electrical Manufacturing Company*, 127 Wis. 22, 33 (1906). On shop management, review *Faerber v. The T. B. Scott Lumber Co. (Ltd.)*, 86 Wis. 226, 233 (1893); *Goff v. The Chippewa River & Menominee Railway Co.*, 86 Wis. 237, 243 (1893); and *Polaski v. Pittsburgh Coal Dock Company*, 134 Wis. 259, 261 (1908).

25. Welke, "Unreasonable Women"; Schlanger, "Injured Women before Common Law Courts, 1860–1930." For women, read *Herold v. Pfister*, 92 Wis. 417 (1896); *Kliegel v. Aitken*, 94 Wis. 432 (1896); *Groth v. Thomann*, 110 Wis. 488 (1901); *Van de Bogart v. Marinette & Menominee Paper Company*, 127 Wis. 104 (1906); *Van de Bogart v. Marinette & Menominee Paper Company*, 132 Wis. 367 (1907); *Kuich v. Milwaukee Bag Company*, 139 Wis. 101 (1909); and *Keena v. American Box Toe Company*, 144 Wis. 231 (1910). On children, see *Luebke v. Berlin Machine Works*, 88 Wis. 442 (1894); *Craven v. Smith*, 89 Wis. 119 (1894); *Casey v. Chicago, St. Paul, Minneapolis & Omaha Railway Company*, 90 Wis. 113 (1895); *Wolski v. The Knapp-Stout & Co. Company*, 90 Wis. 178 (1895); *Disotell v. The Henry Luther Company*, 90 Wis. 635 (1895); *Kucera v. Merrill Lumber Company*, 91 Wis. 637 (1895); *Horn v. La Crosse Box Company*, 123 Wis. 399 (1904); *Schumacher v. Tuttle Press Company*, 142 Wis. 631 (1910); and *Schmolt v. H. W. Wright Lumber Company*, 145 Wis. 577 (1911). About child labor law, see J. Schmidt, "'Restless Movements,'" 338–50.

26. *O'Connor v. Golden Gate Woolen Manufacturing Company*, 135 Cal. 537, 544 (1902); *Quinn v. Electric Laundry Co.*, 155 Cal. 500, 506 (1909); *Berit Jorgensen v. The Johnson Chair Company*, 169 Ill. 429, 430–31 (1897); *Jonathan O. Armour, et al. v. Julianna Golkowski*, 202 Ill. 144, 145–51; *Reaves v. Anniston Knitting Mills*, 154 Ala. 565, 569 (1908); *Lowe Manufacturing Co. v. Payne*, 167 Ala. 245, 248 (1910); *Knisley v. Pratt*, 148 N.Y. 372, 380 (1896); *Bateman v. N.Y.C. & H.R.R.R. Co.*, 178 N.Y. 84, 90–91 (1904); *Kline v. Abraham*, 178 N.Y. 377, 379–81 (1904); *Kirwan v. American Lithographic Co.*, 197 N.Y. 413, 418 (1910).

27. On the vice-principal rule, see *Smith*, 526. On the "wobble" problem, see Friedman, *History of American Law*, 483–84; and Bergstrom, *Courting Danger*, 73–81.

28. WBLIS, *Thirteenth Biennial Report* (1909), 85–86. See also Table 2 in Appendix. On the measure of safety, consult *Baxter v. The Chicago & Northwestern Railway Co.*, 104 Wis. 307, 321–22 (1899); *Hamann v. Milwaukee Bridge Co.*, 127 Wis. 550, 564 (1906). On judges' instructions, review *Suter v. The Park & Nelson Lumber Company*, 90 Wis. 118, 121 (1895); *Raffke v. Patten Paper Company*, 136 Wis. 535, 540 (1908); and *West v. Bayfield Mill Co.*, 144 Wis. 106, 133 (1910). On verdicts, see *Raffke*, 538–40.

On proximate cause, examine *Johnson v. Ashland Water Co.,* 77 Wis. 51, 55 (1890); *Musback v. The Wisconsin Chair Co.,* 108 Wis. 57, 66 (1900); and *Lillis v. Beaver Dam Woolen Mills,* 142 Wis. 128, 134–35 (1910).

29. Just occasionally did Wisconsin judges simply say that employers owed workers "a reasonably safe workplace." See *Nadau v. White River Lumber Co.,* 76 Wis. 120 (1890); and *Hocking v. Windsor Spring Co.,* 125 Wis. 575 (1905). On other state courts, review *Huyck v. McNerney,* 163 Ala. 244, 250 (1909–1910); *Perry v. Rogers,* 151 N.Y. 251, 254–55 (1898); *Vogel v. American Bridge Co.,* 180 N.Y. 373, 376–77 (1905); *Callon v. Bull,* 113 Cal. 593, 603 (1896); and *Brett v. S. H. Frank & Co.,* 153 Cal. 267, 272 (1902). On the insurance firm view, see WLRL, *Testimony, Proceedings, and Report of the Wisconsin Legislature's Special Joint Committee on Industrial Insurance (1909–1911),* 209.

30. *Knudsen v. The La Crosse Stone Company,* 145 Wis. 394, 400, 404–5 (1911). See Friedman and Ladinsky, "Social Change," 67–68.

31. *Guinard v. Knapp-Stout & Co.,* 95 Wis. 482, 487 (1897).

32. Keller, *Affairs of State,* 285–87; Link and McCormick, *Progressivism,* 47–58; E. S. Clemens, *People's Lobby,* 235–71; Dubofsky, *Industrialism and the American Worker,* 72–79.

33. Tomlins, *State and Unions,* 74–77; Forbath, *Law and the American Labor Movement,* 32–135; Fink, "Labor, Liberty, and the Law," 914–18.

34. Aldrich, *Safety First,* 90–91; Witt, *Accidental Republic,* 100–101.

35. E. S. Clemens, *People's Lobby,* 235, 237, 251–52; T. Clark, *Defending Rights,* 25, 125–31; Taft, *Labor Politics,* 6–17, 19; Skocpol, *Protecting Soldiers and Mothers,* 217–45.

36. Gavett, "Labor Movement in Milwaukee," 60; Madden, "Factory Safety," 14, 38, 47–50, 82; Fink, *Workingmen's Democracy,* 23–34; Wisconsin, *Assembly Bills* (1878), Assembly Bill no. 86; (1881), Assembly Bill no. 264; (1882), Assembly Bill no. 335; *Evening Wisconsin* (Milwaukee), February 16, 1883, 2; March 27, 1882, 2; *The Milwaukee Sentinel,* February 14, 1883, 5; March 8, 1883, 4; March 16, 1887, 2; *Wisconsin State Journal,* March 7, 1883, 1; Wisconsin, *Assembly Journal* (1883), 23; (1885), 21, 1251; WBLIS, *First Biennial Report,* 1–5, 169.

Wisconsin created similar bureaus in *Laws of Wisconsin* (1874), Chapter 273 as amended by Chapter 57 (1876); (1876), Chapter 333; (1878), Chapter 214; (1880), Chapter 269; (1882), Chapter 167; (1883), Chapter 319; (1885), Chapter 247; Wisconsin, *Revised Statutes, 1878,* Chapter 87, Sections 1792, 1794, 1795, 1796; Chapter 56, Sections 1404, 1407, 1408.

Regarding other states' labor bureaus, see Knepper, *Ohio and Its People,* 305–7; Beckner, *Labor Legislation in Illinois,* 488; Walker, "Factory Legislation and Inspection," 64–65; Bean, *California: An Interpretive History,* 285–87; Kelly, *Race, Class, and Power,* 108–31; and Grantham, *Southern Progressivism,* 298.

37. Madden, "Factory Safety," 36–41, 51; Gavett, "Labor Movement in Milwaukee," 181–82, 186; *Laws of Wisconsin* (1885), Chapter 245, Section 5; (1887), Chapter 549; (1887), Chapter 453, Section 2.

38. Aldrich, *Safety First*, 67–69; Whiteside, *Regulating Danger*, 57–71; Graebner, *Coal-Mining Safety*, 72–73; Walker, "Factory Legislation and Inspection," 83, 101–2; *Acts of Ohio* (1884), 106–7; Beckner, *Labor Legislation in Illinois*, 228–37; *Statutes and Amendments of California* (1889), Chapter 5; *General Laws of Alabama* (1889), 81; (1907), 338; *Caspar v. Lewin*, 82 Kan. 604, 618 (1910). On early factory laws generally, see Willoughby, "Inspection of Factories"; Nelson, *Managers and Workers*, 123–24; and Aldrich, *Safety First*, 100–101.

39. K. Sklar, *Florence Kelley*, 206–65, 281–85; Beckner, *Labor Legislation in Illinois*, 254–70.

40. *Columbian Biographical Dictionary*, 528, 600–603; *The Milwaukee Sentinel*, February 26, 1887, 4; September 20, 1886, 2; Aikens and Proctor, *Men of Progress: Wisconsin*, 161–62; Quaife, *Wisconsin*, 3:457, 4:151. On Joseph Beck, see Beck to Janie (Beck), July 7, 8, 1901, Beck Papers. On Sherman's appointment, see Walker, "Factory Legislation and Inspection," 117–31.

41. On inspection work generally, see Kagan, "Regulatory Inspectorates." On Wisconsin inspection, see Flower to Henry (Siebers), May 25, 28, 1887, Flower Papers; WBLIS, *First Biennial Report*, 167; *Second Biennial Report*, xl; *Eighth Biennial Report*, ix–xii, 201–2; *Ninth Biennial Report*, 251, xvi–xvii; *Tenth Biennial Report*, 211–12; *Twelfth Biennial Report*, 1166.

42. WBLIS, *Sixth Biennial Report*, 73a, 52a, 81a; *Eighth Biennial Report*, 439; *Ninth Biennial Report*, 840–43, 897–901; *Tenth Biennial Report*, 1098, 1226; Factory Inspector Reports, WBLIS, July–October, 1904, Archives, State Historical Society of Wisconsin.

43. WBLIS, *Second Biennial Report*, xl, 504, 510, 511–12, 513; *Third Biennial Report*, 278; *Fourth Biennial Report*, vi; *Fifth Biennial Report*, 87a; *Seventh Biennial Report*, 374; *Ninth Biennial Report*, 251.

44. On Wisconsin labor developments, see Ozanne, *Labor Movement in Wisconsin*, 36; *The Milwaukee Sentinel*, February 27, 1899, 8; and *Laws of Wisconsin* (1899), Chapters 77, 79, 189, 232. On other states, see Rose, "Design and Expediency," 51–65; *Acts of Ohio* (1898), 36, 155; (1900), 42–43; Yellowitz, *Labor and the Progressive Movement*, 22–23; and Taft, *Labor Politics*, 6–8. On female employment in Wisconsin, compare U.S. Census, 1890, *Manufacturers*, pt. 1, 629, to U.S. Census, 1900, *Manufacturers*, pt. 2, 959. On women's protective law, see Novkov, *Constituting Workers*, 150–68; Tone, *Business of Benevolence*, 24–27; Skocpol, *Protecting Soldiers and Mothers*, 382–400; Kessler-Harris, *Out to Work*, 185–97; and Baer, *Chains of Protection*, 29–39. On mine safety, consult Whiteside, *Regulating Danger*, 97–113; Graebner, *Coal-Mining Safety*, 73–86.

45. *The Milwaukee Sentinel*, March 1, 1899, 2; *Laws of Wisconsin* (1899), Chapter 152; Walker, "Factory Legislation and Inspection," 74–110; Beckner, *Labor Legislation in Illinois*, 268.

46. Korman, *Industrialization*, 73–135; Beckner, *Labor Legislation in Illinois*, 223–28, 240–64; Pegram, *Partisans and Progressives*, 61–84; Rose, "Design and Expediency,"

11–20, 51–65; Knepper, *Ohio and Its People*, 303–12; Nelson, *Managers and Workers*, 128; Aldrich, *Safety First*, 91–92; Witt, *Accidental Republic*, 103–17.

47. MFTC, *Proceedings*, February 20, 1907, May 15, 1907, March 3, 1909, September 1, 1909; WSFL, *Proceedings* (1905): 4, 30–31, 50–51; (1906): 4, 62; (1907): 77; (1908): 6–7; (1909): 23–24; (1910): 6–7. For pending legislation in 1911, see Wisconsin, *Assembly Bills*, 137A, 175A, 423A; and *Senate Bills*, 311S.

48. *Laws of Wisconsin* (1901), Chapter 409; (1905), Chapter 338; WBLIS, *Twelfth Biennial Report*, 1306; Willoughby, "Inspection of Factories," 555–57. Wisconsin factory inspector Rose Perdue reported strong women's lobbying for female inspectors in *The Milwaukee Journal*, July 27, 1911, 1.

49. WLRL, *Testimony of Committee on Industrial Insurance*, 450, 1250; NYSDL, *Second Annual Report*, I14; *Sixth Annual Report*, I30. Even in 1897, W. F. Willoughby had complained that extraneous duties plagued factory inspection nationwide ("Inspection of Factories," 565–66).

50. Early Wisconsin accident tabulations appear in WBLIS, *Fourth Biennial Report*, 127a–129a; *Fifth Biennial Report*, 140a–143a; *Sixth Biennial Report*, 193a–198a; *Eighth Biennial Report*, 209–10; and *Ninth Biennial Report*, xix–xxviii. Wisconsin statutes mandating accident reports were *Laws of Wisconsin* (1905), Chapter 416; and (1911), Chapter 469. For a nationwide survey of accident reporting laws circa 1910, see *American Labor Legislation Review* 1 (1911): 60–71. On New York, see Bergstrom, *Courting Danger*, 47. On Ohio, see *Acts of Ohio* (1888), 99. For early uses of accident statistics, see NYSDL, *Sixth Annual Report*, I32; and WBLIS, *Thirteenth Biennial Report*, 3–56.

51. Bird, "Standardization of Safety," 1022; WBLIS, *Eighth Biennial Report*, 208. See also NYSDL, *Sixth Annual Report*, I35–I39; Sanborn and Sanborn, *Supplement to the Wisconsin Statutes of 1898*, Sections 1021f, 1636–39; and Commons, "Industrial Commission," 1, 4–5.

52. *Caspar v. Lewin*, 613–18. For catalog of machine laws circa 1910, see *American Labor Legislation Review* 1 (1911): 72–89; and Herron, "Factory Inspection." On federalism in the early 1900s, see Graebner's "Federalism in the Progressive Era" and *Coal-Mining Safety*.

53. Ozanne, *Labor Movement in Wisconsin*, 26–33; WSFL, *Proceedings* (1901): 23–24; (1905): 30–31, 50–57; (1906): 20; (1909): 23–24; MFTC, *Proceedings*, December 20, 1905, October 3, 1906, February 2, 1910; Walker, "Factory Legislation and Inspection," 125; NYSDL, *Sixth Annual Report*, I25–I26.

54. Forbath, *Law and the American Labor Movement*, 177–92; Wiecek, *Lost World*, 126–32; Freund, *Police Power*, 99, 133–34. Wisconsin police power cases include *Baker v. State*, 54 Wis. 368 (1882); and *The State v. Heineman*, 80 Wis. 253 (1891). On Wisconsin factory law's basis in police power, see August C. Backus's report in WBLIS, *Ninth Biennial Report*, 844–49.

55. *Thompson v. Johnston Brothers, Co.*, 86 Wis. 576, 582 (1893); *Klatt v. The N.C. Foster Lumber Co.*, 97 Wis. 641, 646 (1897); *Laws of Wisconsin* (1887), Chapter 549.

On dangerous proximity, examine *Thompson v. The Edward P. Allis Company,* 89 Wis. 523, 528 (1895); *Miller v. Kimberly & Clark Co.,* 137 Wis. 138, 142 (1908); and *Gulland v. Northern Coal & Dock Company,* 147 Wis. 391, 395 (1911). On danger, see *Guinard,* 484–85; *Schweikert v. John R. Davis Lumber Co.,* 145 Wis. 632, 637 (1911); and *Koutsky v. Forster-Whitman Lumber Company,* 146 Wis. 425, 431 (1911). On methods of safeguarding, review *Willette v. Rhinelander Paper Company,* 145 Wis. 537, 545 (1911); and *Kuich,* 107.

56. *West,* 117, 128–30; *Willette,* 545–50; *Laws of Wisconsin* (1911), Chapter 396.

57. *Glen Falls P. C. Co. v. Travelers' Insurance Co.,* 161 N.Y. 399, 403–4 (1900). See also *Dillon v. National Tar Co.,* 181 N.Y. 215 (1905); *Wynkoop v. Ludlow Valve Manufacturing Co.,* 196 N.Y. 324 (1909); and *Caspar v. Lewin,* 620–22.

58. Tripp, "Instance of Cooperation," 535–37; *Narramore v. Cleveland, C. C. & St. L. Ry. Co.,* 96 Fed. Rep. 298, 300–301 (1899); *St. Louis, Iron Mountain and Southern Railway Co. v. Taylor,* 210 U.S. 281 (1907); *Chicago, Burlington & Quincy Railway Co. v. U.S.,* 220 U.S. 559 (1910); *Variety Iron & Steel Works Co. v. Poak,* 89 O.S. 297 (1914).

59. Wiecek, *Lost World,* 132–33; Lippmann, *Drift and Mastery,* 157.

60. Kens, "Source of a Myth," 78–98. For studies questioning the courts' conservative image, see Gillman, *Constitution Besieged,* 101–46; Chomsky, "Progressive Judges," 401–40; Urofsky, "State Courts"; Batlon, "Reevaluation of the Court of Appeals"; Witt, *Accidental Republic,* 152–86; and Novak, "Legal Origins," 262, 267–68.

Chapter 2: The Administrative Transformation of Work Safety and Health Law

1. On the "transformation of governance," see Link and McCormick, *Progressivism,* 58–66, 72–84; and Chambers, *Tyranny of Change,* 230–33. For legal changes, consult Horwitz, *Transformation, 1870–1960,* introduction and chap. 8; W. Ross, *Muted Fury,* 56–153; and Wiecek, *Lost World,* 175–97. On growing administrative capacity, examine Skowronek, *New American State;* Teaford, *Rise of the States,* 1–10; and Novak, "Legal Origins," 251–52, 260–64.

2. Friedman and Ladinsky, "Social Change," 69–72; Fishback and Kantor, *Prelude to the Welfare State,* 32–50, 93–112, 120–21; Witt, *Accidental Republic,* 119–85; Rodgers, *Atlantic Crossings,* 246–50; Witt, *Patriots and Cosmopolitans,* 159–70.

3. On Progressive interest-group politics, see McCormick, "Party Period"; Link and McCormick, *Progressivism,* 47–66; McCraw, "Regulation in America," 162–71; Rodgers, "In Search of Progressivism," 114–17; and Keller, *Regulating a New Economy,* 3, 125–37. For the "corporate liberal" and "capture" interpretations, consult Kolko, *Triumph of Conservatism;* Weinstein, *Corporate Ideal;* M. Sklar, *Corporate Reconstruction;* and Caine, *Myth of Progressive Reform.* The politics of "corporate liberalism of the left" are suggested in M. Sklar, *Corporate Reconstruction,* 33–37. On "state corporatism," see Schmitter, "Century of Corporatism"; Wunderlin, *Visions of a New Industrial Order,* 113–29; and Schatz, "From Commons to Dunlap." For a contrasting view, see Hawley, "Herbert Hoover," 117–19; and Hawley, "Discovery and Study."

4. Asher, "Wisconsin Workmen's Compensation Law," 124; Fishback and Kantor, *Prelude to the Welfare State,* 5, 20–21, 88–130.

5. U.S. Census, 1900, *Manufacturers,* pt. 2, 958–59; U.S. Census, 1910, vol. 9, *Manufacturers,* 1356–57; WBLIS, *Thirteenth Biennial Report* (1909), 7, 13; *Fourteenth Biennial Report* (1911), 71–72; Buenker, *The History of Wisconsin,* 81–121; Korman, *Industrialization,* 61–84; Nelson, *Managers and Workers,* 3–100; Ramirez, *When Workers Fight,* 87–99.

6. Buenker, *The History of Wisconsin,* 276–95; Ozanne, *Labor Movement in Wisconsin,* 26–33, 38–45, 125; Margulies, *Decline of the Progressive Movement,* 153–54; Gavett, *Development of the Labor Movement,* 77–89, 94–124; E. S. Clemens, *People's Lobby,* 243–44.

7. On labor-inspired liability reform, see Wisconsin, *Laws of Wisconsin* (1875), Chapter 173; (1880), Chapter 232; (1889), Chapter 438; (1893), Chapter 220; (1903), Chapter 448; (1905), Chapter 303; (1907), Chapter 254; Wisconsin, *Senate Bills* (1907), Senate Bill no. 225; Wisconsin, *Assembly Bills* (1911), Assembly Bill no. 226; Asher, "Wisconsin Workmen's Compensation Law," 125–26; G. Schmidt, "History of Labor Legislation," 46–61; and WSFL, *Proceedings* (1905): 5; (1906): 4; (1907): 6; (1908): 6–7; (1909): 6–7; (1910): 6–7; (1911): 6–7. On worker complaints, consult *Milwaukee Social-Democratic Herald,* January 14, 1911, 1; and Asher, "Industrial Safety," 122. On business safety, see Aldrich, *Safety First,* 90–91; and Witt, *Accidental Republic,* 118–22.

8. Korman, *Industrialization,* 80–83; Asher, "Wisconsin Workmen's Compensation Law," 126–31; Friedman and Ladinsky, "Social Change," 65–69; Fishback and Kantor, *Prelude to the Welfare State,* 93–98; MMAM, *Civics and Commerce,* December 1910, 12.

9. Asher, "Wisconsin Workmen's Compensation Law," 127–40; Buenker, *The History of Wisconsin,* 543–48; Nesbit, *Wisconsin: A History,* 429–30; Reagan, "Ideology of Social Harmony"; Pierce, "Organized Labor," 895–96; Beckner, *Labor Legislation in Illinois,* 431–61; Pegram, *Partisans and Progressives,* 75–78; Wesser, "Conflict and Compromise," 359–71; Mason, "Neither Friends nor Foes," 65–66; Bean, *California: An Interpretative History,* 329–30; Tindal, *Emergence,* 23. See also *Ives v. South Buffalo Railway Co.,* 201 N.Y. 271 (1911).

10. *Laws of Wisconsin* (1911), Chapter 50, Sections 2394-4, 2394-7, 2394-9; Friedman and Ladinsky, "Social Change," 70–72; Fishback and Kantor, *Prelude to the Welfare State,* 56–59, 88–89, 100–130.

11. *Laws of Wisconsin* (1911), Chapter 50, Sections 2394-1, 2394-8; (1913), Chapter 599; (1917), Chapter 624; Asher, "Wisconsin Workmen's Compensation Law," 136–40; Altmeyer, *Industrial Commission,* 26–27. On other states, see Wesser, "Conflict and Compromise," 358–59, 361, 369; Witt, *Accidental Republic,* 161–74, 180, 183; Taft, *Labor Politics,* 54; Reagan, "Ideology of Social Harmony," 324–25; Pierce, "Organized Labor," 896–99; and Fishback and Kantor, *Prelude to the Welfare State,* 103–4.

12. *Laws of Wisconsin* (1911), Chapter 50, Section 2394-4(3); *Statutes and Amendments of California* (1911), Chapter 399, Section 3(3); *Acts of Ohio* (1911), Section 22-2,

p. 529; *Laws of Illinois* (1911), Section 4, pp. 316–17. See also Friedman and Ladinsky, "Social Change," 71; Reagan, "Ideology of Social Harmony," 321; Beckner, *Labor Legislation in Illinois,* 447; and Witt, *Accidental Republic,* 181–82.

13. Asher, "Wisconsin Workmen's Compensation Law," 135–36; *Laws of Wisconsin* (1911), Chapter 50, Sections 2394-13 to 2394-18; *General Laws of Alabama* (1919), Section 21, p. 224; Section 28, p. 227; *Laws of Illinois* (1911), Section 18, p. 315; Grantham, *Southern Progressivism,* 298–301; Beckner, *Labor Legislation in Illinois,* 431–63; Pegram, *Partisans and Progressives,* 70–78; Fishback and Kantor, *Prelude to the Welfare State,* 103–4.

14. Rodgers, *Atlantic Crossings,* 223–25, 246–65; Wesser, "Conflict and Compromise," 366–70; Nash, "Influence of Labor," 250–51; Reagan, "Ideology of Social Harmony," 324–28; Yellowitz, *Labor and the Progressive Movement,* 108–18. See also *Laws of New York* (1914), Chapter 41, Article 5; *Statutes and Amendments of California* (1913), Chapter 175, Sections 36–40; *Acts of Ohio* (1913), 72–92.

15. WSFL, *Proceedings* (1907): 45–48; (1908): 91–107; Asher, "Wisconsin Workmen's Compensation Law," 128–29, 134–35; *The Milwaukee Journal* July 22, 1911, 1; Glad, *History of Wisconsin,* 210–11.

16. Buenker, *The History of Wisconsin,* 593–94; Witt, *Accidental Republic,* 154, 171–72, 176–77; Asher, "Wisconsin Workmen's Compensation Law," 124; Lubove, *Struggle for Social Security,* 45–49, 57–60; Weinstein, *Corporate Ideal,* 40–61.

17. Buenker, *The History of Wisconsin,* 526, 532–35, 543; Margulies, *Decline of the Progressive Movement,* 132–34; Asher, "Wisconsin Workmen's Compensation Law," 126–29, 135–38; E. S. Clemens, *People's Lobby,* 242–44.

18. MMAM, *Civics and Commerce,* January 1909, 2; WBLIS, *Thirteenth Biennial Report,* 107; McEvoy, "Triangle Shirtwaist Factory Fire," 621–25, 643–45; Witt, *Accidental Republic,* 154, 167, 170–74; Higgens-Evenson, "Industrial Police," 365–66.

19. *Laws of Wisconsin* (1903), Chapter 323; (1905), Chapters 147, 250, 296; (1907), Chapters 112, 115; (1909), Chapters 163, 373; Skocpol, *Protecting Soldiers and Mothers,* 400; *The Milwaukee Journal,* July 27, 1911, 1; Whiteside, *Regulating Danger,* 97–113; Graebner, *Coal-Mining Safety,* 73–86.

20. On Wisconsin industrialists' safety work, see Korman, *Industrialization,* 76–108, 120. On the NCF, see Green, *National Civic Federation,* 5–22. On the Wisconsin conferences and legal precedents, consult Bird, "Standardization of Safety," 1021; I. Andrews, "New Spirit," 357; McNeill, "Massachusetts Board of Boiler Rules"; Andrews and Andrews, "Scientific Standards," 131–32; Hatch, "The Prevention of Accidents," 107; and Beckner, *Labor Legislation in Illinois,* 228–39.

21. Margulies, *Decline of the Progressive Movement,* 130–32; Buenker, *The History of Wisconsin,* 519–26; Thelen, *La Follette,* 70–78; Mowry, *The California Progressives,* 90–102; Taft, *Labor Politics,* 42–43; Lower, *Bloc of One,* 21–23, 31–34; Starr, *Inventing the Dream,* 237–38, 254–55; Warner, *Progressivism in Ohio,* 354–57; Cayton, *Ohio,* 227–33; Pierce, "Organized Labor," 900.

22. Reeve, "Our Industrial Juggernaut"; Reeve, "Death Roll of Industry"; Hard, "Making Steel and Killing Men"; Hard, "Law of the Killed."

23. Cebula, *James M. Cox*, 47; McCarthy to McGovern, December 27, 1910, Personal Correspondence, McCarthy Papers; Wisconsin, *Assembly Journal* (1911), 21–22; Teaford, *Rise of the States*, 20; Buenker, *The History of Wisconsin*, 295; Nesbit, *Wisconsin: A History*, 424–26. On New York, see Yellowitz, *Labor and the Progressive Movement*, 128–41, 146–48. On Illinois, consult Beckner, *Labor Legislation in Illinois*, 269–70; Pegram, *Partisans and Progressives*, 63–84. On Alabama, review S. Hackney, *Populism to Progressivism*, 309–16; Kelly, *Race, Class, and Power*, 66–67. On California, see Starr, *Inventing the Dream*, 237–38; Mowry, *The California Progressives*, 142–46; Taft, *Labor Politics*, 43–46; Lower, *Bloc of One*, 31–33; and Mason, "Neither Friends nor Foes," 57–70.

24. Hoeveler, "Social Gospel," 289–98; Margulies, *Decline of the Progressive Movement*, 119, 130–32; Buenker, *The History of Wisconsin*, 580–94; McCarthy, *The Wisconsin Idea*, 11–17; Unger, *Fighting Bob La Follette*, 121–22, 137–38; Fink, *Progressive Intellectuals*, 86–90.

25. Wisconsin, *Assembly Journal* (1911), 21–22, 52–53; Warner, *Progressivism in Ohio*, 425; Mowry, *The California Progressives*, 90–102; Lower, *Bloc of One*, 28–30; Wesser, *Charles Evans Hughes*, 308–10; McCormick, *From Realignment to Reform*, 231–41; Buenker, *The History of Wisconsin*, 527–30.

26. Commons, *Myself*, 154; *Milwaukee Free Press*, March 4, 1911, 1; Wisconsin, *Assembly Journal* (1911), 101, 446, 683.

27. Church, "Economists as Experts," 596–98, 600, 608; Watkins, "Review Essay," 65. On the AALL, see Lubove, *Struggle for Social Security*, 29–32; Yellowitz, *Labor and the Progressive Movement*, 2–3, 55–58, 72–73, 81; and Skocpol, *Protecting Soldiers and Mothers*, 176–89.

28. On McCarthy, see Fink, *Progressive Intellectuals*, 898–90. For Commons on unionism, see Tomlins, *State and Unions*, 79; Fink, "Labor, Liberty, and the Law," 920; Furner, "Knowing Capitalism," 273; and Ernst, "Common Laborers?" 67–68. Compare these to Horne, "Practical Idealism"; Harter, *John R. Commons*, 70–72, and Lubove, *Struggle for Social Security*, 31–34. On "corporatism," see Schatz, "From Commons to Dunlap," 96–100; and Wunderlin, *Visions of a New Industrial Order*, 12–15, 28–34, 66–67, 102, 112. On labor economics, see McNulty, *Labor Economics*, 131–32, 138–40; Fine, *Laissez-Faire*, 178, 186–87, 198–200; Herbst, *German Historical School*, 111, 125, 162–68; and D. Ross, *Origins*, 12–13. On public administration, consult Burgess, *Political Science*, 1:83–89; Goodnow, *Comparative Administrative Law*, iv, 1–4; and Goodnow, *Politics and Administration*, 18–25, 35, 85. See also D. Ross, *Origins*, 74–75; Fine, *Laissez-Faire*, 91–95; Rodgers, *Contested Truths*, 158, 163, 169; McClay, "John W. Burgess"; Ricci, *Tragedy of Political Science*, 77–87; and Novak, "Legal Origins," 270–72. On interest representation, see Commons, *Proportional Representation*, 357–59; Commons, *Labor and Administration*, v–vi, 51–84; Furner, *Advocacy and Objectivity*, 198–203, 273–75; Fink, "'Intellectuals' versus 'Workers,'" 403–5; and Weinstein, *Corporate Ideal*, 189–90. Commons inspired student Francis H. Bird to investigate functional representation in Bird's dissertation, "Advisory Representative Labor Council."

29. Commons, *Myself,* 118–30, 154; Commons, "Wisconsin Public Utilities Law"; Rodgers, *Atlantic Crossings,* 149–52; Novak, "Legal Origins," 267–68.

30. E. W. Clemens, *Economics and Public Utilities,* 14–29, 50–51; Mosher and Crawford, *Public Utility Regulation,* 6–19; Sharfman, "Commission Regulation," 8–9; *Munn v. Illinois,* 94 U.S. 113 (1877); Wiecek, *Lost World,* 110–11, 133–34; Paul, *Conservative Crisis,* 8–10, 39–41; McCraw, *Prophets of Regulation,* 4–21, 57–65; Cushman, *The Independent Regulatory Agencies,* 20–27, 39–44, 65–70, 84–87; Caine, *Myth of Progressive Reform,* 118–20; Rodgers, *Atlantic Crossings,* 151–52.

31. Wisconsin, *Assembly Bills* (1911), Substitute Amendment no. 1A to no. 963A, Section 1, Subsections 1021b-2, 1021b-3, 1021b-10, 1021b-11, 1021b-12; E. S. Clemens, *People's Lobby,* 248–49.

32. On railroad and public utility law, see *Laws of Wisconsin* (1905), Chapter 362, Section 3; (1907), Chapter 499, Section 1797m-3. On the industrial commission bill, consult Wisconsin, *Assembly Bills* (1911), no. 963A, Section 1, Subsection 1021b-14; Substitute Amendment no. 1A to no. 963A, Section 1, Subsections 1021b-1(11), 1021b-8, 1021b-9; Commons, *Organization and Methods,* 6–8; and Altmeyer, *Industrial Commission,* 107–8.

33. Commons, *Myself,* 155; Wisconsin, *Assembly Bills* (1911), Substitute Amendment no. 1A to no. 963A, Section 1, Subsection 1021b-1, 1021b-9(11); Commons, "Industrial Commission," 6. On the OSHAct, see Ashford, *Crisis in the Workplace,* 12; C. Noble, *Liberalism at Work,* 4, 68–69.

34. Wisconsin, *Assembly Bills* (1911), no. 963A, Section 1, Subsections 1021b-17(13) (4), (5), Substitute Amendment no. 1A to no. 963A, Section 1, Subsection 1021b-1(8); Altmeyer, *Industrial Commission,* 106, 129; Buenker, *The History of Wisconsin,* 550.

35. Glaeser, *Public Utilities Economics,* 235–54; Horwitz, *Transformation, 1870–1960,* 216–17, 222–24; R. Stewart, "American Administrative Law," 1071–75; *Minneapolis, St. Paul & Saulte St. Marie Railway Company v. Railroad Commission of Wisconsin,* 136 Wis. 146, 159–64 (1908); McCraw, *Prophets of Regulation,* 213–14; Chase, *American Law School,* 7–14; White, "Allocating Power," 252; White, *Constitution and New Deal,* 96–103; Wiecek, *Lost World,* 146–47.

36. Wisconsin, *Assembly Bills* (1911), Substitute Amendment no. 1A to no. 963A, Section 1, Subsection 1021b-12(1); Commons, *Organization and Methods,* 12–15; Commons, *Myself,* 157; Andrews and Andrews, "Scientific Standards."

37. Breyer and Stewart, *Administrative Law,* 23–26, 34; R. Stewart, "American Administrative Law," 1671–76, 1760–62.

38. Wisconsin, *Assembly Bills* (1911), no. 963A, Section 1, Subsection 1021b-34; Substitute Amendment no. 1A to no. 963A, Section 1, Subsections 1021b-15, 1021b-17, 1021b-19, 1021b-27, 1021b-28, 1021b-29; *Laws of Wisconsin* (1907), Chapter 499, Section 1, Subsections 1797m-67, 1897m-68; (1905), Chapter 362, Section 16. See also R. Stewart, "American Administrative Law," 1074–75; Breyer and Stewart, *Administrative Law,* 25.

39. *Minneapolis Railway,* 164–69; Skowronek, *New American State,* 259–60.

40. On factory inspectors, note *Laws of Wisconsin* (1887), Chapter 453, Section 2. On deputies, see Wisconsin, *Assembly Bills* (1911), Substitute Amendment no. 1A to no. 963A, Section 1, Subsections 1021b-10(3), 1021b-12(1).

41. The original proposal was Wisconsin, *Assembly Bills* (1911), no. 963A, Section 1, Subsection 1021b-20. On the ensuing controversy, see Commons, *Myself,* 102–3; Commons, *Organization and Methods,* 10; *Milwaukee Free Press,* June 4, 1911, 15; June 9, 1911, 10; June 15, 1911, 12; *Madison Democrat,* June 4, 1911, 8; June 29, 1911, 3; Wisconsin, *Assembly Bills* (1911), Substitute Amendment no. 1A to no. 963A, Section 1, Subsection 1021b-12(1); Amendment no. 4A to Substitute Amendment no. 1A to no. 963A; Amendment no. 5A to Substitute Amendment no. 1A to no. 963A; Wisconsin, *Senate Bills* (1911), no. 529S, Amendment no. 1A to no. 529S; Wisconsin, *Assembly Journal* (1911), 1222, 1338; Wisconsin, *Senate Journal* (1911), 1223, 1284.

42. *The Milwaukee Journal,* May 23, 1911, 1; May 27, 1911, 1; Mahon, "Conserving Human Life"; Bird, "Proposed Wisconsin Industrial Commission."

43. *Milwaukee Free Press,* March 4, 1911, 1; June 15, 1911, 12; *Madison Democrat,* June 4, 1911, 8; Wisconsin, *Blue Book* (1911), 342–49; Wisconsin, *Assembly Journal* (1911), 1533, 1578, 1635; Wisconsin, *Senate Journal* (1911), 1179–80, 1255; *Laws of Wisconsin* (1911), Chapter 485, Section 3; Buenker, *The History of Wisconsin,* 549, 523–26, 543.

44. Caine, *Myth of Progressive Reform,* 137–85; Asher, "Wisconsin Workmen's Compensation Law," 129–30, 135; WBLIS, *Fifteenth Biennial Report,* 123–44; Korman, *Industrialization,* 119–20; Buenker, *The History of Wisconsin,* 549–50; Nesbit, *Wisconsin: A History,* 429–30.

45. Buenker, *The History of Wisconsin,* 543; WSFL, *Proceedings* (1910): 42–43, 60, 66; (1911): 47. E. S. Clemens offers different analysis of socialists in *People's Lobby,* 243–44.

46. On McGovern Progressivism, see Margulies, *Decline of the Progressive Movement,* 130–33. On professionals, consult Haber, *Efficiency and Uplift,* 99–116; Larson, *Rise of Professionalism,* xii–xviii; Larson, "Production of Expertise"; Hobson, "Professionals, Progressives, Bureaucratization"; Jacoby, *Employing Bureaucracy,* 126–30.

47. Skowronek, *New American State,* 39–46; Commons, *Organization and Methods,* 4, 6; Crownhart, *The Workmen's Compensation Act,* a pamphlet reprinted by the Industrial Commission of Wisconsin, circa February 1913, 37–38, 40–41, 42–43; Commons, "Industrial Commission," 7–8; ICW, "The Industrial Commission and Its Predecessors," *Bulletin* (August 20, 1912), 159–60.

48. Crownhart, *The Workmen's Compensation Act,* 40–41; Commons, *Organization and Methods,* 1–3, 6, 8–9, 11–12, 17–18, 22, 31; Commons, "Industrial Commission," 4–5.

49. Editions of *Principles* appeared in 1916, 1920, 1927, and 1936. See also J. Andrews, *Administrative Labor Legislation;* Altmeyer, *Industrial Commission;* and Brandeis, "Administration of Labor Laws."

50. Freund, *Police Power,* 2–3, 17–18; McNulty, *Labor Economics,* 139; Commons, *Myself,* 143.

51. Burgess, *Political Science,* 1:83–89; W. Wilson, "The Study of Administration"; Goodnow, *Comparative Administrative Law,* iv, 1–4; Goodnow, *Politics and Administration,* 18–25, 35, 85; Ricci, *Tragedy of Political Science,* 77–87; Commons and Andrews, *Principles of Labor Legislation,* 415–16, 430–43; Novak, "Legal Origins," 270–72.

52. Landis, *The Administrative State,* 1–2, 9–11, 15–16; McCraw, *Prophets of Regulation,* 212–16; Horwitz, *Transformation, 1870–1960,* 213–16; White, *Constitution and New Deal,* 114–16; Commons and Andrews, *Principles of Labor Legislation,* 443–46. For treatments viewing the Industrial Commission as an "expert" agency, see Buenker, *The History of Wisconsin,* 549–50; Thelen, *La Follette,* 109; Dubofsky, *State and Labor,* 55.

53. Commons, *Myself,* 141–43; Commons and Andrews, *Principles of Labor Legislation,* 446–47.

54. Commons and Andrews, *Principles of Labor Legislation,* 447–48, 450–51, 464. On the modern views of administrative representation, see R. Stewart, "American Administrative Law," 1760–70.

55. On "corporatism," see Schatz, "From Commons to Dunlap," 94; and Wunderlin, *Visions of a New Industrial Order,* 28–32. On Commons as a "corporate liberal," see Dubofsky, *State and Labor,* 55. For "industrial pluralism," consult Ernst, "Common Laborers?" 67–68; Tomlins, *State and Unions,* xi, 78–79, 206–13; Tomlins, "New Deal," 19–29; Fink, "Labor, Liberty, and the Law," 914–22; Furner, "Knowing Capitalism," 273; and Stone, "Post-war Paradigm," 1511–14.

56. Commons and Andrews, *Principles of Labor Legislation,* 462–64.

57. Skocpol, *Protecting Soldiers and Mothers,* 299–300; J. Andrews, *Administrative Labor Legislation,* 17–27; Brandeis, "Administration of Labor Laws," 644–54; Altmeyer, *Industrial Commission;* Harter, *John R. Commons,* 112–13.

58. Warner, *Progressivism in Ohio,* 425; *Statutes and Amendments of California* (1913), Chapter 176, Section 36, Subsections 51(1), (2), (5), (8), (9); Section 52, Subsections 57(1), (2); Section 63, Subsection 65(3); Section 72, Chapter 324; *Acts of Ohio* (1913) (Amended Senate Bill no. 137), Subsections 13(1), (6), (11); Section 16, Subsections 22(3), (4), Section 25, pp. 98–99, 101, 104.

59. *Laws of Massachusetts* (1913), Chapter 813; *Laws of Pennsylvania* (1913), Chapter 267; *Laws of New York* (1913), Chapter 145.

60. Other states adopting administrative procedures in work safety law were Colorado (1915), Idaho (1915), Montana (1915), New Hampshire (1917), Utah (1917), Nevada (1919), North Dakota (1919), Washington (1919), Oregon (1920), North Carolina (1921), Tennessee (1923), Arizona (1925), Maryland (1929), and Nebraska (1929). Beyond this, several states extended administrative powers over mine safety and boilers. See J. Andrews, *Administrative Labor Legislation,* 45–52. On the chief mine inspectors' rules, see Graebner, *Coal-Mining Safety,* 78–83.

61. Teaford, *Rise of the States,* 8–10; Graebner, "Federalism in the Progressive Era"; Graebner, *Coal-Mining Safety,* 9–10; Keller, *Regulating a New Economy,* 125, 130; Weir,

Orloff, and Skocpol, *Politics of Social Policy,* 19; Skocpol, *Protecting Soldiers and Mothers,* 206–45; Mettler, *Dividing Citizens,* 1–2, 13–17; Hamm, *Shaping the Eighteenth Amendment,* 6–8.

62. Yellowitz, *Labor and the Progressive Movement,* 146–47; Pegram, *Partisans and Progressives,* 70–84; S. Hackney, *Populism to Progressivism,* 249, 314–23; Kelly, *Race, Class, and Power,* 66–70.

63. Starr, *Inventing the Dream,* 254; Block, *Revising State Theory,* 16.

Chapter 3: Selling the Safety Spirit

1. On the intellectual milieu, see McEvoy, "Triangle Shirtwaist Factory Fire," 621–26, 640–47; Witt, *Accidental Republic,* 110–22; and Bergstrom, *Courting Danger,* 168–96.

2. On managerial change, see Nelson, *Managers and Workers,* 48–53, 59–78, 101–24; Nelson, *Frederick Taylor,* 12–19, 171–76; D. Noble, *America by Design,* 261–65; Jacoby, *Employing Bureaucracy,* 40–44, 49–54; Brody, *Workers in Industrial America,* 9–14, 45–54; Montgomery, *Workers' Control in America,* 113–34; Montgomery, *House of Labor,* 214–56; Ramirez, *When Workers Fight,* 129–37, 147–56; and Witt, *Accidental Republic,* 103–17, 119–22. On welfare work, see Korman, *Industrialization,* 87–108, 110–16; Tone, *Business of Benevolence,* 1–103; Jacoby, *Modern Manors,* 4–19; and Mandell, *Corporation as Family,* 3–9. On safety programs, examine Aldrich, *Safety First,* 91–93, 198–202, 233–39; Bennett and Graebner, "Safety First," 248–56; Whiteside, *Regulating Danger,* 123–125, 142–44; and Kelly, *Race, Class, and Power,* 67.

3. Carpenter, *Forging of Bureaucratic Autonomy,* 13–15; Teaford, *Rise of the States,* 5–10.

4. Wisconsin, *Senate Journal* (1911), 1255–56; *Laws of Wisconsin* (1911), Chapter 485, Section 1021b-31.

5. "Labor Law Administration, New York, 1931," 34, in AALL Papers. See also Yellowitz, *Labor and the Progressive Movement,* 119; New York, *Annual Report of the Industrial Commission* (1915), 9–15, 20–22; and NYSFL, *Proceedings* (1915), 125.

6. Cebula, *James M. Cox,* 48; Reagan, "Ideology of Social Harmony," 325.

7. E. S. Clemens, *People's Lobby,* 112–13, 120–26, 250–51, 253–55; Mason, "Neither Friends nor Foes," 60–64; Lower, *Bloc of One,* 31–36; Starr, *Inventing the Dream,* 237–38; Bean, *California: An Interpretive History,* 329–32; Taft, *Labor Politics,* 54; Nash, "Influence of Labor," 250; Mowry, *The California Progressives,* 90–102, 142–44.

8. On Wisconsin politics, see Margulies, *Decline of the Progressive Movement,* 119–22, 124–63; Nesbit, *Wisconsin: A History,* 430–32; and Buenker, *The History of Wisconsin,* 627–28. On Mahon and Wilcox appointments, see *The Milwaukee Journal,* July 26, 1913, 2; August 8, 1913, 2; August 9, 1913, 1; Milwaukee *Leader,* August 9, 1913, 1; August 29, 1913, 6; and Wisconsin, *Senate Journal* (1913), 1272, 1296–98, 1307–9. On Hambrecht, see *Wisconsin State Journal,* August 13, 1915, 12; Milwaukee *Leader,* August 13, 1915, 2; June 22, 1917, 12; *The Milwaukee Journal,* August 14, 1915, 4; and Wisconsin, *Senate Journal* (1915), 1286. On Beck, consult the Milwaukee *Leader,* June

22, 1917, 12; and Wisconsin, *Senate Journal* (1917), 1099, 1115–16. On Wilcox's failed removal, see *The Milwaukee Journal*, July 10, 1919, 13; Milwaukee *Leader*, July 10, 1919, 2; and Wisconsin, *Senate Journal* (1919), 1402.

9. ICW, "The Industrial Commission and Its Predecessors," *Bulletin* (August 20, 1912), 150; *Laws of Wisconsin* (1911), Chapter 485; (1913), Chapter 675; (1915), Chapter 541; ICW, *Report on Allied Functions* (June 30, 1915), 51; Margulies, *Decline of the Progressive Movement*, 145–48, 168–70. On commission streamlining, examine Altmeyer, *Industrial Commission*, 109–11; ICW, *Biennial Reports* (1924–1926), 58; ICW, *Minutes*, July 24, 1917, January 14, 1918; and ICW, *Report on Allied Functions* (1917), 7; (1918), 5. On administrative reorganization in other states, see Beckner, *History of Labor Legislation*, 500–524; and Teaford, *Rise of the States*, 70–76. On the Wisconsin agency's wartime budget, see "Answer to Interrogatories under the Interpellation Statute," April 10, 1919, 36/C716 ICW; George P. Hambrecht to Emanuel Philipp, May 24, 1918, 33/C661 ICW; and typescript, "Argument upon New Employes and New Departures in the Budget of the Industrial Commission, 1919–1920," 41/C890 ICW.

10. Commons to Gertrude Beeks (National Civic Federation), May 2, 1912, 13/C172 ICW.

11. ICW, *Minutes*, September 18, 1911, January 18, 1912; press release, January 26, 1912, 89/C448 ICW; Altmeyer, *Industrial Commission*, 114–15, 151–52, 178–79, 166; C. W. Price, *How a State Can Promote Safety*, September 23, 1913, pamphlet, ICW General Archives; Assistant to the Commission (Price) to Robert Adair, September 8, 1913, 13/C165 ICW; C. W. Price, "Waste as a Result of Accidents in Industrial Establishments," speech to Wisconsin Buttermakers' Association, Fond du Lac, circa 1915, 6/C1747 ICW. See also Korman, *Industrialization*, 100–102, 120, 125.

12. ICW, *Report on Allied Functions* (1914), 10; C. W. Price, *Cooperation for Safety Between Wisconsin Industrial Commission and Manufacturers and Workmen,* May 19, 1914, 5–6, reprinted by the Industrial Commission of Wisconsin, ICW General Archives. On the safety campaign and welfare capitalism, see Aldrich, *Safety First,* 122–23; and Jacoby, *Modern Manors,* 4–5, 17–18. On foremen, see Nelson, *Managers and Workers,* 42–44, 51.

13. Price, *How a State Can Promote Safety,* 7; Price, *Cooperation for Safety,* 7; *The Milwaukee Journal,* July 25, 1913, 10; "Proceedings at the First Annual Joint Meeting of Liability Insurance Agents and Industrial Commission of Wisconsin," Milwaukee, July 24, 1913, 3–15, 63/C1617 ICW. On welfare work, see Jacoby, *Employing Bureaucracy,* 49–51; Jacoby, *Modern Manors,* 17–18; Mandell, *Corporation as Family,* 7–9; and Nelson, *Managers and Workers,* 101.

14. "Proceedings of Liability Insurance Agents," 4, 10; ICW, *Report on Allied Functions* (1914), 9; (1915), 6–7; *The Milwaukee Journal,* March 6, 1914, 1; Price, "Waste as a Result of Accidents"; Price to C. H. Crownhart, March 7, 1914, 11/C131 ICW.

15. MMAM, *Civics and Commerce,* January 1909, 21–24. On compensation's stimulus, see Aldrich, *Safety First,* 97–99; and Fishback and Kantor, *Prelude to the Welfare State,* 55–56, 77–83. Regarding worker cooperation, examine Korman, *In-*

dustrialization, 120–21; Ramirez, *When Workers Fight,* 129–31; and Jacoby, *Modern Manors,* 13–18.

16. Building Officials Conference (third annual, Washington D.C.), *Proceedings* (Madison: Democrat Printing, 1917), 42; Banner, "Safety for Loggers," ICW General Archives; ICW, *Report on Allied Functions* (1915), 5, 7.

17. Korman, *Industrialization,* 110–20; Commons, *Myself,* 141–43; Commons and Andrews, *Principles of Labor Legislation,* 446–47.

18. Aldrich, *Safety First,* 107; NYSIC, "A Plan for Shop Safety and Health Organization," second draft, November 1918, AALL Papers; NYSIC, *Proceedings of Industrial Safety Congresses* (1916), 23–24; (1917), 152, 154–57; NYSFL, *Twenty-first Proceedings* (1917), 67–68. On union attitudes, see Montgomery, *Workers' Control in America,* 113–34; Montgomery, *House of Labor,* 247–49; and Tone, *Business of Benevolence,* 182–84.

19. IAIABC, *Proceedings,* in USBLS, *Bulletin No. 264* (1919): 29–34.

20. ICO, *Annual Report* (1915–1916), 22.

21. Will J. French, "Milestones of Safety," 3–5, AALL Papers; CIAC, *Report of the [California] Industrial Accident Commission* (1917), 33–34; John R. Brownell, "Results of Accident Prevention Campaigns," CIAC, *California Safety News* 1 (May 1917): 12–14; IAIABC, *Proceedings,* in USBLS, *Bulletin No. 264* (1919): 25–27.

22. Novkov, *Constituting Workers,* 156–64; Hepler, *Women in Labor,* 13–36, 46–50; Kessler-Harris, *Out to Work,* 188–201.

23. Kroes to Beck, February 1914; Kroes to Watrous, March 13, 1914; Beck to Kroes, March 26, 1914; Kroes to Watrous, March 13, 1914; Kroes to [Watrous], circa October 1914; Kroes to Watrous, December 25, 1914; Kroes to Industrial Commission, circa December 1914; Beck to Cameo Cinema Productions, March 30, 1917, in 106/C989 ICW; E. E. Witte to National Association of Manufacturers, October 1, 1918, 72/C12 ICW.

24. Kroes to Watrous, circa November 1914, with attached handwritten note by Beck; Kroes to Industrial Commission, circa December 1914; Kroes to Beck, February 3, 1915, 106/C989 ICW; MMAM, *Civics and Commerce,* June 1914, 10; ICW, *Report on Allied Functions* (1915), 7–8.

25. ICW, *Report on Allied Functions* (1915), 8–9; Price, *Cooperation for Safety,* 6–7; "Minutes of the Safety and Sanitation Committee of the Milwaukee Merchants and Manufacturers Association," October 23, November 4, 1913, 39/C882 ICW.

26. CIAC, "Review of Motion of Picture," *California Safety News* 2 (April 1918): 3–12; IAIABC, *Proceedings* in USBLS, *Bulletin No. 264* (1919): 26–27.

27. *Laws of Wisconsin* (1911), Chapter 485, Section 1021b-12(3); ICW, *Bulletin,* "The Industrial Commission and Its Predecessors," August 20, 1912, 161–64; Altmeyer, *Industrial Commission,* 153; ICW, "General Orders on Safety," in *Bulletin,* May 20, 1912, 5–6; Beck to Seth Low, November 20, 1911, 13/C172 ICW; Price to Will J. French, November 9, 1911, 15/C182 ICW; Price to F. A. Barker, September 5, 1913; Price to Barker, March 8, 1913, 72/C12 ICW; Price to Howard L. Hindley, June 12, 1912, 13/C165 ICW. See also Bennett and Graebner, "Safety First"; and Aldrich, *Safety First,* 109–19.

28. ICW, "Industrial Commission and Its Predecessors," 164; ICW, *Report on Allied Functions* (1914), 6; Kroes to Witte, September 11 and 14, 1917; Kroes to S. J. Williams, October 1, 1917, 106/C989 ICW; newsclippings from Fond du Lac newspaper, circa October 1917, 106/C989 ICW.

29. NYSDL, *Annual Report of the Industrial Commission* (1919), 251.

30. ICO, "Papers on Accident Prevention," *Special Bulletin* 2, no. 3 (March 1915). See pictures on pp. 9, 19, 29, 43.

31. CIAC, "The Safety Museum," *California Safety News* 1 (September 1917): 3–15; *Statutes and Amendments of California* (1913), Chapter 175, Section 65(1).

32. ICW, *Bulletins:* "Jointer Accidents and Their Prevention," February 20, 1913, 61–63; "Metal Burns and Their Prevention," April 20, 1913; "Falls of Workmen and Their Prevention," June 20, 1913; "Infections and Their Prevention," October 20, 1913; "Accidents Caused by Objects Striking Workers," November 20, 1913; "Gear Accidents and Their Prevention," February 20, 1914; "Elevator Accidents and Their Prevention," July 20, 1914; "Emery Wheel Accidents and Their Prevention," January 20, 1915. On "engineering revision," see Aldrich, *Safety First,* 116–18.

33. ICW, *Bulletins,* "General Orders on Safety," May 1912; "General Orders on Sanitation," January 20, 1913; "Jointer Accidents," February 20, 1913, 61–63; "General Orders on Elevators," January 20, 1913; "Emery Wheel Accidents," January 20, 1915, 4, 6; "Wisconsin's Movement for Safety," May 1, 1915, 8.

34. Aldrich, *Safety First,* 105–7; *Laws of Wisconsin* (1911), Chapter 469; transcript, interview with Ben Kuechle, n.d., 10, 17–19, Wausau Archives. On the science of managing production processes, see D. Noble, *America by Design,* 26–34; Montgomery, *House of Labor,* 230–31; Aldrich, *Safety First,* 116–20.

35. Bulletin list, NYSDL, *Annual Report of the Industrial Commission* (1919), 243; index to articles for 1919, CIAC, *California Safety News* 4 (January 1920): 13–15; ICO, *Annual Report* (1913–1914), 27–28; (1916–1917), 23. See also history of Factory Inspection Department in IDL, *Eighth Annual Report* (1924–1925), 137–51.

36. ICW, *Minutes,* February 4, 1918; Altmeyer, *Industrial Commission,* 152–53; ICW, *Report on Allied Functions* (1914), 7–8; (1918), 6–7; ICW, *Wisconsin Safety Review,* June 1918, page inside front cover, February, June 1919, January, February, April–May 1922, and July 1923; Kuechle interview, 18; ICW, *Bulletins,* "Eye Injuries and Their Prevention," March 20, 1913, page inside front cover; "Shop Organization for Safety," February 20, 1913; A. L. Kaems and R. T. Solensten, "Safeguarding the Punch Press," ICW, *Wisconsin Safety Review,* October 1918, 25. On managerial change, see Jacoby, *Employing Bureaucracy,* 99–105; D. Noble, *America by Design,* 263–65; and Aldrich, *Safety First,* 157–58.

37. ICW, *Wisconsin Safety Review,* June 1918, 3–4; "Punch Press," 19–21; J. S. Herbert, "Safety and the Foreman," ICW, *Wisconsin Safety Review,* June 1918, 20–21; John C. White, "Safety in the Power Plant," ICW, *Wisconsin Safety Review,* February 1922, 4, 6. On foremen, see Nelson, *Managers and Workers,* 34–54, 77–78; Jacoby, *Employing Bureaucracy,* 48; and Montgomery, *House of Labor,* 225.

38. Memos, Price to all deputies, September 20 and 30, 1915; Sidney Williams to all deputies, October 12, 1917; Price to all deputies, June 22 and July 2, 1915, 34/C661 ICW; Price, "How to Promote Safety," 3; Beck to D. D. Evans, June 14, 1912, 78/C76.2 ICW; memos, Beck to deputies, July 7 and October 30, 1915, 34/C661 ICW; George P. Hambrecht, address of January 1, 1918, 5–7, *Addresses of George P. Hambrecht,* ICW General Archives.

39. Al Kroes, "Report of Inspection Work," circa 1913, 108/C1598 ICW.

40. Henry Schreiber, "Report on Inspection Work," circa 1913, August Kaems, "Report on Inspection Work," 1913; Kroes, "Report on Inspection Work," 1913, 108/C1598 ICW.

41. A. R. Week (John Weeks Lumber Co.) to Beck, January 17, 1917; J. J. Lingle (Westboro Lumber Co.) to Beck, January 18, 1917; Beck to John Weeks Lumber Co., January 19, 1917; Beck to Lingle, January 19, 1917; A. Quickhass (Menominee Indian Mills) to Industrial Commission, January 15, 1917; W. B. Clubine to Beck, January 8, 1917, 39/C850 ICW.

42. E. S. Hammond to Beck, January 8 and 16, 1917; Beck to Hammond, January 9, 1917, 39/C850 ICW.

43. Hurst, *Law and Economic Growth,* 490–96; Nesbit, *Wisconsin: A History,* 296–310; Glad, *History of Wisconsin,* 210–11. On West Coast lumbermen, see Tripp, "Instance of Cooperation," 530–50; and Tripp, "Law and Social Control," 466–74.

44. IAIABC, *Proceedings,* in USBLS, *Bulletin No. 264* (1919): 18; French, "Milestones of Safety," 3.

45. Commons and Andrews, *Principles of Labor Legislation,* 430–40, 446–48; Carpenter, *Forging of Bureaucratic Autonomy,* 14–15; Teaford, *Rise of the States,* 5–6.

46. Witt, *Accidental Republic,* 187; Fishback and Kantor, *Prelude to the Welfare State,* 54–56; Aldrich, *Safety First,* 91–114; Nelson, *Managers and Workers,* 137–38.

47. Kelly, *Race, Class, and Power,* 73–80; Grantham, *Southern Progressivism,* 290–301; S. Hackney, *Populism to Progressivism,* 309–23; IDL, *Second Annual Report* (1918–1919), 48–55; Beckner, *Labor Legislation in Illinois,* 269–71; Pegram, *Partisans and Progressives,* 61–84; Aldrich, *Safety First,* 91–93, 107–11.

48. NYSIC, *Proceedings of the Industrial Safety Congresses* (1916), 26–41; (1917), 182–213; (1918), 21–33, 91–107.

49. Reagan, "Ideology of Social Harmony," 324–31; Knepper, *Ohio and Its People,* 293–95, 327–39.

50. E. S. Clemens, *People's Lobby,* 252–55; Nash, "Influence of Labor," 246.

Chapter 4: The First Safety Codes

1. Horwitz, *Transformation, 1870–1960,* 213–46; Benedict, "Law and Regulation," 249–53; White, *Constitution and New Deal,* 100–103, 116–20; Wiecek, *Lost World,* 146–47; Chase, *American Law School,* 13–14; Breyer and Stewart, *Administrative Law,* 23–28; White, "Allocating Power," 241; B. Schwartz, *Administrative Law,* 17–22; R. Stewart, "American Administrative Law," 1671–81.

2. Urofsky, "State Courts"; Chomsky, "Progressive Judges," 426–31; Kens, *Lochner v. New York*; Kens, "Source of a Myth," 83–91; Gillman, *Constitution Besieged*, 122–40; Witt, *Accidental Republic*, 152–74.

3. M. Sklar, *Corporate Reconstruction*, 14–33; Montgomery, *House of Labor*, 180–84, 230–34; Hays, "Political Choice in Regulation"; Hays, "The New Organizational Society," 244–63; Ramirez, *When Workers Fight*, 87–99, 129–31; Chandler, *Visible Hand*, 240–44, 345–48, 370–72; D. Noble, *America by Design*, xxiii–xxvi; Nelson, *Managers and Workers*, 11–78.

4. Comparative state economic statistics show Wisconsin industry's relatively transitional character in the 1910s. See U.S. Census, 1910, vol. 9, *Manufacturers*, 32–33, 100–101, 292–93, 878–82, 984–89, 1356–59; and U.S. Census, 1920, vol. 9, *Manufacturers*, 40–43, 112–19, 348–59, 1038–51, 1186–93, 1638–43. See also Buenker, *The History of Wisconsin*, 80–125; Cayton, *Ohio*, 180–81; Knepper, *Ohio and Its People*, 303–4; and Howard, *Illinois*, 391–92.

5. Tomlins, *State and Unions*, 204–13; Ernst, "Common Laborers?" 66–68, 80–82; Furner, "Knowing Capitalism," 273.

6. Wisconsin, *Assembly Bills* (1913), no. 311A; *The Milwaukee Journal*, February 20, 1913, 2.

7. Beck to Fred D. Schwedtman, November 21, 1913; Schwedtman to Beck, November 24, 1913, 72/C12 ICW.

8. Milwaukee *Leader*, February 22, 1913, 4.

9. *The Milwaukee Journal*, February 20, 1913, 2; Wisconsin, *Assembly Journal* (1913), 246.

10. Charles McCarthy to Robert La Follette, April 25, 1913; Commons to La Follette, April 24, 1913, Special Correspondence, La Follette Papers; *The Milwaukee Journal*, April 29, 1913, 2. On evolution of Wisconsin Progressivism, see Margulies, *Decline of the Progressive Movement*, 119; and Thelen, *La Follette*, 32–78. On rule making for women's hours and wages, refer to Skocpol, *Protecting Soldiers and Mothers*, 387–89, 402–6; Brandeis, "Women's Hours Legislation," 482; Brandeis, "Minimum Wage Legislation," 513; and "Legislation for Women," 369, 381, 392–93.

11. Howard, *Illinois*, 447–51; Beckner, *Labor Legislation in Illinois*, 500–504; Pegram, *Partisans and Progressives*, 196–202.

12. Yellowitz, *Labor and the Progressive Movement*, 119–20, 138–41, 147; Wesser, *Response to Progressivism*, 99–166; Wesser, "Conflict and Compromise," 364–72; "Labor Law Administration, New York 1931," 34–37, AALL Papers; *Laws of New York* (1913), Chapter 145, Sections 51, 52; (1915), Chapter 674, Sections 40a, 51a, 52, 52b.

13. *State v. Lange Canning Company*, 164 Wis. 228, 235–42 (1916); *Wisconsin Telephone Company v. Railroad Commission of Wisconsin*, 162 Wis. 383, 405–6 (1916); *Chicago & Northwestern Railway Company v. Railroad Commission*, 162 Wis. 91, 93 (1916); *Klein v. Barry*, 182 Wis. 255, 274 (1923); *Wisconsin-Minnesota Light & Power Company v. Railroad Commission of Wisconsin*, 183 Wis. 96, 103–4 (1924); *Kreutzer v. Westfahl*, 187 Wis. 463, 485–86 (1925); *State ex. rel. Owen v. Stevenson*, 164 Wis. 569,

277–79 (1917); J. Andrews, *Administrative Labor Legislation*, 174–79; Wiecek, *Lost World*, 133–34, 146–47; Chomsky, "Progressive Judges," 426–28; Gillman, *Constitution Besieged*, 201; Kens, "Source of a Myth"; Shapiro, "Supreme Court's 'Return.'"

14. *Laws of Wisconsin* (1911), Chapter 485, Sections 1021b-1(1), (2), (3), (4), (5), (11), 1021b-8; ICW, *Minutes*, February 14, 1922, May 16, 1922; Altmeyer, *Industrial Commission*, 107–8; Commons, "Industrial Commission," 8.

15. *Acts of Ohio* (1913), Sections 13(1), (2), (11), pp. 98–99; *Statutes and Amendments of California* (1913), Chapter 176, Sections 51(1), (2), (8); *Laws of New York* (1913), Chapter 145, Section 3 (adding Section 20–b); (1915), Chapter 674, Section 51–a; *Ursprung v. Winter Garden Co., Inc.*, 183 N.Y. App. Div. 718, 728 (1918); *John Wilks v. United Marine Contracting Corporation*, 199 N.Y. App. Div. 788, 793 (1922).

16. *Szeliwicki v. Connor Lumber and Land Company*, 163 Wis. 20, 23 (1916); *Sparrow v. Menasha Paper Company*, 154 Wis. 459, 463–64 (1913); Reuss, "Thirty Years," 337; *Rosholt v. Worden-Allen Co.*, 155 Wis. 163, 174–76 (1913); *Sadowski v. Thomas Furniture Company*, 157 Wis 443, 448–49 (1914). On the new intellectual milieu, see McEvoy, "Triangle Shirtwaist Factory Fire," 643–45; and Bergstrom, *Courting Danger*, 181–84.

17. *American Woodenware Manufacturing Company v. Schorling*, 96 O.S. 305, 310–59 (1917).

18. *Olson v. Whitney Brothers Company*, 160 Wis. 606, 609–13, 616–21 (1915); *Frank Helme v. Great Western Milling Company*, 43 Cal. App. 416, 428–24 (1919); Reuss, "Thirty Years," 344, 357–61.

19. Mashaw, *Due Process*, 61–69, 73–79; Benedict, "Law and Regulation," 232–34, 250–51; Chase, *American Law School*, 3–22; White, "Allocating Power," 241; Breyer and Stewart, *Administrative Law*, 20–28; R. Stewart, "American Administrative Law," 1671–76, 1711–61.

20. *Laws of Wisconsin* (1911), Chapter 485, Sections 1021b-12(1), (3), (4), (7), 1021b-13, 1021b-16, 1021b-17; Altmeyer, *Industrial Commission*, 125; *Borgnis v. Falk*, 363.

21. *Statutes and Amendments of California* (1913), Chapter 176, Sections 57 and 65(3).

22. *Laws of New York* (1915), Chapter 674, Section 52.

23. *Laws of Wisconsin* (1911), Chapter 485, Sections 1021b-19, 1021b-28; *Acts of Ohio* (1913), Sections 38, 39, 40, pp. 107–8; *Statutes and Amendments of California* (1913), Chapter 176, Sections 84(b), (c); *Laws of New York* (1915), Chapter 674, Sections 52–b, 52–c; J. Andrews, *Administrative Labor Legislation*, 181.

24. Altmeyer, *Industrial Commission*, 117, 120; ICW, *Report on Allied Functions* (1914), 4–5, 28–29; Commons and Andrews, *Principles of Labor Legislation*, 446–47; Commons, "Industrial Commission," 5–6; ICW, *Report on Allied Functions* (1914), 4–5, 28–29; ICW, "Wisconsin's Movement for Safety," *Bulletin*, May 1, 1915, 3; ICW, *Biennial Report* (1924–1926), 13.

25. Commons and Andrews, *Principles of Labor Legislation*, 443–49; J. Andrews, *Administrative Labor Legislation*, 74–77; Brandeis, "Administration of Labor Laws,"

652–53; Brandeis, "Women's Hours Legislation," 487–92; Brandeis, "Minimum Wage Legislation," 513–15; "Legislation for Women," 372, 381, 392–94.

26. See Table 3 in Appendix. Fred M. Wilcox, "Work of the Industrial Commission of Wisconsin in the Field of Safety and Sanitation," ICW, *Wisconsin Safety Review* 3 (1922): 4; Altmeyer, *Industrial Commission,* 125, 171; ICW, "General Orders on Safety," *Bulletin,* no. 1 (May 20, 1912), 3–4; ICW, "Proposed Orders on Safety in Building Construction," ICW, *General Orders* (1914), 2–3; ICW, "General Orders on Safety—Elevators," ICW, *General Orders* (1914), 1–2; ICW, *Code of Boiler Rules* (1914), 1–2; ICW, *Building Code* (1914), 2–3; Beck to George Mutter, December 12, 1914, 39/C882 ICW.

27. NYSDL, *Report of the Industrial Commission* (1915), 20–22; ICO, *Safety Bulletin* 2, no. 3 (March 1, 1915): 77–78; CIAC, *Report of the [California] Industrial Accident Commission* (1916), 31–32; (1917), 30–32; (1918), 30–31; CSFL, *Proceedings* (1915): 81.

28. ICW, "General Orders on Safety," *Bulletin,* May 20, 1912, 3–4; WMA, *Classified Directory,* 51, 82, 92, 101, 109.

29. Typescript, "Meeting of the Committee on Safety and Sanitary Standards," February 9, 1912, 39/C882 ICW; Caine, *Myth of Progressive Reform,* 137–62, 187–88; Hays, "Political Choice in Regulation," 128.

30. Ira Lockney to Beck, August 28, 1911; Evans to Beck, August 25, 1911; August Lehnhoff to Wisconsin Industrial Commission, September 4, 1911, 39/C882 ICW; Evans to Beck, February 28, 1915, 78/C76.2 ICW; ICW, *General Orders on Safety,* April 20, 1915, 2–3; August 1, 1915, 2; Altmeyer, *Industrial Commission,* 122; typescripts, "Minutes of Meeting of Committee on Book of Standards," Milwaukee, November 25, 1911; "Meeting of the Committee on Safety and Sanitary Standards," January 26, May 8, 1912; "Minutes of the Meeting of the Committee on Sanitation," May 8, December 24, 1914, all in 39/C882 ICW; Beck to Barker (NAM), September 16, 1913, 72/C12 ICW; Price memo to J. E. Vallier, A. L. Kaems, August Lehnhoff, and Ira Lockney, April 11, 1914, 24/C661 ICW; Price, memo to all deputies, November 5, 1914; 34/C661 ICW; Price to Committee on Safety and Sanitation, December 24, 1914, 39/C882 ICW.

31. Typescript, "Minutes of Meeting of Committee on Safety and Sanitation," December 9, 1911, 39/C882 ICW; Beck to Frank Weber, December 14, 1912, 15/C191 ICW; Price to members of the Large Committee of Safety and Sanitation, December 21, 1912, 39/C882 ICW.

32. *Laws of Wisconsin* (1911), Chapter 485, Sections 1021b-12(3), (4), (5), 1021b-16, 1021b-17; *Statutes and Amendments of California* (1913), Chapter 176, Section 57; *Laws of New York* (1915), Chapter 674, Section 52; *Acts of Ohio* (1913), Sections 871-22(3), (4), (5), 871-26, 871-27, pp. 101–5; J. Andrews, *Administrative Labor Legislation,* 82–83; Altmeyer, *Industrial Commission,* 125.

33. Hearing transcript, "Special Committee on Safety and Sanitation Standards," January 27, 1912, 3–13, 39/C882 ICW.

34. Ibid., 42–50.

35. Ibid., 62–63.

36. Ibid., 93–100, 102–3.

37. Ibid., 90.

38. ICW, "General Orders on Safety," *Bulletin* 1, no. 1 (May 20, 1912): 4–5; ICW, *Labor Laws of the State of Wisconsin and Orders of the Industrial Commission* (Madison, 1938), 15; ICW, *Hearing on Factory Rules,* January 2, 1912, 3–12.

39. Weber to Beck, December 10, 1912, 15/C191 ICW; *The Milwaukee Journal,* July 25, 1913, 10; James P. Holland, "Safety from the Viewpoint of Organized Employees," in NYSIC, *Proceedings of the Fourth Industrial Safety Congress* (December 1–4, 1919), 24.

40. On Wilson, see Dubofsky, *State and Labor,* 53–60.

41. *Addresses of George P. Hambrecht,* September 1, 1917, 5, ICW General Archives.

42. Altmeyer, *Industrial Commission,* 132; ICO, *Report* (1916–1917), 17; NYSIC, *Annual Reports* (1918), 223; (1919), 249.

43. Link and McCormick, *Progressivism,* 58–66; Teaford, *Rise of the States,* 1–10.

44. J. Andrews, *Administrative Labor Legislation,* 45–51.

45. Ibid., 122–49; "Labor Law Administration, Massachusetts, 1931," 9–10, 28–30, AALL Papers.

46. ICO, *Report* (1916–1917), 17, 22; J. Andrews, *Administrative Labor Legislation,* 138–39; Reagan, "Ideology of Social Harmony," 326–31; Rose, "Design and Expediency," 94–103.

47. CIAC, *Report of Industrial Accident Commission* (1916), 31–37; (1917), 30–35; (1918), 28–31; (1919), 25–26; (1920), 35–36; Smith, "Chronological Development," 7–17; U.S. Census, 1910, vol. 9, *Manufacturers,* 100–101; U.S. Census, 1920, vol. 9, *Manufacturers,* 112–19; E. S. Clemens, *People's Lobby,* 251–54.

48. NYSDL, *Report of the Industrial Commission* (1918), 217–23; (1919), 243–49; Smith, "Chronological Development," 89, 100. On the New York economy, see U.S. Census, 1910, vol. 9, *Manufacturers,* 878–83; U.S. Census, 1920, vol. 9, *Manufacturers,* 1038–51; and Ellis et al., *New York State,* 509–32, 541–46.

49. J. Andrews, *Administrative Labor Legislation,* 146–48; Smith, "Chronological Development," 155–61.

50. On Wisconsin industry, see U.S. Census, 1910, vol. 9, *Manufacturers,* 1356–59; U.S. Census, 1920, vol. 9, *Manufacturers,* 1638–43; Brownlee, *Progressivism and Economic Growth,* 90–94; Korman, *Industrialization,* 110–34.

Chapter 5: The Club of the Law

1. Korman, *Industrialization,* 118; Commons and Andrews, *Principles of Labor Legislation,* 462; Rodgers, *Atlantic Crossings,* 225.

2. Moss, *Socializing Security,* 59–72; Witt, *Accidental Republic,* 138–49; Commons, *Myself,* 142; Commons and Andrews, *Principles of Labor Legislation,* 454, 462–63. See also Brandeis, "Administration of Labor Laws," 626–54.

3. Aldrich, *Safety First*, 91–120, 123–30; Fishback and Kantor, *Prelude to the Welfare State*, 54–55, 57–63, 77–83; Witt, *Accidental Republic*, 123–30.

4. On insurance generally, see Grant, *Insurance Reform*, 3–133; McDowell, *Deregulation and Competition*, 14–16; and Kolko, *Triumph of Conservatism*, 89–97. On compensation insurance, see Fishback and Kantor, *Prelude to the Welfare State*, 148–67; and Rodgers, *Atlantic Crossings*, 250.

5. Witt, *Accidental Republic*, 143–49; Glenn, "Risk, Insurance, and Obligation," 296–313; Baker and Simon, *Embracing Risk*, 1–13; Baker, "Risk, Insurance, and Responsibility," 35–45; Heimer, "Insuring More," 119–20; Aldrich, *Safety First*, 99–100; J. Hackney, "Intellectual Origins," 444–45, 454–58, 462–78; McDowell, *Deregulation and Competition*, 52; Kimball, *Insurance and Public Policy*, 5.

6. ICW, *Workmen's Compensation: Fourth Annual Report* (1915), 1; speech by C. H. Crownhart at meeting of the state commissioners at Lansing, Michigan, April 15, 1914, 5, 107/1040 ICW; Beck to Herman Ekern, May 6, 1914, 6/C82 ICW. On modern appraisals of compensation's impact, see Aldrich, *Safety First*, 97–99; Fishback and Kantor, *Prelude to the Welfare State*, 55, 77–81; and Witt, *Accidental Republic*, 187–88.

7. Aldrich, *Safety First*, 99; Fishback and Kantor, *Prelude to the Welfare State*, 148–67; Asher, "Wisconsin Workmen's Compensation Law," 134; Wesser, "Conflict and Compromise," 369–70; Reagan, "Ideology of Social Harmony," 321; Nash, "Influence of Labor," 250–51. See also *Statutes and Amendments of California* (1913), Chapter 176, Sections 34, 35, 36; *Laws of New York* (1914), Chapter 41, Article 3, Section 50; *Acts of Ohio* (1911), Sections 20, 21, pp. 528–30; Wisconsin Compensation Insurance Board, *Insurance Experience under Compensation Act*, January 3, 1919, 7–8; ICW, *Workmen's Compensation: Second Annual Report* (1913), 2; Beck to Ekern, May 6, 1914, 6/C82 ICW; and *Laws of Wisconsin* (1911), Chapter 50, Sections 2394-26, 2394-27, 2394-28.

8. *Laws of Illinois* (1911), Sections 1, 3, 10, 15, 18, 23½, pp. 314–26; (1913), Section 27, pp. 354–55; Fishback and Kantor, *Prelude to the Welfare State*, 126–27; Beckner, *Labor Legislation in Illinois*, 431–72; Howard, *Illinois*, 424; Pegram, *Partisans and Progressives*, 74–78; "Labor Law Administration, Illinois, 1930," 34–35, AALL Papers.

9. *General Laws of Alabama* (1919), 229–31; Grantham, *Southern Progressivism*, 298–301.

10. Wesser, "Conflict and Compromise," 367–70; NYSDL, *Report of the Industrial Commission* (1915), 10–24; Yellowitz, *Labor and the Progressive Movement*, 119–20; Wesser, *Response to Progressivism*, 162–66.

11. *Laws of New York* (1911), Chapter 460; (1912), Chapter 175; (1913), Chapter 26; (1914), Chapter 16; (1914), Chapter 41, Section 95; NYSDL, *Report of the Industrial Commission* (1915), 25–26; "Labor Law Administration, New York, 1931," 76d–76f, AALL Papers. On New York fire insurance, see Grant, *Insurance Reform*, 125–28.

12. *Statutes and Amendments of California* (1913), Chapter 176, Sections 2(b), 34, 36, 37, 38, 40; (1915), Chapter 642; CIAC, *Report of the California Industrial Accident Commission* (1918), 27; Nonet, *Administrative Justice*, 20–36, 49–62, 66–70.

13. Fishback and Kantor, *Prelude to the Welfare State*, 160–67; Reagan, "Ideology of Social Harmony," 320–28.

14. ICW, *Workmen's Compensation, Second Annual Report*, 2–3; E. H. Downey to Beck, March 17, 1915, 78/C76.2 ICW; typescript, IAIABC (at Topeka, Kansas), "Report of Committee on Administration and Procedure," September 21, 1936, 44/C1040 ICW; Dan Hagge, interview by the author, May 19, 1981, Wausau, Wisconsin.

15. *Wisconsin State Journal*, August 13, 1915, 12; Milwaukee *Leader*, August 13, 1915, 2; *The Milwaukee Journal*, August 14, 1915, 4; William A. Fricke to Beck, May 10, 1915; Beck to Crownhart, August 16, 19, 1915, Box 3, Folder 5, Crownhart Family Papers.

16. *Laws of Wisconsin* (1913), Chapter 599; (1917), Chapter 637; (1919), Chapter 136; ICW, *Bulletin: Workmen's Compensation Insurance*, June 1, 1915, 7–10, 13–16; Crownhart to Schwedtman, August 21, 1912, 72/C12 ICW. On life and fire insurance in Wisconsin, see Grant, *Insurance Reform*, 60–63, 128–31. On Wisconsin's "service state," see Buenker, *The History of Wisconsin*, 593–94.

17. *Laws of Wisconsin* (1917), Chapter 637, Section 1921-7; *Statutes and Amendments of California* (1913), Chapter 176, Section 40a; (1915), Chapter 642, pp. 1269–70; ICW, *Bulletin: Workmen's Compensation Insurance* (1915), 12–13; Wisconsin Compensation Insurance Board, *1930 Report of the Compensation Insurance Board*, 4; ICO, *The Ohio State Insurance Manual* (1915), 4; Emile E. Watson, "Merit Rating in Workmen's Compensation Insurance," IAIABC *Proceedings*, in USBLS, *Bulletin*, no. 210 (1917): 58–75; *Laws of New York* (1914), Chapters 16 and 41: Aldrich, *Safety First*, 99–100; ICW, *Wisconsin Labor Statistics*, no. 7 (March 30, 1928), 1; WSFL, *Proceedings* (1914): 42.

18. *Ohio State Insurance Manual*, 3–7; ICO, *Bulletin 2*, no. 1 (January 1, 1915): 9–11; Watson, "Merit Rating," 68–75; Aldrich, *Safety First*, 99–100; Heimer, "Insuring More," 136–38; Nonet, *Administrative Justice*, 68.

19. Kimball, *Insurance and Public Policy*, 106. On the Blaine era's politics, see Margulies, *Decline of the Progressive Movement*, 244–82.

20. *Laws of Illinois* (1911), Section 3, pp. 316–17; (1913), Section 6, p. 340; Fishback and Kantor, *Prelude to the Welfare State*, 126–27; Beckner, *Labor Legislation in Illinois*, 447–63.

21. *Acts of Ohio* (1911), Sections 20-2, pp. 529–30; (1913), Section 29, p. 84; Pierce, "Organized Labor," 895–96; Fishback and Kantor, *Prelude to the Welfare State*, 122–25.

22. *Laws of Wisconsin* (1913), Chapter 599, Section 2394-9(5); (1917), Chapter 624, Section 2394-9(7); *Addresses of George P. Hambrecht*, January 1, 1918, 8, ICW General Archives; *Employers Mutual News Bulletin*, no. 10 (June 1920), 4, Wausau Archives; Fred Braun, interview by the author, June 3, 1981, Wausau, Wisconsin; Altmeyer, *Industrial Commission*, 168.

23. *Statutes and Amendments of California* (1911), Chapter 399, Section 3(3); (1917), Chapter 586, Sections 6(a)(4), 6(b), 31(b).

24. Graebner, *Coal-Mining Safety*, 86–100; Brandeis, "Administration of Labor

Laws," 632–54; I. Andrews, "New Spirit"; Commons and Andrews, *Principles of Labor Legislation,* 454, 462–63; *Laws of Wisconsin* (1911), Chapter 485, Sections 1021b-1(6), 1021b-10(3), 1021b-26, 1021b-30; Fred M. Wilcox, "Work of the Industrial Commission of Wisconsin in the Field of Safety and Sanitation," ICW, *Wisconsin Safety Review* 3 (March 1922): 5; ICW, *Biennial Report* (1918–1920), 84–85; ICW, *Report on Allied Functions* (1914), 9 (emphasis in the original); (1917), 8; Altmeyer, *Industrial Commission,* 119–20, 152, 321.

25. Voyta Wrabetz to Ernest Kelly, March 14, 1940, 2/C52 ICW; Commons and Andrews, *Principles of Labor Legislation,* 450–52; *Wisconsin Statutes* (1911), Chapter 110a, Section 2394-52(1); Wisconsin Civil Service Commission, *State Civil Service: Manual of Competitive Examinations* (Madison, 1913), 22; typescript, "Proposed Examination Questions for Male Deputy of the Industrial Commission," 34/C661.4 ICW. On various deputies' qualifications, see John A. Hoeveler to Fred M. Wilcox, December 17, 1916, 78/C76.2 ICW; Wilcox to L. J. Parrish, December 6, 1927, 78/C76.2 ICW; ICW, *Minutes,* November 3, 1920; June 17, November 1 and 22, 1921; February 11, 1922.

26. Beck to Kroes, October 5, November 13 and 21, 1916, 106/C989 ICW; typescript, "Instructions to Deputies," n.d., 34/C661.1 ICW; Robert McA. Keown to State Board of Labor and Industries, February 4, 1920, 10/C99 ICW; Witte, memo, "Re: Mr. Conklin's meeting with Commissioners," August 25, 1920; George P. Hambrecht to I. N. Conklin, August 31, 1920; Wilcox to Conklin, March 20, 1923, 78/C76.2 ICW; Crownhart to Harry P. Peterson, September 5, 1911, and attached correspondence, 22/C309 ICW.

27. *Addresses of George Hambrecht,* January 1, 1918, 4, ICW General Archives; Price to Bird, October 19, 1914, 68/C1770 ICW; Beck to Bird, December 21, 1914, 69/C1770 ICW; Evans to Price, November 16, 1911, 34/C661 ICW; Beck to Evans, May 21, 1917, 78/C76.2 ICW.

28. *Acts of Ohio* (1913), Section 13(6), p. 98; ICO, *Report No. 26,* "Inspection of Workshops, Factories, and Public Buildings in Ohio," (1915); ICO, *Reports* (1913–1914), 19; (1916–1917), 16; (1917–1918), 52–53.

29. CIAC, *Reports of the California Industrial Accident Commission* (1915) 33; (1916) 33; (1917) 32–33; F. A. Short, "Enforcement of Safety Orders," CIAC, *California Safety News* 4, no. 9 (September 1920): 4; Warren H. Pillsbury, "The Law of Safety in California," CIAC, *California Safety News* 9, no. 1 (January 1925): 19; *Statutes and Amendments of California* (1913), Chapter 176, Sections 7(5), 72(b); (1919), Chapter 471, Section 46½; Aldrich, *Safety First,* 100.

30. J. J. McSherry, "General Inspection" (ca. 1917), 5; John J. Hanlon, "The Best Method of Establishing Co-operation between Employer, Employee, and Inspector" (ca. 1917), 2, AALL Papers. NYSIC, *Proceedings of the Second Industrial Safety Congress* (1917): 263; NYSDL, *Report of the Industrial Commission* (1915): 22–23; NYSDL, *Annual Report* (1925): 67; *Laws of New York* (1921), Chapter 50, Section 256(3).

31. Beckner, *Labor Legislation in Illinois,* 495–96, 501–2; "Labor Law Administration, Illinois, 1930," 42–47, 67–73, 75–76.

32. Buenker, *The History of Wisconsin*, 82–121, 275–76; Nesbit, *Wisconsin: A History*, 323–34; Altmeyer, *Industrial Commission*, 110–14; ICW, *Report on Allied Functions* (1917), pages inside front cover; J. Andrews, *Labor Laws in Action*, 75–76; U.S. Census, 1920, vol. 9, *Manufacturers*, 1638–43.

33. *General Laws of Alabama* (1907), 335, 338–39; (1911), 356–65; Kelly, *Race, Class, and Power*, 66–67; S. Hackney, *Populism to Progressivism*, 307.

34. *Laws of Illinois* (1911), 326–27; IDL, *Second Annual Report* (1918–1919), 48–57; Beckner, *Labor Legislation in Illinois*, 496; Pegram, *Partisans and Progressives*, 67–73; U.S. Census, 1920, vol. 9, *Manufacturers*, 40–43, 348–59.

35. T. Clark, *Defending Rights*, 20–23; CIAC, *Reports* (1916), 33; (1917) 33; *Statutes and Amendments of California* (1913), Chapter 324; E. S. Clemens, *People's Lobby*, 253–54; U.S. Census, 1920, vol. 9, *Manufacturers*, 112–19.

36. Knepper, *Ohio and Its People*, 293–305; ICO, "Inspection of Workshops, Factories and Public Buildings in Ohio," *Report No. 26* (1915): 5; U.S. Census, 1920, vol. 9, *Manufacturers*, 1186–93.

37. Ellis et al., *New York State*, 510–32; *Laws of New York* (1913), Chapter 145, Article 4; (1915), Chapter 674, Sections 40–a(4), 42; NYSDL, *Annual Report* (1925), 64; U.S. Census, 1920, vol. 9, *Manufacturers*, 1038–51; J. Andrews, *Labor Laws in Action*, 75–76; Walker, "Factory Legislation and Inspection," 74–145; Yellowitz, *Labor and the Progressive Movement*, 146–57.

38. ICW, *Reports on Allied Functions* (1914), 20–25; (1915), 9–19; (1917), 8–13; (1918), 12–16; U.S. Census, 1910, vol. 9, *Manufacturers*, 1356; Altmeyer, *Industrial Commission*, 134–35, 157–59.

39. IDL, *Second Annual Report* (1918–1919), 48–55.

40. ICO, *Report No. 26*, 6–25.

41. Kroes, "Report on Inspection Work," circa 1913, 108/C1598 ICW; Beck, memos to deputies, January 23, March 28, 1914, 34/C661 ICW; Kaems to Price, November 18, 1914; Lockney to Price, November 30, 1914; John Humphrey to Price, November 22, 1914; C. J. Kremer to Industrial Commission, December 18, 1916; Maud Sweet to Beck, December 17, 1916, 34/C661.1 ICW; Kroes to Beck, March 31, 1914, 106/C989 ICW.

42. Lehnhoff to Industrial Commission of Wisconsin, November 28, 1911, 34/C661 ICW; Evans to Price, August 1, 1915, 78/C76.2 ICW; Humphrey to Price, November 15, 1911, 34/C661 ICW; correspondence from deputies Lockney, Kroes, Kremer, Evans, and Humphrey to Beck, 34/C661.1 ICW.

43. Typescript, Horace Secrist, "An Exposition and Criticism of the Statistical Methods and Output of the Industrial Commission of Wisconsin," October 22, 1914, 19–20, 60/C1427 ICW; memos, Beck to deputies, October 7, 1913; Beck to inspectors, June 9, 1914; Beck to deputies, August 16, 1916, 34/C661.1 ICW; Lockney to Beck, April 14, 18, 1913, 46/C1235 ICW.

44. Memo, Price to all deputies, May 19, 1915, 34/C661 ICW; Evans to S. J. Williams, November 3, 1917, 78/C76.2 ICW; memo, Beck to inspectors, February 3, 1913, 34/C661 ICW; Altmeyer, *Industrial Commission* 63–64, 145–46.

45. F. J. Emery to Industrial Commission, January 8, 1918; Williams to Emery, January 10, 1918; Beck to American Saw Mill Machinery Co., March 5, 1913; W. F. & John Barnes Co. to Beck, March 14, 1913; Beck to Barnes Drill Co., April 2, 1913; C. G. Richardson to Industrial Commission of Wisconsin, August 22, 1918; R. J. Solensten to Parks & Woolsen Machinery Company, October 16, 1918, 46/C1235 ICW; Aldrich, *Safety First,* 131.

46. Beck to W. F. & John Barnes Co., March 21, 1913, 46/C1235 ICW; Wilcox to Williams, August 9, 1929, 90/C589 ICW; Price to inspectors, June 17, 1914, 45/C1175 ICW.

47. Wilcox to Conklin, May 9, 1922, 78/C76.2 ICW; Altmeyer, *Industrial Commission,* 144, 168–69, 178; memo, Beck to deputies, September 20, 1916, 34/C661.1 ICW.

48. ICW, *Minutes,* December 16, 1919, February 14, 1920, February 25, 1921, August 17, 1920, and June 25, 1921; transcript, "Hearing before the Wisconsin Railroad and Industrial Commissions," March 16, 1917, 10, 57/C1407.3 ICW; Williams to M. A. Edgar, May 22, 1918, 68/C1801 ICW; transcript, "Suspension of Enforcement of General Order #4509," September 27, 1918, 35/C661.17 ICW.

49. Evans to Beck, August 11, 1912, 78/C76.2 ICW; Vallier to Watrous, May 29, 1914, 34/C661.1 ICW.

50. ICW, *Report on Allied Functions* (1914), 92–94; (1915), 49; (1917), 8–9, 49–52.

51. Kaems to Beck, May 5, 1913, 78/C76.2 ICW; ICW, *Minutes,* December 30, 1918, February 6, March 24, 1919, February 27, September 21, 1920, August 23, 1921, May 11 and 20, 1925; A. J. Altmeyer to John W. Reynolds, April 10, 1928; Reynolds to Altmeyer, January 22, 1932; Altmeyer to Reynolds, March 31, 1932; Reynolds to Altmeyer, April 4, 1932, 110/C2 ICW.

52. Pillsbury, "Law of Safety in California," 19 (see n. 29, this chapter); IDL, *Annual Reports* (1918–1919), 55; (1920–1921) 102–3; ICO, *Reports* (1916–1917), 17; (1917–1918), 52–53; ICO, *Report No. 26,* 28.

53. NYSDL, *Report* (1925), 66–67.

54. August Kaems, "Report on Inspection Work," 1913, 108/C1598 ICW; Henry Schreiber, "Report on Inspection Work," 1913; Al Kroes, "Report on Inspection Work," 1913, 108/C1598 ICW; Kroes to Industrial Commission of Wisconsin, November 22, 1913; Kroes to Beck, February 6, 1914, 106/C989 ICW; ICW, *Report on Allied Functions* (1914), 2, 6; Crownhart to Industrial Accident Board, April 15, 1913, 15/C182 ICW; *The Milwaukee Journal,* August 22, 1913, 2; March 6, 1914, 1; ICW, *First Aid: A Handbook for Use in Shops,* 1915, 8.

55. ICO, *Report* (1916–1917), 16; Aldrich, *Safety First,* 91–114, 164–66, 298–302, 309–11; Nelson, *Managers and Workers,* 138–39.

56. ICW, *Reports on Allied Functions* (1914), 24; (1915), 14; (1917), 9; (1918), 12; ICO, *Reports* (1913–1914), 19; (1916–1917), 16; (1917–1918), 53–54; (1921–1922), 30; (1922–1923), 24.

57. Aldrich, *Safety First,* 55–62, 240–41, 300–301, 164; ICW, *Wisconsin Safety Review*

(October 1921): 5–7; CIAC, *Reports* (1915), 48; (1916), 42–43; (1917), 43; (1919), 43; (1920), 53; ICO, *Reports* (1913–1914), 8; (1915–1916), 17; (1916–1917), 13; (1917–1918), 46; Curran, *Dead Laws for Dead Men,* 64–66; Whiteside, *Regulating Danger,* 115–17.

58. ICW, *Wisconsin Safety Review* (October 1921): 5–9.

59. NYSIC, *Special Bulletin No. 75* (March 1916): 1–22.

60. CIAC, *Reports* (1916), 42–43; ICO, *Reports* (1913–1914), 8; (1915–1916), 17; (1916–1917), 13; (1917–1918), 46.

Chapter 6: Politics and Work Safety Education in the Interwar Economy

1. Hawley, *Great War,* 100–104.

2. Kolko, *Triumph of Conservatism,* 2–4; Weinstein, *Corporate Ideal,* ix–xv; M. Sklar, *Corporate Reconstruction,* 1–40; Caine, *Myth of Progressive Reform,* 202; Link and McCormick, *Progressivism,* 58–66; Wiebe, *Search for Order,* 222–23, 295–96; Keller, *Regulating a New Society,* 6–7, 197–99, 202, 214–15.

3. Keller, *Regulating a New Society,* 6–7, 214–15; Patterson, *New Deal and the States,* 3–11; Teaford, *Rise of the States,* 96–118.

4. ICW, *Report on Allied Functions* (1917), 8, 53; (1918), 4, 8–9; ICW, *Biennial Report* (1918–1920), 87; ICW, *Minutes,* May 8, June 3, 1918, January 6, September 15, 1919; M. A. Edgar to Industrial Commission, May 26, 1917, 68/C1801 ICW; Robert McA. Keown to Kaems, March 21, 1919, 38/C76.2 ICW; Williams to all deputies, December 18, 1917, 34/C661.1 ICW; Edgar to Industrial Commission, August 29, 1919, 68/C1801 ICW; Hambrecht to Emanuel L. Philipp, May 24, 1918; Philipp to Hambrecht, June 5, 1918, 33/C661 ICW; Hambrecht to C. F. Otto, October 12, 1920, 79/C76.2 ICW.

5. Margulies, *Decline of the Progressive Movement,* 214–50; ICW, *Minutes,* November 11, 1918, July 22, 1919; petition, March 28, 1919, and Resolution of the Wisconsin State Federation of Labor, March 3, 1919, 36/C716 ICW; "Answer to Interrogatories under the Interpellation Statute," April 10, 1919, 36/C716 ICW; "Argument upon New Employees and New Departures in the Budget of the Industrial Commission for 1919–1921," April 10, 1919, 41/C890 ICW; *The Milwaukee Journal,* June 11, 1919, 5; *Laws of Wisconsin* (1919), Chapter 599; ICW, *Report on Allied Functions* (1917), 53; ICW, *Biennial Report* (1920–1922), 56; (1922–1924), 55; (1918–1920), 4, 7–8.

6. Nesbit, *Wisconsin: A History,* 464; Glad, *History of Wisconsin,* 300; Wisconsin, *Senate Journal* (1921), 1240, 1269, 1283, 1290; *The Milwaukee Journal,* June 22, 1921, 2; July 7, 1921, 21; Milwaukee *Leader,* June 25, 1921, 10; *Wisconsin State Journal,* July 3, 1921, 1.

7. Nesbit, *Wisconsin: A History,* 460–61, 463–65, 469; Glad, *History of Wisconsin,* 232–34, 239–46, 301–11; Ozanne, *Labor Movement in Wisconsin,* 128–31. On the budget, see *The Milwaukee Journal,* March 31, 1921, 1; *Laws of Wisconsin* (1921), Chapter 314; and Wisconsin, *Senate Journal* (1925), 23–24. On Tarrell and Wilcox, see Wisconsin, *Senate Journal* (1923), 74, 198; (1925), 1178, 1293.

8. On cutbacks, see ICW, *Biennial Report* (1922–1924), 6; (1924–1926), 54; and ICW, *Minutes,* March 22, 1921. On salaries, consult ICW, *Minutes,* July 7, 1921, September 6, 1921. On travel, see Altmeyer to Hoeveler, February 4, 1925, 78/C76.2 ICW; and ICW, *Minutes,* September 11, 1923. On expense accounts, review Altmeyer to Traveling Employees of the Commission, February 18, 1925, 34/C661.1 ICW; and ICW, *Minutes,* October 21, 1921. About new agency duties, see ICW, *Biennial Report* (1924–1926) 3–4, 6. On the budget generally, see ICW, *Biennial Reports* (1922–1924), 7–8; (1924–1926), 3–6; typescript, "Budget Report to Governor," 1925, in *Clippings: Labor Regulation and Regulatory Agencies in Wisconsin,* WLRL; and Wilcox to Weber, March 12, 1926, 15/C191 ICW.

9. On politics, see Nesbit, *Wisconsin: A History,* 468–69; Glad, *History of Wisconsin,* 313–16; and Wisconsin, *Senate Journal* (1927), 27–31, 352, 781, 856; (1929), 18–21, 979, 1037. On appropriations, see ICW, *Biennial Report* (1926–1928), 3–4; (1928–1930), 4, 17; *Laws of Wisconsin* (1927), Chapter 389; (1929), Chapter 480; and Altmeyer to Walter J. Kohler, September 9, 1929, 80/C181 ICW. About the education post, consult ICW, *Biennial Report* (1928–1930), 3–4; ICW, *Minutes,* July 26, 1928; and Wisconsin, *Assembly Bills* (1929), 841A.

10. NYSDL, *Annual Report of the Industrial Commissioner* (1925), 63–65; "Labor Law Administration, New York, 1931," 7, 12–14, 34–37, AALL Papers; NYSFL, *Proceedings* (1921): 5–9; (1924): 29, 31; Ellis et al., *New York State,* 393–403. On safety meetings, see NYSDL, *Proceedings of Industrial Safety Congresses.*

11. NYSDL, *Proceedings of Industrial Safety Congresses* (1920–1927); NYSDL, *Special Bulletins* 101 (1920); 102 (1921); 104 (1921); 108 (1921); 116 (1923); 127 (1924); 128 (1924); 129 (1924); 130 (1924); 131 (1924); 134 (1924); 139 (1925); 144 (1926); 150 (1927). On the compensation fund, see NYSDL, *Proceedings of the Tenth Industrial Safety Congress* (1926): 222–24.

12. ODIR, *Annual Report* (1921–1922), 3–4, 30; (1922–1923), 24; Knepper, *Ohio and Its People,* 355–56; Cayton, *Ohio,* 306–8; Joint Resolution Amending Article II, Section 35, of the Ohio Constitution, *Acts of Ohio* (1923), 631–32; *Acts of Ohio* (1925), Section 1465-89a, pp. 226–27; OSFL, *Proceedings* (1927): 118–20.

13. ODIR, *Annual Reports* (1923–1924), 26–27, 45; (1924–1925), 29, 48; (1925–1926), 26; Thomas P. Kearns, "Safety Education in Industry Under State Supervision as a Means to Prevent Accidents," in IAIABC, *Proceedings,* in USBLS, *Bulletin No. 485* (1929): 175–80.

14. OSFL, *Proceedings* (1927): 109, 115–16, 120.

15. Bean, *California: An Interpretive History,* 362–64; "Labor Law Administration, California, 1940," 27–30, 46, AALL Papers; *Statutes and Amendments of California* (1913), Chapter 176, Sections 65, 69, Chapter 178; (1917), Chapter 586, Section 51; (1921), Chapter 604; (1927), Chapter 761; CSFL, *Proceedings* (1922): 32–33; (1923): 23; Taft, *Labor Politics,* 84–85; Nonet, *Administrative Justice,* 126.

16. Dubofsky, *State and Labor,* 83–102; Brody, *Workers in Industrial America,* 57–58; Tomlins, *State and Unions,* 90–94; I. Bernstein, *Lean Years,* 190–220; Gordon, *New*

Deals, 87–91; Forbath, *Law and the American Labor Movement,* 59–127; Jacoby, *Employing Bureaucracy,* 167–74; Hawley, *Great War,* 48–49; Montgomery, *House of Labor,* 420–38.

17. Hawley, *Great War,* 53–54, 81–82, 84, 90, 91–92; D. Noble, *America by Design,* 259–76, 290–91, 301–19; Gordon, *New Deals,* 35–37, 128–62; Brody, *Workers in Industrial America,* 50–55; Jacoby, *Employing Bureaucracy,* 180–96; Jacoby, *Modern Manors,* 20–34; Tone, *Business of Benevolence,* 226–44; Mandell, *Corporation as Family,* 128–29; Cohen, *Making a New Deal,* 161–211.

18. Aldrich, *Safety First,* 123–57, 236–37; Whiteside, *Regulating Danger,* 142–44.

19. U.S. Census, 1920, vol. 9, *Manufacturers,* 40–43; U.S. Census, 1930, *Manufacturers,* vol. 3, *Reports by States,* 48–49; Tindall, *Emergence of the New South,* 331–38, 523–27; Rogers et al., *Alabama,* 79–80.

20. U.S. Census, 1920, vol. 9, *Manufacturers,* 112–19; U.S. Census, 1930, *Manufacturers,* vol. 3, *Reports by States,* 66–69; Bean, *California: An Interpretative History,* 361–62; Taft, *Labor Politics,* 79–85.

21. U.S. Census, 1920, vol. 9, *Manufacturers,* 348–59; U.S. Census, 1930, *Manufacturers,* vol. 3, *Reports by States,* 146–49; Howard, *Illinois,* 467–69; Cohen, *Making a New Deal,* 161–82.

22. U.S. Census, 1920, vol. 9, *Manufacturers,* 1038–51; U.S. Census, 1930, *Manufacturers,* vol. 3, *Reports by States,* 358–61; Ellis et al., *New York State,* 547–49; Montgomery, *House of Labor,* 438–56.

23. U.S. Census, 1920, vol. 9, *Manufacturers,* 1186–93; U.S. Census, 1930, *Manufacturers,* vol. 3, *Reports by States,* 404–6; Knepper, *Ohio and Its People,* 357–59; Cayton, *Ohio,* 306–10.

24. U.S. Census, 1920, vol. 9, *Manufacturers,* 1638–43; U.S. Census, 1930, *Manufacturers,* vol. 3, *Reports by States,* 566–67; Nesbit, *Wisconsin: A History,* 460–61; Glad, *History of Wisconsin,* 136–50, 234–46.

25. Hawley, *Great War,* 53–54, 68–70, 90–93, 100–104; Gordon, *New Deals,* 130–60; Aldrich, *Safety First,* 105. For business technology, see Lorin Frankel to Williams, January 8, 1918, 46/C1235 ICW. On eye protectors, see MMAM, *Civics and Commerce,* June 1916, 30. On ventilation, consult booklet, Clark Automatic Dust Collecting Company, 64/C1661 ICW. On optical service, see E. W. Duperrault to Industrial Commission, March 19, 1925; and R. W. Brinsor (Kindly Optical Company) to Industrial Commission, September 16, 1926, 10/C100 ICW. About the bribe, see William G. Clark to state factory inspectors, October 29, 1913; Beck to Clark Automatic Dust Collecting Co., November 4, 1913; W. G. Clark to Beck, November 6, 1913; Beck to Clark Automatic Dust Collecting Co., November 8, 1913; D. McCool (Cement Co.) to Beck, November 17, 1913; and Beck to Clark Automatic Dust Collecting Company, November 19, 1913, 64/C1661 ICW.

26. Aldrich, *Safety First,* 111–13; Fred Braun, interview by the author, June 3, 1981, Wausau, Wisconsin. See also William A. Fricke, "The Value and Need of Cooperation Under the Workmen's Compensation Act," address delivered at the First Annual Joint

Meeting of Liability Insurance Agents and the Industrial Commission of Wisconsin, July 24–25, 1913; H. J. Hagge, "Motives and Methods," *Coverage*, September 1935, 1–2; transcripts of interviews with J. C. Youmans (January 1974), 3; Ben Kuechle (n.d.), 5; Fred Braun (July 1973), 4, 8; *Employers Mutual News Bulletin*, September 1919, 2; and articles in *Employers Mutual News Bulletin* and *The Foreman*, Wausau Archives.

27. MMAM, *Civics and Commerce*, November 1913, 16; December 1913, 12; March 1914, 21; April 1914, 12; May 1914, 14; June 1914, 10; December 1914, 14; March 1915, 10; May 1915, 13; ICW, *Report on Allied Functions* (1915), 5–6; (1918), 7; speech of January 1, 1918, 8–9, in *Addresses of George P. Hambrecht*, ICW General Archives; ICW, *Minutes*, January 20, 1919, February 10, November 26, 1920, December 13, 1921, February 29, 1922; ICW, *Biennial Reports* (1922–1924), 6; (1930–1932), 9; (1934–1936), 11; (1936–1938), 2; Altmeyer, *Industrial Commission*, 154; programs for foremen safety schools and regional safety conferences, 1929–1931, 59/C1417 ICW; report, "Safety Schools in Wisconsin—1940," 1–4, 59/C1417 ICW; Aldrich, *Safety First*, 150–54.

28. ICW, *Biennial Reports* (1930–1932), 9; (1926–1928), 8; (1928–1930), 9; Altmeyer, *Industrial Commission*, 156; ICW, *Minutes*, January 22, 1930; Keown to Altmeyer, July 23, 1932, 59/C1417 ICW; Clarence J. Muth to Altmeyer, December 21, 1931, 59/C1417 ICW; Henry J. Bell to Wrabetz, January 18, 1930, 59/C1417 ICW. On the Bureau of Mines program, consult Price to Joseph A. Holmes, July 21, 1914; Witte to Van H. Manning, November 12, 1918, 45/C1171 ICW; A. H. Findeisen to Industrial Commission, March 30, 1922, 37/C807 ICW; circular, "Train Quarrymen to Extend First Aid," circa 1922, 37/C807 ICW; ICW, *Biennial Report* (1926–1928), 21.

29. Title pages, ICW, *Biennial Reports* (1918–1920 to 1930–1932); "Labor Law Administration, Wisconsin, 1931," 11, AALL Papers; ODIR, *Annual Reports* (1921–1922 to 1931–1932); "Labor Law Administration, New York, 1931," 34–37; Teaford, *Rise of the States*, 87–95.

30. IAIABC, *Proceedings* (1920–1928), published as USBLS, *Bulletins*, nos. 281, 304, 333, 359, 385, 406, 432, 456, 485; Aldrich, *Safety First*, 103; Jacoby, *Employing Bureaucracy*, 126–31; Larson, *Rise of Professionalism*, xii–xvii; Hobson, "Professionals, Progressives, Bureaucratization"; Hays, "The New Organizational Society," 245–59.

31. Hagglund, "Some Factors," 7–9, 10, 13, 22–23; Aldrich, *Safety First*, 150–52, 157–59; D. Noble, *America by Design*, 259–76, 290–91, 301–19; Tone, *Business of Benevolence*, 229–31.

32. Fred W. Braun, "The Psychology of Safety," *Coverage*, December 1937, 2, 11; Fred W. Braun, "Man Failure—or Was It Our Alibi?" *Coverage*, May 1940, 2, Wausau Archives; C. B. Auel, "Is There an Accident-Prone Employee?" paper presented at Pennsylvania Safety Conference, 1930, AALL Papers; IAIABC, *Proceedings*, in USBLS, *Bulletin No. 333* (1923): 199–205; NYSDL, *Proceedings of the Tenth Industrial Safety Congress* (1926): 21; Aldrich, *Safety First*, 137–42.

33. ICW, *Biennial Report* (1936–1938), 20; Fred Wilcox, typescript, "How Far Have We Come?" address to the Foremen's Safety School, Wisconsin Rapids, December 5, 1930, 9, 44/C1040 ICW; Voyta Wrabetz, typescript, "Accident Prevention in Machine

Shops," circa 1939, 6–7, 44/C1040 ICW; ICW, *Wisconsin Labor Statistics,* January–February 1924, 4; Aldrich, *Safety First,* 151.

34. Voyta Wrabetz, typescript, "Accident Prevention in Public Utilities," November 11, 1935, 21/C297.36 ICW; Wilcox, "How Far Have We Come?" 15; Wrabetz, "Accident Prevention in Machine Shops," 2, 4–6; ICW, *Wisconsin Labor Statistics,* September 5, 1929, 1; NYSDL, *Proceedings of the Ninth Industrial Safety Congress* (1925): 42–43; D. J. Harris, "Safeguards Alone Do Not Eliminate Accidents," CIAC, *California Safety News* 9 (January 1925): 14.

35. Typescript, Harry A. Nelson, "Wisconsin's Program of Pre-employment and Medical Examination Procedure," circa 1939, 2–4; 75/C24 ICW; Voyta Wrabetz, typescript, "Address before the Rock River Safety Conference," Beloit, April 27, 1939, 4, 44/C1040 ICW; ICW, "Wisconsin's Movement for Safety," 4–5; Wrabetz, "Accident Prevention in Machine Shops," 7.

36. Mandell, *Corporation as Family,* 128–29; Tone, *Business of Benevolence,* 182–98; Cohen, *Making a New Deal,* 184–211; Montgomery, *House of Labor,* 421–22; John Sullivan, "Additional Building Inspectors—a Needed Remedy for a Bad Situation," NYSDL, *Proceedings of the Tenth Industrial Safety Congress* (1926): 24–25; OSFL, *Proceedings* (1927): 109, 113–14.

37. OSFL, *Proceedings* (1927): 109, 110, 114–15, 119–20, 122; Sullivan, "Additional Building Inspectors," 25.

38. Sullivan, "Additional Building Inspectors," 25; OSFL, *Proceedings* (1927): 119, 122.

39. OSFL, *Proceedings* (1927): 116; NYSFL, *Proceedings* (1929): 17.

40. Teaford, *Rise of the States,* 120–52; Patterson, *New Deal and the States,* 26–37, 39–49, 50–73, 121–24, 132–44; Dubofsky, *State and Labor,* 103–67; Tomlins, *State and Unions,* 99–241; Hamby, *Survival of Democracy,* 314.

41. Rogers et al., *Alabama,* 480–504; *General Laws of Alabama* (1935), 911–13; (1939), 234; *Laws of Illinois* (1934–1936), 33–40; NYSFL, *Proceedings* (1932): 19; ICO, Division of Safety and Hygiene, *Annual Statistical Reports* (1934–1940); "Labor Law Administration, California, 1940," 40–49; Nesbit, *Wisconsin: A History,* 476; Glad, *History of Wisconsin,* 356–57, 365–72; Gordon, *New Deals,* 160–64.

42. Kasparek, *Fighting Son,* 118–22; Nesbit, *Wisconsin: A History,* 486–88; Glad, *History of Wisconsin,* 378–82, 391–96; *Laws of Wisconsin* (1931), Chapter 67; *Wisconsin State Journal,* July 10, 1931, 2; August 27, 1931, 1, August 9, 1933, 4; *The Capital Times,* August 27, 1931, 8.

43. Teaford, *Rise of the States,* 121–31; Patterson, *New Deal and the States,* 26–49, 132–65; Howard, *Illinois,* 505–13; Knepper, *Ohio and Its People,* 368–76; Bean, *California: An Interpretive History,* 410–20; Ellis et al., *New York State,* 416–30.

44. *Laws of Wisconsin* (1933), Chapter 140. On budgets, see ICW, *Biennial Report* (1930–1932), 54; (1932–1934), 86. On Schmedeman, see Wisconsin, *Senate Journal* (1933), 172, 202; and Wisconsin, *Senate Bills* (1933), 64S, Amendment no. 1S to no. 64S. About administrative demands, review ICW, *Minutes,* circa 1933–1934; and ICW,

Biennial Reports (1930–1932), 4–5; (1932–1934), 3–7, 11, 16; (1936–1938), 3, 19. On Ohio, examine forewords in ODIR *Annual Reports* through the 1930s.

45. *Wisconsin State Journal,* May 5, 1933, 3; May 11, 1933, 3; May 26, 1933, 3; July 15, 1933, 3; *The Capital Times,* May 24, 1933, 18; May 31, 1933, 1; *The Milwaukee Journal,* May 24, 1933, 13; *Wisconsin State Journal,* May 24, 1933, 1; Wisconsin, *Senate Journal* (1933), 1408, 1474, 1912, 1988; Fact Sheet, Harry R. McLogan, 63/C1605 ICW; Nesbit, *Wisconsin: A History,* 488–89.

46. Rogers et al., *Alabama,* 481–85; Tindall, *Emergence of the New South,* 505–8, 523–27, 531–32; Howard, *Illinois,* 513–14; Cayton, *Ohio,* 315–19; Knepper, *Ohio and Its People,* 377–81; Ozanne, *Labor Movement in Wisconsin,* 59–69, 78–89, 137–38; Kasparek, *Fighting Son,* 202; Glad, *History of Wisconsin,* 426–35, 510–12, 534–36; Nesbit, *Wisconsin: A History,* 490–93; Jacoby, *Modern Manors,* 35–56; Cohen, *Making a New Deal,* 238–46, 253, 285–86; Dubofsky, *State and Labor,* 105–67; Gordon, *New Deals,* 214–16, 220–22, 234–35.

47. Nesbit, *Wisconsin: A History,* 490; Glad, *History of Wisconsin,* 528–30; Wisconsin, *Senate Journal* (1931), 138–39; (1937), 524, 824; Wisconsin, *Legislative Journals* (for the 1937 special session), 258, 262; Wisconsin, *Assembly Bills* (1933), 32A; (1935), 62A; Wisconsin, *Senate Bills* (1935), 246S; (1937), 443S; Peter A. Napiecinski to Philip La Follette, April 24, 1937, 13/C181 ICW; *The Capital Times,* November 24, 1938, 1; January 27, 1955, 2; WMA, *Weekly Bulletin,* May 28, 1937, 5, WMA Papers; WSFL, *Proceedings* (1933): 123; Brownlee, *Progressivism and Economic Growth,* 88–89; Teaford, *Rise of the States,* 132–38.

48. ICW, *Biennial Reports* (1934–1936), 96; (1936–1938), 2–3, 11, 17, 20, 27, 30, 74; (1938–1940), 5–6, 52.

49. Nesbit, *Wisconsin: A History,* 493–94, 526–27; Raney, *Wisconsin,* 369–70; Wisconsin, *Senate Journal* (1939), 234–37, 1521–25; WMA, *Weekly Bulletin,* February 3, 1939, WMA Papers; *The Milwaukee Journal,* January 1, 1939, 1; June 7, 1939, 2; July 18, 1939, 1; *Laws of Wisconsin* (1939), Chapter 142; Julius P. Heil to Wrabetz, July 18, 1939, 13/C181 ICW. Compare expenditures in ICW, *Biennial Report* (1936–1938), 74; (1938–1940), 52; (1940–1942), 52. See also ICW, *Biennial Report* (1938–1940), 2, 6, 11, 21; Wrabetz to Victor H. Tousley, November 1, 1939, 56/C1407.2 ICW; Harry E. Pike and K. B. Stevens to Wrabetz, January 24, 1940, 71/C1929 ICW; and Wrabetz to Pike, January 31, 1940, 71/C1929 ICW.

Chapter 7: The Technocrats Take Command

1. J. Andrews, *Administrative Labor Legislation,* 122–47; NYSDL, *Annual Reports* (1923), 137–39; (1925), 132; "Labor Law Administration, California, 1940," chart opposite p. 18, AALL Papers; ICO, *Annual Report* (1916–1917), 22; ODIR, *Annual Reports* (1922–1923), 25; (1923–1924), 25, 27; (1924–1925), 29–30; (1925–1926), 26; Rosner and Markowitz, *Dying for Work,* xvi; Keller, *Regulating a New Society,* 6–7, 198–99; Aldrich, *Safety First,* 107–8; Whiteside, *Regulating Danger,* 151–52.

2. On business concentration, see Chandler, *Visible Hand*, 484–500; M. Sklar, *Corporate Reconstruction*, 4–33; Hawley, *Great War*, 53–55, 81–82, 92; Gordon, *New Deals*, 18–29, 35–85; Keller, *Regulating a New Economy*, 3–5; Alchon, *Invisible Hand of Planning*, 306; and Wiebe, *Search for Order*, 293–302. On the standardization movement, see Graebner, *Coal-Mining Safety*, 100–111; E. B. Rosa, "The Organization and Work of the National Safety Code Committee," offprint by USBLS, 1920; R. McA. Keown, "Uniform Safety Standards," offprint by USBLS, 1921; American Engineering Standards Committee, "The Status of the National Safety Codes," September 15, 1927, 1, all in AALL Papers; Aldrich, *Safety First*, 103; and C. Noble, *Liberalism at Work*, 43–46. On technocratic authority, see Hays, "The New Organizational Society," 244–63; Hays, "Political Choice in Regulation," 133–53; and D. Noble, *America by Design*, 74–75.

3. R. Stewart, "American Administrative Law," 1671–76; White, *Constitution and New Deal*, 98–103; Wiecek, *Lost World*, 146–47; Hays, "The New Organizational Society," 245–50, 255–63; Hays, "Political Choice in Regulation," 147–50; M. Sklar, *Corporate Reconstruction*, 16–23.

4. White, *Constitution and New Deal*, 94–127; Zeppos, "Legal Profession"; Shepherd, "Fierce Compromise"; Shamir, *Managing Legal Uncertainty*; Horwitz, *Transformation, 1870–1960*, 225–35; Verkuil, "Emerging Concept"; Fetner, *Ordered Liberty*, 62–64; B. Schwartz, *Administrative Law*, 18–22; Davis, *Administrative Law Text*, 7–9; Kagan, *Adversarial Legalism*, 189; Burke, *Lawyers, Lawsuits, and Legal Rights*, 8–12; Witt, *Patriots and Cosmopolitans*, 209–10, 228–33.

5. Note comparison of mechanical systems to electricity in hearing transcript, "Hearing before the Wisconsin Railroad and Industrial Commissions on the Construction and Operation of Electrical Systems," Milwaukee, March 16, 1917, 62, 57C1407.3 ICW. On electrical code development, see Lewis E. Gettle to Price, July 19, 1913, 1/C10 ICW; ICW, *Report of Allied Functions* (1915), 3–4; (1917), 6; (1918), 11; ICW, *State Electrical Code* (1917), 1–2; (1922), 3–5; (1924), 6–7; (1930), 5–6; (1934), 9–10; ICW, *Biennial Reports* (1918–1920), 3; (1920–1922), 21–22; (1922–1924), 25; ICW, *Labor Laws of the State of Wisconsin and Orders of the Industrial Commission* (1938), 23.

6. ICW, *Electrical Safety Code* (1922), 3–5: ICW, *Minutes*, July 29, 1919; CIAC, *Report* (1918): 30.

7. Price to Wilcox, October 16, 1915, 12/C15.5 ICW; transcript, "Hearing before the Wisconsin Railroad and Industrial Commissions," March 16, 1917, 3–6, 8, 17, 57/C1407.3 ICW; typescript, "Minutes of Meeting of Electrical Committee," October 18, 1916, 55/C1407 ICW; D. Noble, *America by Design*, 74–75; ICW, *Biennial Report* (1920–1922), 21–22; ICW, *Wisconsin Safety Review* 2 (October 1921): 9; ICW, *Minutes*, November 19, 1919; memo, Hoeveler to Industrial Commission, November 15, 1918, 58/C1407.4 ICW; transcript, "Electrical Code Hearing," April 16, 1924, 15–16, 55/C1407 ICW; ICW, *State Electrical Code* (1924), 6–7; S. W. Bratton to Wilcox, March 31, 1922, 56/C1407.1 ICW; Hoeveler to Electrical Code Advisory Committee, January 1, 1924, 55/C1407 ICW; Benjamin, *Administrative Adjudications*, 39–47.

8. Hoeveler to P. E. Barnum, circa 1924; Hoeveler to Peter C. Fabere, March 7, 1923, 56/C1407.1 ICW; Benjamin, *Administrative Adjudications,* 51–54; transcript, "Hearing before the Wisconsin Railroad and Industrial Commission on the Construction and Operation of Electrical Systems," March 16, 1917, 11, 46–49, 56, 61–62, 57/C1407.3 ICW.

9. Transcript, "Electrical Code Hearing," Milwaukee, April 16, 1924, 3–5, 9, 55/C1407 ICW.

10. Ibid., 7, 14, 17–18, 20.

11. Ibid., 10–12, 14; Hoeveler to Martin Fenneberg, May 4, 1923, along with attached correspondence, 56/C1407.1c ICW. Wisconsin's Industrial Commission granted twenty-four out of twenty-seven requested electrical-code waivers in the 1920s, and sixty-three of sixty-four in the 1930s, according to various agency *Minutes.*

12. "Electrical Code Hearing," April 16, 1924, 16.

13. Circular, "Bill 630A"; Wilcox to W. L. Smith, June 28, 1927, 55/C1407.1a ICW; Wisconsin, *Assembly Journal* (1927), 1399, 1665, 1724, 1779, 1878, 2099; ICW, *State Electrical Code* (1930), Orders 1305.07, 1306.11.

14. Williams to Geuder, Paeschke & Frey Co., August 15, 1917, 39/C857 ICW; "Explosion of Oxy-Acetylene Apparatus in California," CIAC, *California Safety News* 2, no. 7 (1918): 17–19; ICW, *Biennial Reports* (1918–1920), 23; (1920–1922), 15; ICW, *Minutes,* January 20 and 27, February 17 and 24, May 19, September 20, October 20, November 4, 1919, and March 11 and May 4, 1920; ICW, *Labor Laws and Orders,* 30; Wisconsin, *Assembly Bills* (1919), no. 423A; Wisconsin, *Assembly Journal* (1919), 419, 1226. See also Industrial Commission ruling, "In the Matter of the Petition of the Alexander Milburn Company and Others for a Modification of General Orders 4600 and 4635," May 2, 1921, 1–3, 81/C248.2 ICW.

15. Petition by the Alexander Milburn Company, "Re: General Orders on Gas Hazards," December 3, 1920, 81/C248.2 ICW.

16. Witte to O. H. Skinner, January 12, 1921; Witte to [Members of the Advisory Committee on Gas Hazards], April 4, 1921, 81/C248.2 ICW; ICW, *Minutes,* January 26, March 8, April 1, 19, 1921; ICW, "Ruling on Alexander Milburn Case," 3–10, 81/C248.2 ICW; Williams to Witte, January 24, 1921, 82/C248.3 ICW; Fred H. Haggerson to Industrial Commission of Wisconsin, March 22, 1921; Witte to the American Welding Society, April 12, 1921; H. S. Smith to Witte, April 18, 1921, 81/C248.2 ICW.

17. News clipping, "Acetylene Combine Grips State, Charge," 82/C248.3 ICW; *Alexander Milburn Co. v. Union Carbide & Carbon Corporation, et al.,* 15 F. (2d) 678 (1926); D. Noble, *America by Design,* 70–71, 84–109.

18. WSFL, *Proceedings* (1920): 48; Price to Henry Schreiber et al., March 23, 1915, 34/C661.1 ICW; Price to Vallier, April 11, 1916, 36/C793 ICW; Beck to Weber, March 8, 1917, 15/C191 ICW. On the 1921 legislation consult ICW, *Labor Laws and Orders,* 25; ICW, *Minutes,* February 26, 1921; E. H. Kiefer to Witte, March 1, 1921; Witte to Kiefer, February 28, 1921; Keown to W. M. Matthews & Brothers, Inc., September 15, 1921; Witte to Frederick J. Peterson, March 2, 1921, 28/C576 ICW; Wisconsin, *Assembly*

Bills (1921), 89A; and Wisconsin, *Assembly Journal* (1921), 148, 593, 638–39, 663, 674, 720, 721, 890.

19. NYSFL, *Proceedings* (1921): 81, 83; (1922): 33, 35; (1923): 87; Nugent, "Organizing Trade Unions," 433–36.

20. Wisconsin, *Assembly Journal* (1923), 257, 533; Wisconsin, *Assembly Bills* (1923), 331A; ICW, *Labor Laws and Orders,* 25; *The Milwaukee Journal,* November 22, 1923, 3; Milwaukee *Leader,* November 21, 1923, 2; Altmeyer, *Industrial Commission,* 125; ICW, *General Orders on Spray Coating* (1924), 3–11; WSFL, *Proceedings* (1924): 74–75. On labor's view, note letter and petition, Brotherhood of Painters, Decorators, and Paperhangers of America (Kenosha) to Wilcox, March 21, 1924, 28/C576 ICW. On the business view, see W. J. Mishella to Wilcox, November 30, 1923; and Howard G. Bartlay to Wilcox, November 27, 1923, 28/C576 ICW.

21. Orrin Fried to Kaems, December 19, 1923; Wilcox to Workers Health Bureau, October 19, 1926; Wilcox to Waubesa Lodge no. 470, December 22, 1923, 28/C576 ICW.

22. Wilcox to Brotherhood of Painters, Decorators, and Paperhangers of America, March 24, 1924, 28/C576 ICW.

23. ICW, *General Orders on Spray Coating* (1925), 2; (1929), 2–3; (1939), 7–8; ICW, *Minutes,* September 12, 1925, September 30, 1929, October 9, 1939; ICW, *Biennial Report* (1928–1930), 8; Wisconsin, *Assembly Bills* (1925), 298A; (1927), 515A; WSFL, *Proceedings* (1925): 54, 97; (1929): 58; (1933): 83; Wilcox to W. H. Cameron, March 28, 1927, 28/C576 ICW.

24. Benjamin, *Administrative Adjunctions,* 33–39; NYSDL, *Report* (1925), 134; *Russell v. Department of Labor of the State of New York,* 135 N.Y. Misc. 199 (1929); John J. Sheridan, "How to Get Safety in Small Plants," NYSDL, *Proceedings of the Eleventh Industrial Safety Congress* (1927): 160–61.

25. Sullivan, "Additional Building Inspectors," 25 (see chap. 6, n. 36); OSFL, *Proceedings* (1927): 110; Nugent, "Organizing Trade Unions," 443; CSFL, *Proceedings* (1927): 64.

26. Smith, "Chronological Development," 17–18, 100, 113, 154–62; J. Andrews, *Administrative Labor Legislation,* 147–48.

27. Tomlins, *State and Unions,* 60–94; Fink, "Labor, Liberty, and the Law," 918–22; Dubofsky, *State and Labor,* 104–46; Glad, *History of Wisconsin,* 510–12; Nesbit, *Wisconsin: A History,* 492; Ozanne, *Labor Movement in Wisconsin,* 61–69, 136–38.

28. J. Andrews, *Administrative Labor Legislation,* 146–47; USBLS, "Safety Codes and Standard Safe Practices," 306–9; WSFL, *Proceedings* (1935): 209; W. C. Muehlstein, "Meeting of American Society for Testing Materials, Chicago, June 25–26, 1931," 60/C1428 ICW; ICW, *Biennial Report* (1934–1936): 27; ICW, *Minutes,* August 5, 1940; J. E. Wise to E. A. Luedke, March 23, 1931, 55/C1407 ICW; transcript, "Hearing before the Wisconsin Public Service and Industrial Commissions on the Electrical Code," August 28, 1934, p. 45, 58/C1407.8 ICW; C. Noble, *Liberalism at Work,* 44–46; C. H. Fry, "The Relationship Between Division of Workmen's Compensation and, Factory Inspection, Safety and Health Promotion," 1935, pp. 6–7, AALL Papers.

29. ICW, *General Orders,* March 13, 1932, 7; memo, Altmeyer to Muehlstein, May 27, 1930, 41/C882.7 ICW; ICW, *General Orders on Safety in Construction* (1933), 6; ICW, *Electrical Safety Code* (1934), 9–10.

30. *Laws of Illinois* (1935–1936), 34; *General Laws of Alabama* (1939), 234; Benjamin, *Administrative Adjudications,* 33–39.

31. Wiecek, *Lost World,* 146–47; White, *Constitution and New Deal,* 103–4; J. Andrews, *Administrative Labor Legislation,* 174–79.

32. *The Capital Times,* October 8, 1925, 7; *The Milwaukee Journal,* October 9, 1925, 8; *The Milwaukee Sentinel,* October 9, 1925, 1; Nesbit, *Wisconsin: A History,* 304–10; Hurst, *Law and Economic Growth,* 496; Dicey, *Law of the Constitution;* Horwitz, *Transformation, 1870–1960,* 225–30; Davis, *Administrative Law Text,* 15–17; B. Schwartz, "Administrative Agency," 266–68.

33. Shamir, *Managing Legal Uncertainty,* 99–100; Chase, *American Law School,* 27–28, 39–40, 50–70, 82; Auerbach, *Unequal Justice,* 130–43, 191; Fetner, *Ordered Liberty,* 34–46; Dickinson, *Administrative Justice,* 32–38, 41–42, 84–136; White, *Constitution and New Deal,* 103–7; Witt, *Patriots and Cosmopolitans,* 209–10.

34. *Van Zandt's, Inc. v. Department of labor of the State of New York,* 129 N.Y. Misc. 747 (1927); *Russell v. Department of Labor of the State of New York,* 135 N.Y. Misc. 199 (1929); *Acme Steel and Malleable Iron Works, Inc. v. Department of Labor of the State of New York,* 255 N.Y. 555 (1930); *Schumer v. Caplin,* 241 N.Y. 346 (1925).

35. On Ohio, consult *Clemmer & Johnson Co. v. Industrial Commission of Ohio,* 112 O.S. 421 (1925); and *Slatmeyer v. Industrial Commission of Ohio,* 115 O.S. 654 (1926). Notably, Ohio's newly enlightened supreme court now acknowledged that the state's safe place statute raised legal safety standards. See *The Ohio Automatic Sprinkler Co. v Fender,* 108 O.S. 149 (1923). On California, consult *Hoffman v. Department of Industrial Relations,* 209 Cal. 383 (1930). On Wisconsin, review Altmeyer, *Industrial Commission,* 134, 149–51; *Bentley Brothers, Inc. v. Industrial Commission,* 194 Wis. 610, 614–15 (1928); *Wenzel & Henoch v. Industrial Commission,* 202 Wis. 595, 598–602 (1930); and *Saxe Operating Corporation v. Industrial Commission,* 197 Wis. 552, 553–54 (1929).

36. *Miller v. Paine* 202 Wis. 77, 82, 89–90, 92 (1930); Altmeyer, *Industrial Commission,* 107–8.

37. *Baker v. Janesville Traction Company,* 204 Wis. 452, 455–57, 461–63 (1931); *Popowich v. American Steel & Wire Co.,* 13 F. (2d) 381, 382 (1926).

38. *Fritschler v. Industrial Commission* 209 Wis. 558, 592 (1932); Benedict, "Law and Regulation," 261–62.

39. Wisconsin, *Senate Bills* (1929), 394S, 406S, 424S, Substitute 1S to 424S; Wisconsin, *Assembly Bills* (1929) 374A; *Laws of New York* (1927), Chapter 166, Sections 10, 29.

40. Fred M. Wilcox, typescript, "Paper Read to the State Bar Meeting," 1929, 4–5, 7, 9, 44/C1040 ICW; *The Capital Times,* April 22, 1929, main section, 20. Failed bills included *Assembly Bills* (1931), 202A; and *Senate Bills* (1931), 168S; (1933), 167S.

41. SBAW, *Bulletin* (July 1930): 144; (July 1931): 132–36; (July 1933): 133; Shepherd, "Fierce Compromise," 1565–66; CBA, *State Bar Journal* (April 1931): 94–96, 105–6; *Ohio State Bar Association Report* (January 24, 1929): 1, 6–8; (July 22, 1930): 222–25; Shamir, *Managing Legal Uncertainty*, 95.

42. Dubofsky, *State and Labor,* 103–45; Leuchtenburg, *Supreme Court Reborn,* 213–36; Hamby, *Liberalism and Its Challengers,* 32–35; Teaford, *Rise of the States,* 120–31; Benedict, "Law and Regulation," 258–59; Wiecek, *Lost World,* 218–45; Auerbach, *Unequal Justice,* 158–224; Chase, *American Law School,* 142–49; Fetner, *Ordered Liberty,* 59–64; Horwitz, *Transformation, 1870–1960,* 230–33, 237–40; B. Schwartz, *Administrative Law,* 21; Davis, *Administrative Law Text,* 8–9; White, *Constitution and New Deal,* 114–20; Verkuil, "Emerging Concept," 261–79; Shamir, *Managing Legal Uncertainty,* 53–54, 99–113; Shepherd, "Fierce Compromise," 1559–1678; Zeppos, "Legal Profession," 1124–56; Witt, *Patriots and Cosmopolitans,* 228–33.

43. McGill, "Alabama Administrative Regulatory Agencies," 75–90; McGill, "Alabama Supreme Court"; *General Laws of Alabama* (1981), 1534–56; *Chicago Bar Association Record* 20 (1938–1939): 42, 101, 165–66, 199–200; 21 (1939–1940): 259–60; 22 (1940–1941): 339–42; *Laws of Illinois* (1945), 1144–48; (1951), 327–28; (1975), 3311–19; Shamir, *Managing Legal Uncertainty,* 79, 108; Galie, *New York State Constitution,* 107–8; Galie, *Ordered Liberty,* 243; *Laws of New York* (1975), Chapter 167; *Ohio Bar Association Report* 8 (1935): 481; 12 (1939): 1–2; 13 (1940): 191–92; 14 (1941): 70–71, 155–63, 389; 15 (1943) 581–85; *Acts of Ohio* (1943), 358–405; CBA, *State Bar Journal* 13 (1938): 1–6; 14 (1939): 46–50; 15 (1940): 61; 16 (1941): 214–15; 16 (1941): pt. 2, 32–40; 18 (1943): 422–28; 19 (1944): 244–86; 20 (1945): 124–29, 198–201; 22 (1947): 391–96; *Statutes and Amendments of California* (1941), Chapter 628; (1945), Chapter 867; (1947), Chapter 1425.

44. SBAW, *Bulletin* (August 1935): 154–56; (November 1936): 251–52; (February 1937): 5–6, 65–67; (August 1937): 189–90; (August 1938): 2; (November 1939): 174–76; letters and memos attached to letter, Wrabetz to Ray A. Brown, October 20, 1938, 14/C181.1 ICW.

45. Nesbit, *Wisconsin: A History,* 544; WMA, *Weekly Bulletin,* February 9, 1940, Box 11, Folder 3, WMA Papers; SBAW, *Bulletin* (May 1940): 103; (August 1940): 135; Harry A. Nelson to E. Blythe Statson, May 18, 1943, 70/C1893 ICW; WSFL, *Proceedings* (1939): 226; (1941): 297; Wisconsin, *Senate Bills* (1939), 421S; *The Milwaukee Journal,* May 25, 1939, 2, August 24, 1939, 1; *Laws of Wisconsin* (1943), Chapter 375; Hoyt, "Wisconsin Administrative Procedure Act," 214–16.

46. *Laws of Wisconsin* (1943), Chapter 375, Sections 227.15–227.21; Hoyt, "Wisconsin Administrative Procedure Act," 217–18; Leuchtenburg, *Supreme Court Reborn,* 216–36; Benedict, "Law and Regulation," 261–62; Shapiro, "Supreme Court's 'Return,'" 91–116.

47. *Robert A. Johnston v. Industrial Commission,* 242 Wis. 299, 301–5 (1943).

48. R. Stewart, "American Administrative Law," 1711–1813.

Chapter 8: The Limits of Law Enforcement

1. Lubove, *Struggle for Social Security,* 49–63; Ashford and Caldart, *Technology,* 8–9; D. Berman, *Death on the Job,* ix–x, 56–73; Kolko, *Triumph of Conservatism,* 89–97; McDowell, *Deregulation and Competition,* 14–19; Fishback and Kantor, *Prelude to the Welfare State,* 82–83; Bale, "Compensation Crisis," 269; Keller, *Regulating a New Society,* 197–99; Rosner and Markowitz, *Dying for Work,* xvi–xvii; Aldrich, *Safety First,* 97–108, 114, 164–66.

2. Hawley, *Great War,* 68–70, 100–104; Wiebe, *Search for Order,* 298; Lubove, *Struggle for Social Security,* 59–65; Link and McCormick, *Progressivism,* 66; Gordon, "Does the Ruling Class Rule?" 292.

3. D. Berman, *Death on the Job,* 60–61; Hawley, *Great War,* 53–54, 91–92; Brody, *Workers in Industrial America,* 50; Gordon, *New Deals,* 18–19, 128–35; Alchon, *Invisible Hand of Planning,* 3–6; Baker and Simon, *Embracing Risk,* 17, 119.

4. D. Noble, *America by Design,* 259–76, 290–91, 301–19; Jacoby, *Employing Bureaucracy,* 177–89; Montgomery, *House of Labor,* 411–14, 438–52; Aldrich, *Safety First,* 99–100, 165–67; Nonet, *Administrative Justice,* 66–75.

5. Hawley, *Great War,* 53–54; Gordon, *New Deals,* 18–23, 35; William Leslie, "Factors Used in Rate Making for Compensation Insurance—Their Explanation and Illustration," in IAIABC, *Proceedings,* in USBLS, *Bulletin No. 406* (1926): 158–75; Leonard W. Hatch, "Method of Rate Making of the New York State Insurance Fund," in IAIABC, *Proceedings,* in USBLS, *Bulletin No. 385* (1925): 152–53; *Laws of New York* (1922), Chapter 660, Section 141b; *Laws of Wisconsin* (1917), Chapter 637, Sections 1921-12, 1921-23.

6. IAIABC, *Proceedings,* in USBLS, *Bulletin No. 406* (1926): 176–81; "Labor Law Administration, New York, 1931," pp. 76d–76e, and "Labor Law Administration, Illinois, 1930," p. 34, both in AALL Papers; Hatch, "Method of Rate Making," 151–53. On the WMA view, consult WMA, *Weekly Bulletin,* August 31, 1931, in Box 10, Folder 1; June 21, 1938, in Box 10, Folder 6; and August 4, 1938, in Box 10, Folder 6; and Earl F. Cheit, "Confidential Report: Analysis of Wisconsin's Workmen's Compensation Insurance Experience" (circa 1952), pp. 52–54, Box 8, Folder 2, WMA Papers.

7. ICW, *Bulletin: Workmen's Compensation Insurance,* June 1, 1915, 12–13; ICW, *Minutes,* November 25, 1919; Witte to Carl Young, October 11, 1921, 16/C244 ICW. See also ICW, *Wisconsin Labor Statistics,* March 30, 1928, 1; Leon S. Senior, "Schedule Rating Versus Experience Rating: A Critical Review of the Systems for Rating Workmen's Compensation Risks," in IAIABC, *Proceedings,* in USBLS, *Bulletin No. 333* (1923): 252–64; "Labor Law Administration, New York, 1931," 76d–76f; and George A. Kingston, "Merit Rating," in IAIABC, *Proceedings,* in USBLS, *Bulletin No. 304* (1922): 26–27.

8. Senior, "Schedule Rating," 264–67; Nonet, *Administrative Justice,* 68. For a different view, see Aldrich, *Safety First,* 99–100, 153, 214–15, 280–81.

9. Wisconsin, *Senate Bills* (1921) 203S; *The Milwaukee Journal,* May 7, 1924 17; May

8, 1924, 2; May 9, 1924, 1 and 2; May 10, 1924, 2. On the 1925 law, see *Laws of Wisconsin* (1925) Chapter 399; and Kimball, *Insurance and Public Policy*, 160. On subsequent legislation, consult *Laws of Wisconsin* (1927), Chapter 45; (1933), Chapter 230; Chapter 353, Section 205.24; Chapter 487, Section 182; Chapter 489, Section 25; and *Wisconsin Compensation Rating & Inspection Bureau v. Mortensen*, 227 Wis. 335 (1938).

10. Sydney D. Pinney to Wilcox, October 6, 1925, 7/C82.1 ICW; "Memorandum Regarding Experience Rating," circa 1925, 7/82.1 ICW; Altmeyer to John J. Blaine, October 9, 1925, 80/C181 ICW; E. W. Kitzrow to Henry Meigs, May 14, 1931; F. M. Wylie to Wisconsin Compensation Insurance Board, May 22, 1931; W. H. Burhop to Kitzrow, June 30, 1931, 108/C1062 ICW; IAIABC, *Proceedings*, in USDLS, *Bulletin No. 2* (1935): 214; Wisconsin, Compensation Insurance Board, *Report* (1930), 9–10; *Laws of New York* (1926), Chapter 533, Section 9; Chapter 545, Section 67; *Statutes and Amendments of California* (1935), Chapter 145, Section 11730. See also Aldrich, *Safety First*, 100.

11. Aldrich, *Safety First*, 99–100; Albert W. Whitney, "Accident Prevention Work from the Viewpoint of Stock Insurance Companies," NYSDL, *Proceedings of the Tenth Industrial Safety Congress* (1926): 211–12; S. Bruce Black, typescript, "Recent Experience in Casualty Insurance," May 4, 1933, 7, AALL Papers; Kingston, "Merit Rating," 25–35; T. J. Duffy, "Method of Rate Making in an Exclusive State-Fund State," in IAIABC, *Proceedings*, in USBLS, *Bulletin No. 385* (1925): 148–51, 158–59; E. I. Evans, "Merit Rating, an Incentive for Accident Prevention," in IAIABC, *Proceedings*, in USDLS, *Bulletin No. 2* (1935): 209–13.

12. Harry J. Aldrich, "Self-Insurance Plan Nearest the Ideal for Selling Safety," NYSDL, *Proceedings of the Ninth Industrial Safety Congress* (1925): 64–65; Charles G. Smith, "Accident Prevention Work of the State Insurance Fund," NYSDL, *Proceedings of the Tenth Industrial Safety Congress* (1926): 226–27.

13. Wisconsin, *Assembly Bills* (1923), 393A; (1925), 174A; (1927), 1A, 199A; (1931), 395A; Milwaukee *Leader*, November 27, 1930, 4; Fishback and Kantor, *Prelude to the Welfare State*, 173–76.

14. *Laws of Wisconsin* (1913), Chapter 599, Subsections 2394-9(5)(a) to 2394-9(5) (d); (1917), Chapter 624, Section 2394-9(7); Evans, "Merit Rating," 214.

15. ICW, *Progress of Work Report* (January 1922), 5; *Addresses by George P. Hambrecht*, January 3, 1921, both in ICW General Archives; Altmeyer, *Industrial Commission*, 146, 168; *Milwaukee Corrugated Company v. Industrial Commission of Wisconsin*, 197 Wis. 414, 419–20 (1928); ICW, *Wisconsin Labor Statistics*, no. 39 (May 12, 1932): 10; *Builders Mutual Casualty Company v. Industrial Commission*, 210 Wis. 311, 312–14 (1933).

16. ICW, *Wisconsin Labor Statistics*, no. 39 (May 12, 1932): 1, 10–11; ICW, *Workmen's Compensation Reports* (1934), 5; (1936), 5; (1938), 5; (1940), 6; (1942), 6.

17. *Statutes and Amendments of California* (1911), Chapter 399, Section 3(3); (1917), Chapter 586, Sections 6(b), 31(b); *E. Clemens Horst Company v. Industrial Accident Commission of the State of California*, 184 Cal. 180, 189–90 (1920); Pillsbury, "Law of

Safety in California," 19 (see chap. 5, n. 29); Warren H. Pillsbury, "The Law of Safety in California," CIAC, *California Safety News* 9 (February 1925): 13; *Leo Hoffman v. Department of Industrial Relations of the State of California*, 209 Cal. 383, 384–91 (1930); *Ethel D. Company v. Industrial Accident Commission*, 219 Cal. 699, 703–4 (1934).

18. *Acts of Ohio* (1911), Section 21-2, p. 529; (1913), Section 29, p. 84; (1923), 631; (1925), 218; *Constitution of the State of Ohio* (1913), Article II, Section 35; OSFL, *Proceedings of the 44th Annual Convention* (1927): 118–19; *The American Woodenware Manufacturing Company v. Schorling*, 96 O.S. 305 (1917); *The Toledo Cooker Co. v. Sniegowski*, 105 O.S. 161 (1922); *The Ohio Automatic Sprinkler Co. v. Fender*, 108 O.S. 149 (1923).

19. OSFL, *Proceedings* (1927): 118–21.

20. *Slatmeyer v. Industrial Commission of Ohio*, 115 O.S. 654, 656–57 (1926); *State ex rel. Rudd v. Industrial Commission of Ohio*, 116 O.S. 67, 69–73, 76–78 (1927); *Swift & Co. v. See*, 30 Ohio App. 127, 129–31 (1928); *State ex rel. Stuber v. Industrial Commission of Ohio*, 127 O.S. 325, 329–30 (1933); *State ex rel. Rae v. Industrial Commission of Ohio*, 136 O.S. 168, 170–73 (1939).

21. Altmeyer, *Industrial Commission*, 110–15, 134–45; ICW, *Report on Allied Functions* (1917): page inside front cover; ICW, *Biennial Reports* (1918–1920), 26–28; (1920–1922), 13, 18; (1922–1924), 7; (1926–1928), 9; (1928–1930), 50; report, "Work of the Industrial Commission of Wisconsin in the Field of Safety and Sanitation," circa 1921, 33/C661 ICW; "Labor Law Administration, Wisconsin, 1931," 72–101, AALL Papers.

22. "Labor Law Administration, Wisconsin, 1931," 40–57, 141–54, 169; Altmeyer, *Industrial Commission*, 119–20, 321; ICW, *Biennial Reports* (1918–1920), 10; (1924–1926), 6–7; (1928–1930), 9; (1932–1934), 4, 11.

23. ICW, *Biennial Reports* (1922–1924), 8; (1924–1926), 3–4; (1926–1928), 3–4; typescript, "Industrial Commission," circa 1924, *Clippings: Labor Regulation and Regulatory Agencies in Wisconsin*, WLRL; Altmeyer, *Industrial Commission*, 134–45, 175; Alfred Briggs to Altmeyer, March 11, 1931, 9/C86 ICW; "Labor Law Administration, Wisconsin, 1931," 113, 126.

24. ICW, *Biennial Reports* 1920–1922 to 1938–1940; Altmeyer, *Industrial Commission*, 157–61, 176; "Labor Law Administration, Wisconsin, 1931," 126; Witte to Commons, February 14, 1921, 16/C215 ICW; ICW, *Minutes*, March 27, October 5, 1920, December 20, 1921, February 28, 1922; ICW, *Progress of Works Report* (February 1922), 8; (March 1922), 12.

25. "Labor Law Administration, Wisconsin, 1931," 158; Altmeyer, *Industrial Commission*, 162; ICW, *Biennial Reports* (1920–1922), 55; (1922–1924), 54; (1924–1926), 55; (1926–1928), 59; (1928–1930), 52; (1930–1932), 56–57; ICW, *Minutes*, 1921–1940; *Laws of Wisconsin* (1913), Chapter 599; (1917), Chapter 624; (1931), Chapter 87. See also Table 4 in Appendix.

26. Aldrich, *Safety First*, 153. On boiler inspections, see Altmeyer, *Industrial Commission*, 139–40; ICW, *Biennial Report* (1938–1940), 11; (1934–1936), 20; ICW, *Minutes*, June 16, November 28, 1931, May 4, August 13, 1932, May 19, 1933, April 17, June 4, July 9, August 6, October 4, December 4, 1934. See Table 5 in Appendix.

27. Altmeyer, *Industrial Commission,* 64–65, 162. See Table 5 in Appendix.

28. *General Laws of Alabama* (1939), 234.

29. "Labor Law Administration, Illinois, 1930," 14–15, 42–47, 50–61, 67–73, 75–76; IDL, *Annual Reports* (1920–1921), 95–103; (1921–1922), 49; (1922–1923), 91–94; (1923–1924), 125–28; (1924–1925), 152–55; (1925–1926), 75–80; (1926–1927), 63–67; (1927–1928), 57–61; (1928–1929), 53–57; (1929–1930), 63–71.

30. ODIR, *Annual Reports* (1921–1922), 3–4; (1922–1923), 24; (1923–1924), 25; (1924–1925), 28; (1925–1926), 25; (1926–1927), 23; (1927–1928), 26; (1928–1929), 25–26; (1930–1931), 26; (1931–1932), 28; IAIABC, *Proceedings,* in USBLS, *Bulletin No. 304* (1922): 94–96.

31. C. H. Fry, "The Relationship between Division of Workmen's Compensation, Factory Inspection, Safety, and Health Promotion," paper delivered to the convention of the International Association of Industrial Accident Boards and Commission (1935), 5; "Labor Law Administration, California, 1940," 41–43, both in AALL Papers. See also Pillsbury, "Law of Safety," 19 (see chap. 5, n. 29); *Statutes and Amendments of California* (1919), Chapter 471, Section 12; F. A. Short, "Enforcement of Safety Orders," CIAC, *California Safety News* 4 (1920): 4.

32. NYSDL, *Annual Report* (1925), 63–65; NYSFL, *Proceedings* (1924): 29–31; Ellis et al., *New York State,* 395–405; *Laws of New York* (1921), Chapter 50, Section 18; (1923), Chapters 607, 884; (1924), Chapter 464; (1927), Chapter 166; (1929), Chapter 399; "Labor Law Administration, New York, 1931," 1–4, 28, 38, 63–65, 75–76, 76a.

33. "Labor Law Administration, New York, 1931," 43–44, 57–59, 71–75, 77–96, 99; NYSFL, *Proceedings* (1928): 31.

34. NYSDL, *Annual Report* (1925), 66–68; "Labor Law Administration, New York, 1931," 99, 101–5, 110–14; *Lown v. The Department of Labor of the State of New York,* 216 N.Y. App. Div. 474 (1930).

35. ICW, *Biennial Reports* (1932–1934), 3–7, 11; (1936–1938), 3, 19; *General Laws of Alabama* (1935), 146, 911–13; (1939), 232, 234; Rogers et al., *Alabama,* 500–501; Table 5 in Appendix; ICW, *Minutes,* 1931–1940.

36. Leonard W. Hatch, "What Is the Industrial Hazard in New York State?—Is it Increasing?" in NYSDL, *Proceedings of the Ninth Industrial Safety Congress* (1925): 39; Eugene S. Patton, "Are Accident Statistics Entirely Satisfactory?" IAIABC, *Proceedings,* in USBLS, *Bulletin No. 564* (1932): 239–41, 244; E. Stewart, "Are Accidents Increasing?" 161; "Statistics of Industrial Accidents to 1927," 1–3, 6–10, 44–67; Aldrich, *Safety First,* 164, 240–44; Curran, *Dead Laws for Dead Men,* 69–86.

37. USBLS, "Statistics of Industrial Accidents to 1927," 118–37, 158–66; E. Stewart, "Are Accidents Increasing?" 162; ICW, *Wisconsin Safety Review* (October 1921): 6–7; Hatch, "Industrial Hazard," 39–43; "Industrial Accidents, 1926 to 1932," *Monthly Labor Review,* 1388–98; "Industrial Accidents During 1939," *Monthly Labor Review,* 104–8; ICO, Division of Safety and Hygiene, *Annual Statistical Reports* (1934), 5; (1937), 6; (1938), 6; (1939), 6; (1940), 7; Aldrich, *Safety First,* 211, 298–314.

38. Hatch, "Industrial Hazard," 42–43; Aldrich, *Safety First,* 268–71, 317–18; ICW, *Wisconsin Labor Statistics,* no. 20 (September 5,1929): 1, and 21 (October 1929): 2.

39. "Industrial Accidents to 1927," 119–20; ICO, Division of Safety and Hygiene, *Annual Statistical Reports* (1934), 17; (1937), 17; (1939), 17; (1940), 17; Altmeyer, *Industrial Commission,* 137.

40. ICW, *Biennial Report* (1924–1926), 7; NYSDL, *Annual Report* (1925), 66–67, 72; ODIR, *Reports* (1920–1921), 30; (1921–1922), 24; (1922–1923), 26; (1924–1925), 29; (1925–1926), 26; (1926–1927), 23–24; (1927–1928), 26–27; (1928–1929), 25–26; (1929–1930), 25–26; (1930–1931), 25–29; (1931–1932), 28–29; (1932–1933), 83–84; ICW, *Wisconsin Labor Statistics,* no. 26 (May 20, 1930): 1; no. 39 (May 12, 1932): 10; Altmeyer, *Industrial Commission,* 138–43; Aldrich, *Safety First,* 104–14, 122–67.

41. "Industrial Accidents to 1927," 1–2, 29–112, 138–39.

42. Aldrich, *Safety First,* 153–54. On technomanagerial modernization and welfare work, see Jacoby, *Employing Bureaucracy,* 190–93; Gordon, *New Deals,* 35–165; Tone, *Business of Benevolence,* 52–61; and Jacoby, *Modern Manors,* 26–31. On productivity's impact, see E. Stewart, "Are Accidents Increasing?" 164–65; "Are Accidents Increasing?" *American Labor Legislative Review,* 240; and Aldrich, *Safety First,* 240–44.

43. ICW, *Wisconsin Safety Review,* 2 (October 1920): 6; Hatch, "What the Accident Record Shows," 173; Aldrich, *Safety First,* 164–65; D. J. Harris, "Safeguards Alone Do Not Eliminate Accidents," CIAC, *California Safety News* 9 (January 1925): 14; Hatch, "Industrial Hazard," 43; ICO, Division of Safety and Hygiene, *Annual Statistical Report* (1937), 17; Wrabetz, "Accident Prevention in Machine Shops," 4–6 (see chap. 6, n. 33).

44. "Industrial Injuries in the United States during 1939," 86–108; ICO, Division of Safety and Hygiene, *Annual Statistical Report* (1940), 7, 9. See Table 5 in Appendix.

Chapter 9: The Troubled Campaign against Occupational Disease

1. Nelson to John A. Robb, December 30, 1938, 2/C52 ICW.

2. The term "corporate/industrial/scientific system" is Graebner's in "Private Power," 17. See also Bale, "Compensation Crisis"; Graebner, "Radium Girls, Corporate Boys"; Sellers, *Hazards of the Job;* C. Clark, "Managing an Epidemic"; C. Clark, *Radium Girls;* Hepler, *Women in Labor;* Warren, *Brush with Death;* Rosner and Markowitz, *Deadly Dust;* Derickson, *Black Lung;* Derickson, *Workers' Health;* Rosner and Markowitz, *Dying for Work;* and Cherniack, *Hawk's Nest Incident.*

3. Sellers, *Hazards of the Job,* 2–185; Hepler, *Women in Labor,* 32–65; Crandall and Crandall, "Revisiting the Hawks Nest Tunnel Incident," 276; C. Clark, *Radium Girls,* 6; Cherniack, *Hawk's Nest Incident,* 17–19, 52–53; Warren, *Brush with Death,* 94–99; Rosner and Markowitz, *Deadly Dust,* 13–48, 70; Derickson, *Black Lung,* 60–86.

4. USBLS, "Problems of Workmen's Compensation Legislation," 198.

5. Bale, "Compensation Crisis," 416–25; Bale, "Assuming the Risks," 505–8; Sellers, *Hazards of the Job,* 18–36; *Kliegel v. Aitken,* 94 Wis. 432 (1896); *Segall v. Padlasky,* 123 Wis. 207 (1904); Rosner and Markowitz, *Deadly Dust,* 13–21; Derickson, *Black Lung,* 60–64; Stern, "Industrial Structure," 425–38.

6. *Laws of Wisconsin* (1883), Chapter 319, Section 2; (1885), Chapter 247, Sections 4, 5; (1887), Chapter 549, Section 1; Willoughby, "Inspection of Factories," 551–63. Willoughby overlooked Chapter 5, *Statutes and Amendments of California* (1889). See also WBLIS, *Biennial Reports* (1886), 492, 501–2, 503; (1888), 284; (1892), 4a, 63a, 66a, 69a; (1894), 51a, 81a; Sellers, *Hazards of the Job*, 36–37; Kens, *Lochner v. New York*, 49–66; K. Sklar, *Florence Kelley*, 207–36; C. Clark, *Radium Girls*, 185–88; Beckner, *Labor Legislation in Illinois*, 227, 240, 254–64.

7. *The Milwaukee Sentinel*, September 13, 1898, 1; January 10, 1899, 3; February 27, 1899, 8; March 2, 1899, 3; *Laws of Wisconsin* (1899), Chapters 79, 189, 232; (1901), Chapter 239; (1903), Chapters 230, 323; (1905), Chapter 147; (1907), Chapters 115, 486; (1911), Chapter 170.

8. *Laws of Wisconsin* (1911), Chapter 50, Section 2394-4; *Laws of Illinois* (1913), 337–38; *Labanoski v. Hoyt Metal Co.*, 292 Ill. 218 (1920); *Laws of New York* (1914), Chapter 41, Section 10; *General Laws of Alabama* (1919), 206; *Acts of Ohio* (1911), 529; (1913), 79; *Statutes and Amendments of California* (1911), Chapter 399, Section 3; *Madden's Case*, 222 Mass. 487 (1916); Bale, "Compensation Crisis," 467–69, 476–96.

9. Beckner, *Labor Legislation in Illinois*, 273–81; Hamilton, *Exploring the Dangerous Trades*, 114–26; IDL, *Second Annual Report* (1918–1919), 55–57; C. Clark, *Radium Girls*, 74–75; *Laws of New York* (1913), Chapter 145, Sections 51(4), 60, Chapter 199; NYSDL, *Annual Reports* (1910), 39–40; (1918), 221–22; (1919), 247–48; Sellers, *Hazards of the Job*, 81–87; *Laws of Wisconsin* (1911), Chapter 485, Sections 1021b-1(11), 1021b-8, 1021b-12(3), (4); *Statutes and Amendments of California* (1913), Chapter 176, Sections 51(8), 52, 57(1), (2).

10. ICW, "General Orders on Sanitation," *Bulletin* 2, no. 1 (January 20, 1913); typescript, "Standards for Shop Sanitation," circa 1912, 39/C882 ICW; Beck to Weber, October 1, 1912, 15/C191 ICW; ICW Beck to Anderson & Johnson Brothers, December 6, 1912; Beck to Mills & Co., December 13, 1912, 17/C256 ICW.

11. Francis Bird, typescript, "Lead Poisoning in Milwaukee," 36/C793 ICW; Price to Vallier, April 11, 1916; J. A. Norris to Price, December 6, 1916; Witte to F. C. Rusener, November 14, 1917, 36/C793 ICW; Price to all deputies, March 23, 1915, 34/C661.1 ICW; MMAM, *Civics and Commerce*, n.s., no. 53 (November 1914); n.s., no. 59 (May 1915); n.s., no. 66 (December 1915); ICW, "Shop Bulletin No. 5, Infections and Their Prevention," *Bulletin* 2, no. 11 (October 20, 1913): 279; ICW, *First Aid: A Handbook for Use in Shops*, 1915, 1–2; Smith, "Chronological Development," 154–60; Brandeis, "Women's Hours Legislation," 490–92; Hepler, *Women in Labor*, 22–26.

12. WSFL, *Proceedings* (1914): 118–19, 122–23; (1915): 99; (1916): 95, 105–7; (1917), 103; Watrous to Beck, September 27, 1912, 15/C191 ICW; Beck to Kaems, August 11, 1913; Beck to Allington-Curtis Mfg. Co., September 29, 1913; Lockney to Beck, September 20, 1913; Price to Evans et al., June 14, 1916, 64/C1661 ICW; typescript, "Minutes of Committee on Exhaust Systems for Buffing and Polishing Wheels," December 30, 1915, January 22, June 3, 1916, 64/C1661 ICW; Beck to Mills & Co., December 13, 1912; Pete Peterson to Marathon Granite Co., November 26, 1915, 17/C256 ICW; circular,

Price to all deputies, October 30, 1915, 34/C661 ICW; Beck to deputies, February 12, 1917, 34/C661.1 ICW.

13. J. W. Schereschewsky to Beck, June 16, 1916; Beck to R. M. La Follette, June 19, 1916; R. La Follette to Beck, July 7, 1916; Schereschewsky to Beck, August 28, 1918; Beck to Schereschewsky, August 28, 1916, 15/C183 ICW; Beck to George H. Willmer, June 20, 1917, 64/C1661 ICW; Williams to Committee on Safety and Sanitation, May 5, 1917, 39/C882 ICW; ICW, *Biennial Report* (1918–1920), 34–35.

14. Bale, "Compensation Crisis," 480–510, 568; *Rist v. Larkin & Sangster*, 171 N.Y. App. Div. 71 (1916); *Hiers v. Hull*, 178 N.Y. App. Div. 350 (1917); *Richardson v. Greenberg*, 188 N.Y. App. Div. 248 (1919); *Industrial Commission of Ohio v. Brown*, 92 O.S. 309 (1915); *Industrial Commission of Ohio v. Roth*, 98 O.S. 34 (1918); *Labonoski v. Hoyt Metal Company*, 292 Ill. 218 (1920); *Heilman v. Industrial Commission*, 161 Wis. 46 (1915); *Vennen v. New Dells Lumber Company*, 161 Wis. 370 (1915); IAIABC, *Proceedings*, in USBLS, *Bulletin No. 248* (1919): 251–57, 263–68; Warren, *Brush with Death*, 79–82; Rosner and Markowitz, *Deadly Dust*, 22–44; Sellers, *Hazards of the Job*, 107–40; Derickson, *Black Lung*, 36–38.

15. Bale, "Compensation Crisis," 519–20, 570–73, 582–83; C. Clark, *Radium Girls*, 141; Rosner and Markowitz, *Deadly Dust*, 84; C. Clark, "Managing an Epidemic," 19; E. S. Clemens, *People's Lobby*, 244; *Statutes and Amendments of California* (1915), Chapter 607, Section 12; (1917), Chapter 586, Section 3(4); *Laws of Wisconsin* (1919), Chapters 457, 668; IAIABC, *Proceedings*, in USBLS, *Bulletin No. 248* (1919): 264; L. A. Tarrell to Bloodgood, Kemper & Bloodgood, Attys., May 28, 1917, 17/C266 ICW; Witte to Rusener, November 14, 1917, 36/C793 ICW; *Wisconsin State Journal*, July 31, 1918, 1; ICW, *Wisconsin Labor Statistics*, no. 3 (March 1924): 1.

16. Bale, "Compensation Crisis," 560–68; *Laws of New York* (1920), Chapter 538, Article 2–A; (1922), Chapter 615, Articles 1, 2, 3; *Laws of Illinois* (1923), 351; Beckner, *Labor Legislation in Illinois*, 282; C. Clark, *Radium Girls*, 187–88; *Acts of Ohio* (1921), 181–85; (1923); OSFL, *Proceedings* (1917): 70.

17. Bale, "Compensation Crisis," 602–15; C. Clark, *Radium Girls*, 112; Rosner and Markowitz, *Deadly Dust*, 82–87; Derickson, *Black Lung*, 60–85; Emery Hayhurst, "Industry as a Source of Disease," paper delivered at Midwest Conference on Occupational Disease, 1933, 5–9, AALL Papers.

18. Warren, *Brush with Death*, 93–94; Sellers, *Hazards of the Job*, 145–47; Rosner and Markowitz, *Deadly Dust*, 53–65; C. Clark, *Radium Girls*, 1–4; Derickson, *Black Lung*, 55–56; Seager, "Barre, Vermont," 63–67; Stern, "Industrial Structure," 439–41; Nugent, "Fit for Work," 592–93; Glad, *History of Wisconsin*, 238–39.

19. Fred M. Wilcox, "Duty of the State to Include Disease of Occupation Under Workmen's Compensation," circa 1921, 12–13, 44/C1040 ICW; Fred M. Wilcox to Ethelbert Stewart, January 5, 1931, 42/C1031 ICW; Altmeyer to John L. Baer, February 9, 1931, 16/C244 ICW; ICW, *Wisconsin Labor Statistics*, no. 24 (March 14, 1930): 1; no. 30 (September 15, 1930): 3; Hayhurst, "Industry as a Source of Disease," 10; Bale, "Compensation Crisis," 519–20, 560–67, 573–74.

20. ICW, *Minutes,* October 20 and 21, 1919; ICW, *Biennial Reports* (1918–1920), 34–35; ICW, *Labor Laws of the State of Wisconsin and Orders of the Industrial Commission* (1938), 24–27; ICW, *General Orders on Sanitation* (1921), 3–4; IAIABC, *Proceedings,* in USBLS, *Bulletins No. 281* (1921): 411–30; *No. 304* (1922): 162–68; *No. 333* (1923): 151–57, 217–24; *No. 406* (1926): 62–76, 89–97; and *No. 432* (1927): 52–60.

21. ICW, *Heating and Ventilation Code* (1925), 5–7; ICW, *Labor Laws of the State of Wisconsin and Orders of the Industrial Commission* (1938), 30; ICW, *Biennial Reports* (1922–1924), 13–14; (1924–1926), 10–11; C. W. Keniston to (Photo Engraving Companies), November 5, 1919, typescript, "Minutes of Meeting of Photo Engravers," November 12, 1919; memorandum, Keniston to Industrial Commission, January 31, 1920, 67/C1704 ICW; Keniston to R. A. Sanborn, July 21, 1920, 78/C76.2 ICW; ICW, *Minutes,* February 3, 1919, and March 21, 1922; ICW, *Progress of Work Report* (January 1922), 2; (February 1922), 9–10; (March 1922), 11, in ICW General Archives; R. A. Small to Industrial Commission, May 21, 1923, and Hearing Transcripts, "Before the Industrial Commission," July 6, 1923, August 18, 1924, 40/C882.4 ICW.

22. Report of Legislative Representatives of the Brotherhood of Locomotive Engineers, Brotherhood of Locomotive Firemen and Enginemen, Brotherhood of Railway Trainmen, and Order of Railway Conductors, 1921, p. 29; Witte to Henry Killilea, May 25, 1921; C. M. & St. Paul General Manager to Witte, June 10, 1921; Witte to C. M. & St. Paul Railway Co., June 27, 1921; John Humphrey to Mr. (Robert McA.) Keown, July 6, 1921; J. A. Tyman to Martin Comford, January 24, 1923; C. F. Schweigert to R. G. Knutson, March 16, 1923; Knutson to Fred C. Schweigert, August 1, 1923, 400/C882.2 ICW; ICW, *General Orders on Spray Coating* (1924), 1–3; Smith, "Chronological Development," 158–60.

23. Howard L. Boyd, "Occupational Diseases," CIAC, *California Safety News* 1 (December 1917): 3–9; L. H. Duschak, "Lead Poisoning," CIAC, *California Safety News* 6 (October 1922): 10–11; *Statutes and Amendments of California* (1889), Chapter 5; (1921), Chapter 244; CSFL, *Proceedings* (1920): 26; (1922), 38, 40; (1923), 23, 24; (1926), 57; (1927), 64; (1928), 63; J. Andrews, *Administrative Labor Legislation,* 123–24; "Labor Law Administration, California, 1940," insert after p. 52, in AALL Papers; *Joseph Blanchard v. Industrial Accident Commission* 68 Cal. App. 65 (1924); Smith, "Chronological Development," 10–13.

24. OSFL, *Proceedings* (1924), 50; (1927), 109; (1930), 78; J. Andrews, *Administrative Labor Legislation,* 138–39; Stern, "Industrial Structure," 441; *Ewers, Admx. v. The Buckeye Clay Pot Company,* 29 Ohio App. 396 (1928).

25. IDL, *Annual Reports* (1918–1919), 56; (1920–1921), 101; (1921–1922), 48; (1922–1923), 95; (1923–1924), 122–23, 128; (1924–1925), 148–49, 155; (1925–1926), 81; (1926–1927), 69; (1928–1929), 59; Graebner, "Private Power," 27; *Jannusch v. Weber Brothers Metal Works,* 249 Ill. App. 1 (1928); *May v. Belleville Enameling & Stamping Co.,* 247 Ill. App. 275 (1928); C. Clark, *Radium Girls,* 187–89.

26. *Matter of Connelly v. Hunt Furniture Co.,* 240 N.Y. 83 (1925); *Emma H. Scheerens v. E. W. Edwards & Sons,* 133 N.Y. Misc. 616 (1929); NYSDL, *Annual Reports* (1919),

243–47; (1921), 206; (1923), 137, 140–43; (1925), 69, 75–77, 132–36, 140–65; NYSDL, *Proceedings of Industrial Safety Congresses* (1924): 194–95; (1925): 208–45; (1926): 53–81; NYSDL, "The Health of the Working Child," *Special Bulletin No. 134* (1924), and "Chronic Benzol Poisoning among Women Industrial Workers," *Special Bulletin No. 150* (1927); J. Andrews, *Administrative Labor Legislation*, 136–37.

27. Bale, "Compensation Crisis," 602; J. Andrews, *Administrative Labor Legislation*, 122–49; USBLS, "Workmen's Compensation Legislation of the United States and Canada," foldouts after p. 50; Hayhurst, "Industry as a Source of Disease," 10; ICW, *Wisconsin Labor Statistics*, no. 24 (March 1930): 6–8; no. 30 (September 1930): 3; Warren, *Brush with Death*, 96–99; Rosner and Markowitz, *Deadly Dust*, 24–73; Sellers, *Hazards of the Job*, 141–85; C. Clark, *Radium Girls*, 1–8, 77–164; IAIABC, *Proceedings*, in USBLS, *Bulletin No. 485* (1929): 125–56, 180–92; NYSFL, *Proceedings* (1929): 13–17; ICW, *Biennial Reports* (1928–1930), 4; ICW, *General Orders on Dusts, Fumes, Vapors, and Gases* (March 18, 1932), 5–7.

28. Jones, "Silicosis: An Occupational Hazard," 1–4; Bale, "Compensation Crisis," 666–68; Rosner and Markowitz, *Deadly Dust*, 4–5, 82–91; Sellers, *Hazards of the Job*, 203–5.

29. Voyta Wrabetz, typescript, "Occupational Diseases—Legislative Aspects," May 8, 1936, 15, 44/C1040 ICW; *The Milwaukee Journal*, February 18, 1940, Local section, p. 2; Press Release, August 27, 1932, 89/C448 ICW; *Wausau Daily Record-Herald*, September 1, 1961 (Souvenir Edition), 18; Rosner and Markowitz, *Deadly Dust*, 75–82; Sellers, *Hazards of the Job*, 189–90, 203–4; Crandall and Crandall, "Revisiting the Hawks Nest Tunnel Incident," 271; Nonet, *Administrative Justice*, 106; handwritten memo, O. A. F(ried) to Mr. (Voyta) Wrabetz, April 13, 1936, 65/C1666 ICW; *The Milwaukee Journal*, November 9, 1933, 7. See also S. Bruce Black, "Recent Experience in Casualty Insurance," 1933, 5–6; Thomas N. Barlett, "Workmen's Compensation Legislation: Occupational Diseases," 1935; and Frederick S. Kellogg, "Silicosis Claims—a New Problem in the Insurance Field," 1934, AALL Papers.

30. *Burns v. Industrial Commission*, 356 Ill. 602, 606–7 (1934); IAIABC, *Proceedings*, in USDLS, *Bulletin No. 2* (1935): 82; Employers Mutual to Wrabetz, April 10, 1936, 65/C1666 ICW; USBLS, "Problems of Workmen's Compensation Legislation," 61; C. Clark, *Radium Girls*, 189.

31. *Laws of Wisconsin* (1911), Chapter 50, Section 2394-4(2); Wrabetz to Tom Duncan, April 19, 1935, 13/C181 ICW; *Wisconsin Granite Company v. Industrial Commission*, 208 Wis. 270 (1932); *Wisconsin State Journal*, April 6, 1932, 2.

32. Glad, *History of Wisconsin*, 378–82; *Laws of Wisconsin* (1931), Chapter 403, Section 102.01(2); Wrabetz to Marshall Dawson, February 6, 1942, 25/C412.1 ICW; *North End Foundry Co. v. Industrial Commission*, 217 Wis. 363 (1935).

33. Wisconsin, *Assembly Bills* (1933), 623A; Wisconsin, *Assembly Journal* (1933), 2022–24; *The Milwaukee Journal*, May 29, 1933, 6, June 12, 1933, 1; *The Capital Times*, May 31, 1933, 14; WSFL, *Proceedings* (1933): 125; transcript, interview with Ben Kuechle, 1976, 13, Wausau Archives; Dan Hagge, interview by the author, May 19, 1981,

Wausau, Wisconsin; Wrabetz to Duncan, April 19, 1935, 13/C181 ICW; Rosner and Markowitz, *Deadly Dust,* 5.

34. *Laws of Wisconsin* (1933), Chapters 314, 402, Section 102.01(2); A. T. Flint to Mine Safety Appliances, June 24, 1935, 17/C258 ICW; Rosner and Markowitz, *Deadly Dust,* 90–91.

35. *General Laws of Alabama* (1935), 911–13; (1939), 235, 244–45; *American Mutual Liability Insurance Co. v. Agricola Furnace Co.,* 236 Ala. 535 (1938); *Associated Indemnity Corporation v. Industrial Accident Commission* 124 Cal. App. 378 (1932); NYSFL, *Proceedings* (1925): 143; (1928): 117; OSFL, *Proceedings* (1930): 78; (1934): 72; Rosner and Markowitz, *Deadly Dust,* 96–98; Bale, "Compensation Crisis," 666–68; Crandall and Crandall, "Revisiting the Hawks Nest Tunnel Incident," 270–72; Cherniack, *Hawk's Nest Incident,* 81–88.

36. *Laws of New York* (1935), Chapter 254; (1936), Chapters 887, 888, 889; NYSFL, *Proceedings* (1935): 65–67; (1936): 59–61, 80–82, 105–6; Rosner and Markowitz, *Deadly Dust,* 91–95.

37. *Jannusch v. Weber Bros. Metal Works,* 249 Ill. App. 1 (1928); *May v. Belleville Enameling and Stamping Co.,* 247 Ill. 275 (1928); *The First National Bank of Ottawa v. The Wedron Silica Co.,* 351 Ill. 560 (1933); *Madison v. Wedron Silica Company,* 352 Ill. 60 (1933); *Parks v. Libby-Owens-Ford Glass Co.,* 360 Ill. 130 (1935); *Boshuizen v. Thompson & Taylor Co.,* 360 Ill. 160 (1935); *Vallat v. Radium Dial Co.,* 360 Ill. 407 (1935); *Laws of Illinois* (1936), 34, 40–42; (1937), 563. See also C. Clark, *Radium Girls,* 188–90.

38. *Acts of Ohio* (1937), 268; (1939), 425; OSFL, *Proceedings* (1930): 78; (1934): 72; (1935): 99; Shaw, "Practice and Procedure," 184–85, 190–91.

39. USBLS, "Problems of Workmen's Compensation Legislation," 58–59, 198; Seager, "Barre, Vermont," 76–79; Rosner and Markowitz, *Deadly Dust,* 135–38; Derickson, *Black Lung,* 89–111.

40. USBLS, "Problems of Workmen's Compensation Legislation," 59–64; "Silicosis Compensation Blunder," 166–67; Rosner and Markowitz, *Deadly Dust,* 95.

41. NYSFL, *Proceedings* (1935): 66; Rosner and Markowitz, *Deadly Dust,* 82–168; Graebner, "Private Power," 26–53; Sellers, *Hazards of the Job,* 196–224; C. Clark, *Radium Girls,* 149–69, 189–93; Derickson, *Black Lung,* 89–105, Crandall and Crandall, "Revisiting the Hawks Nest Tunnel Incident," 273–76.

42. ICW, *Labor Laws of the State of Wisconsin and Orders of the Industrial Commission* (1938), 24–25; ICW, *Biennial Reports* (1926–1928), 6–7; (1928–1930), 4, 8; (1930–1932), 8; (1932–1934), 15; ICW, *General Orders on Dusts, Fumes, Vapors and Gases* (March 1932), 5–7; WMA, *Weekly Bulletin,* March 23, 1931 in WMA Papers.

43. ICW, *General Orders on Dust, Vapors, Fumes, and Gases* (March 1932), 5–7, 11; Rosner and Markowitz, *Deadly Dust,* 112–13, 155; Sellers, *Hazards of the Job,* 196–201; Graebner, "Private Power," 53–55.

44. E. G. Meiter, "Detection and Control of Occupational Diseases," *Coverage* (July 1935): 1–3, in Wausau Archives; *The Milwaukee Journal,* February 18, 1940, 2; ICW,

Proceedings of Conference Concerning Effects of Dust upon the Respiratory System, November 1932 (1933); typescript, "Report of the Interim Committee Created under Joint Resolution 155A of the 1933 Regular Session to Study Labor Conditions in the Stone Cutting Industry with Particular Reference to the Disease of Silicosis," 42/ C1031 ICW; *Wausau Daily Record-Herald,* September 1, 1961 (Souvenir Edition), 18; typescript, "Summary of Special Committee on Silicosis," 1–10, 66/C1666 ICW.

45. Reuel C. Stratton (Travelers), "The Engineering Control of Occupational Diseases by Plant Equipment and Operation," May 12–13, 1932, and Metropolitan Life Insurance Company, "Silicosis," circa 1934, AALL Papers; David Stuart Beyer (Liberty Mutual), "The Use of Exhaust Systems for the Protection of Workers Exposed to Dust Vapors and Fumes," in IAIABC, *Proceedings,* in USDLS, *Bulletin No. 2* (1935): 242–51.

46. *Laws of Wisconsin* (1935), Chapter 488; *The Milwaukee Journal,* February 18, 1940, Local section, p. 2; Nelson to Henry Paulus, October 25, 1935, 17/C256 ICW; Harry A. Nelson, typescript, "Silicosis Problem Solved in Wisconsin," circa 1936, 4–9, 44/C1040 ICW; ICW, *News Releases, 1917–1940,* February 3, 1938; ICW, *Biennial Report* (1936–1938), 8; *Wisconsin Compensation Rating & Inspection Bureau v. Mortensen,* 227 Wis. 335 (1938).

47. ICW, *News Releases, 1917–1940,* February 3, 1938; Paul A. Brehm, "Proposed Cooperative Program of the Industrial Hygiene Unit with the Industrial Commission," circa 1937, 41/C882.11 ICW; "Labor Law Administration, California, 1940," 31; USBLS, "Labor Laws and Their Administration" (1940), 32–33; USBLS, "Labor Laws and Their Administration," (1941), 151–52; Rosner and Markowitz, "Research or Advocacy?" in *Dying for Work,* edited by Rosner and Markowitz, 83–99.

48. Typescript, Summary of Testimony of the Special Committee on Silicosis, 3–7, 66/C1666 ICW; "Good Housecleaning at Falks"; Wrabetz, "Occupational Diseases— Legislative Aspects," 11–13 (see n. 29, this chapter); Harry A. Nelson, "Silicosis Problem Solved," 3–5; ICW, *Biennial Report* (1934–1936), 29; WMA, *Weekly Bulletin,* May 2, 1938; *The Milwaukee Journal,* February 18, 1940, Local section, p. 2; Wrabetz to Thomas P. Kearns, April 1, 1943, 15/C186 ICW; Kuechle, "Occupational Disease Liabilities," 18.

49. "Labor Law Administration, California, 1940," 18; *Laws of Illinois* (1935–1936), 34; *Laws of New York* (1936), Chapter 887, Section 65(2); Benjamin, *Administrative Adjudications,* 38–39; Greenburg, "Industrial Hygiene and Worker's Health," 30. See also Rosner and Markowitz, *Deadly Dust,* 98–162, 178–81.

50. Hayhurst, "Occupational Disease Considerations," 90–91.

51. *Laws of Wisconsin* (1937), Chapter 180, Section 3.

52. Edwin W. DeLeon, "The Relation of Medical Examination of Employees to Insurance under Workmen's Compensation Laws," 1914, AALL Papers; Beck to J. W. Schereschewsky, August 7, 1916, 15/C183 ICW; S. J. Williams to Wilcox, July 16, 1918, 90/C589 ICW; Wilcox to Williams, July 25, 1918, 90/C589 ICW; Beck to Rupert Blue, September 9, 1916, 15/C183 ICW; ICO, *Report* (1913–1914), 47; ICO, "Physical

Examination of Wage Earners in Ohio in 1914," *Report No. 18* (1915); Sellers, *Hazards of the Job,* 107–15; Nugent, "Fit for Work," 578–91.

53. Sellers, *Hazards of the Job,* 111–12, 118–20; Nugent, "Fit for Work," 580, 595; Rosner and Markowitz, *Deadly Dust,* 45; Warren, *Brush with Death,* 102–4; WSFL, *Proceedings* (1929): 55–56; OSFL, *Proceedings* (1924): 48; NYSFL, *Proceedings* (1928): 157–58.

54. *Laws of Wisconsin* (1925), Chapter 399; Altmeyer to John B. Andrews, April 26, 1932, 9/C86 ICW.

55. Sellers, *Hazards of the Job,* 200–218; Rosner and Markowitz, *Deadly Dust,* 112–14, 155; C. Clark, *Radium Girls,* 149–64.

56. Nelson to Paulus, May 5, 1933, 17/C256 ICW; Wrabetz to Lester C. Weisse, February 25, 1936, 65/C1666 ICW; A. T. Flint to Harry Lippart, June 29, 1937, 24/C397 ICW; Nelson to Seton S. Williams, February 12, 1941, 65/C1666 ICW; Wrabetz to (N.J. State) Senator I. Grant Scott, October 19, 1943, 65/C1666 ICW; *Laws of Wisconsin* (1935), Chapter 488.

57. ICW, *Physical Examinations of Industrial Workers, Report of the Joint Advisory Committee* (Madison, 1939), 5, 7, 21–22; typescript, "History of Proceedings re Medical Examinations of Employes," circa January 1939, 76/C24 ICW; Voyta Wrabetz, typescript, "Silicosis—Legislation and Administration," June 19, 1939, 71/C1929 ICW; Wrabetz to Andrews, February 21, 1938, 76/C24 ICW; Nelson to C. R. Alt, January 23, 1940, 75/C24 ICW; circular letter (and responses), Nelson to All Self-Insurers under Compensation Act, May 15, 1937, 50/C1130.8 ICW.

58. ICW, *Physical Examinations,* 9–21; Paul A. Brehm, typescript, "Health in Industry," May 19, 1939, 2–5, 44/C1040 ICW; Nelson to C. W. Dickey, December 18, 1939; Nelson to Kimberly-Clark Corporation, January 2, 1940; Nelson to Alt, January 11 and 23, 1940, 75/C24 ICW; Nelson to Gus Kneuse, May 12, 1941, 2/C24 ICW.

59. ICW, *Physical Examinations,* 22; Nelson to B. E. Kuechle, May 8, 1941; Nelson to William MacInnes, June 2, 1943, 2/C24 ICW.

60. *Wisconsin Medical Journal* 39 (1940): 865; 40 (1941): 53–54; Ralph M. Carter to Wrabetz, October 28, 1935, 105/C817 ICW; M. J. Reuter to Kuechle, August 28, 1939; Chester W. Long (M.D.) to Kuechle, August 26, 1939; George Crownhart to Nelson, October 13, 1939; Nelson to J. G. Crownhart, October 25, 1939; Kuechle to Wrabetz, January 18, 1940; Kuechle to Nelson, January 24, 1940, 75/C24 ICW. On doctors' role in medical screening, see Nugent, "Fit for Work."

61. Kuechle to Harold (Falk), November 15, 1937; Kuechle to Joseph A. Padway, December 16, 1937; Kuechle to Wrabetz, December 23 and 31, 1937, 75/C24 ICW; Kuechle to John Enright, January 13, 1938; Wrabetz to Herman Seide, January 10, 1938; Kuechle to Wrabetz, January 24, 1938, 65/C1666 ICW; memorandum for file, Conference between Mr. Wrabetz, Mr. Nelson, and Mr. Kuechle, by Nelson, May 11, 1938, 76/C24 ICW; Kuechle to Wrabetz, October 13, 1939, May 18, 1940; Wrabetz to J. F. Friedrich, June 27, 1940; Charles Elsby to Kuechle, August 26, 1940; Wrabetz to Friedrich, November 16, 1940; Wrabetz to Kuechle, October 29, 1940; Joe (Padway)

to Kuechle, November 9, 1940; Kuechle to Wrabetz, November 18, 1940, 75/C24 ICW; interoffice memoranda (of Employers Mutual); Elsby to Kuechle, August 21 and 28, 1940, 75/C24 ICW; Kuechle to Wrabetz, February 12, 1941, December 3, 1942, 2/C24 ICW.

62. WMA, *Weekly Bulletin,* August 12, 1939, WMA Papers; circular, Nelson to All Self-Insurers under the Workmen's Compensation Act, November 20, 1939, 75/C24 ICW; J. H. Wendel to Nelson, December 9, 1939; Jessel S. Whyte to Nelson, December 13, 1939; Alt to Industrial Commission of Wisconsin, January 17, 1940; Alt to Nelson, January 26, 1940, 75/C24 ICW; Work Sheet, November 20, 1939, 75/C24 ICW.

63. Kuechle to Wrabetz, August 29, 1940, 75/C24 ICW; Nelson to Kuechle, May 8, 1941, 2/C24 ICW.

64. USBLS, "Problems of Workmen's Compensation Legislation," 57–58, 198; Hayhurst, "Occupational Disease Considerations," 90–93; "Silicosis Compensation Blunder," 166–68; Derickson, *Black Lung,* 100–105; Rosner and Markowitz, *Deadly Dust,* 95.

65. Falasz, "Factory Inspection and Safety," 134–39; "Industrial Hygiene Bill"; Rosner and Markowitz, "Research or Advocacy?" 87–99. On the fragmented character of industrial health programs, see USBLS, "Federal and State Agencies Concerned with Health."

Epilogue: The Road to OSHA

1. Curran, *Dead Laws for Dead Men,* 90–129; Whiteside, *Regulating Danger,* 175–95; Derickson, *Black Lung,* 112–82.

2. Markowitz and Rosner, *Deceit and Denial,* 156–64; Ashford, *Crisis in the Workplace,* 52–57; Kelman, "Occupational Safety and Health Administration," 239–41; MacLaury, "Job Safety Law," 21–24; C. Noble, *Liberalism at Work,* 69–91; Mendeloff, *Regulating Safety,* 17–22; W. Berman, *America's Right Turn,* 12–13.

3. Kelman, "Occupational Safety and Health Administration," 242; C. Noble, *Liberalism at Work,* 69–86; Vogel, "'New' Social Regulation," 157–59, 169–75; Eisner, "Discovering Patterns," 175–76; W. Berman, *America's Right Turn,* 11–12.

4. Mintz, *OSHA,* 450–63, 717, 720; Morey, "General Duty Clause"; C. Noble, *Liberalism at Work,* 94; Ashford, *Crisis in the Workplace,* 142–43; Ashford and Caldart, *Technology,* 183.

5. Vogel, "'New' Social Regulation," 172–73; Eisner, "Discovering Patterns," 177–78; C. Noble, *Liberalism at Work,* 95–96; R. Stewart, "American Administrative Law," 1711–56; Ashford, *Crisis in the Workplace,* 144–46, 150–51; Kelman, *Regulating America,* 9–16; Mintz, *OSHA,* 63–65; Bardach and Kagan, *Going by the Book,* 44–57.

6. Ashford, *Crisis in the Workplace,* 147–52; C. Noble, *Liberalism at Work,* 95; Mintz, *OSHA,* 9, 337–64, 513, 539, 558–64, 725–34; Barnako, "Enforcing Job Safety," 38; Kelman, *Regulating America,* 176–82; Seymour, "Federal Role," 28; Mendeloff, *Regulating Safety,* 18–19.

7. Szasz, "Industrial Resistance," 106–13; Markowitz and Rosner, *Deceit and Denial,* 8, 226, 231; W. Berman, *America's Right Turn,* 69, 98–99; Schaller, *Right Turn,* 30–34, 128; Berkowitz, *Something Happened,* 84.

8. Markowitz and Rosner, *Deceit and Denial,* 119–34, 195–99, 219–26; Wahl and Gunkel, "Due Process," 591–612; Szasz, "Industrial Resistance," 109–11; Mintz, *OSHA,* 39–59, 61–65, 82–86; Kelman, *Regulating America,* 18–93; C. Noble, *Liberalism at Work,* 96, 139–67; Schaller, *Right Turn,* 34, 38–39.

9. Markowitz and Rosner, *Deceit and Denial,* 220–25; Ashford and Caldart, *Technology,* 117–23, 125–47; Mintz, *OSHA,* 68, 81, 168; C. Noble, *Liberalism at Work,* 106–7, 168–70; Kelman, *Regulating America,* 70–75; Shapiro, "Supreme Court's 'Return,'" 123–26.

10. Mintz, *OSHA,* 7–8, 718, 735–36; Ashford, *Crisis in the Workplace,* 209–11; C. Noble, *Liberalism at Work,* 55–57, 59–66; Seymour, "Federal Role," 28–29.

11. Mintz, *OSHA,* 617–63, 735; Seymour, "Federal Role," 29–30; Ashford, *Crisis in the Workplace,* 210–31; Simson, "Occupational Safety and Health Act," 602; *OSHR* 4, no. 43 (March 27, 1975): 1407–8; 6, no. 2 (June 10, 1976): 40–41.

12. Vogel, "'New' Social Regulation," 165–79; C. Noble, *Liberalism at Work,* 101–5; Mintz, *OSHA,* 619; *OSHR* 4, no. 35 (January 30, 1975): 1042; 4, no. 40 (March 6, 1975); 1319–20; 5, no. 10 (August 7, 1975): 318.

13. *The Capital Times,* April 15, 1975, 1, May 8, 1975, 30; *The Milwaukee Sentinel,* May 8, 1975, 12; *The Milwaukee Journal,* March 9, 1974, 24; *OSHR* 4, no. 51 (May 22, 1975): 1699.

14. Ashford, *Crisis in the Workplace,* 57–61; Vogel, "'New' Social Regulation," 169; *The Capital Times,* April 15, 1975, 1, July 25, 1975, 36; *Wisconsin State Journal,* September, 13, 1973, 4, March 21, 1974, Section 4, 4, June 25, 1974, 8, December 19, 1974, 2; *The Milwaukee Journal,* March 15, 1974, Accent section, 12.

15. *Wisconsin State Journal,* September 13, 1973, 4, June 25, 1974, 8; *The Milwaukee Sentinel,* October 6, 1973, Part 2, 27, May 8, 1975, 12, May 13, 1975, 8; *The Milwaukee Journal,* March 15, 1974, Accent section 12; *The Capital Times,* April 15, 1975, 1, May 8, 1975, 30; *New York Times,* September 21, 1975, Section 3, 1; *Chicago Tribune,* May 6, 1973, Section 2, 4; *OSHR* 4, no. 51 (May 22, 1975): 1699; 4, no. 39 (February 27, 1975): 1277; 4, no. 46 (April 17, 1975): 1505; Ashford, *Crisis in the Workplace,* 52–61, 228–29; Sheehan, "Job Safety Plans"; D. Berman, *Death on the Job,* 117–20, 168–71.

16. *Wisconsin State Journal,* September 13, 1973, 4; *OSHR* 4, no. 40 (March 6, 1975): 1319; 4, no. 51 (May 22, 1975): 1699.

17. Ashford, *Crisis in the Workplace,* 398–402; "Occupational Safety and Health Act," *Hearings before the United States Senate,* 257–324.

18. *OSHR* 4, no. 38 (February 20, 1975): 1260–61.

19. *OSHR* 5, no. 17 (September 25, 1975): 515; 5, no. 38 (February 19, 1976): 1268; Seymour, "Federal Role," 30.

20. *OSHR* 4, no. 28 (December 12, 1974): 820–21; 4, no. 31 (January 2, 1975): 906; 4, no. 36 (February 6, 1975): 1188; 4, no. 39 (February 27, 1975); 1277; 4, no. 40 (March 6,

1975): 1321–22; 5, no. 2 (June 12, 1975): 45; *Chicago Tribune,* November 15, 1973, sec. 1, p. 26; December 4, 1974, sec. 4, p. 17; December 22, 1974, sec. 1, p. 27; December 24, 1974, sec. 3, p. 2; January 30, 1975, sec. 3, p. 2; *The Wall Street Journal,* December 31, 1974, 1.

21. *New York Times,* February 16, 1975, 26; March 14, 1975, 77; March 16, 1975, sec. 1, 51, September 21, 1975, sec. 3, 1; *OSHR* 4, no. 35 (January 30, 1975): 1042, 1048–49; 4, no. 39 (February 27, 1975): 1277; 4, no. 41 (March 13, 1975): 1358; 4, no. 42 (March 20, 1975): 1385–86.

22. *San Francisco Chronicle,* January 13, 1972, 8; February 3, 1972, 9; February 11, 1972, 14; February 23, 1972, 5; January 5, 1973, 41; January 24, 1973, 57; June 5, 1973, 41; September 14, 1973, 41; *Los Angeles Times,* January 14, 1972, pt. 1, p. 3; February 3, 1972, pt. 1, p. 3; February 11, 1972, pt. 1, p. 3; February 23, 1972, pt. 1, p. 3; June 5, 1973, pt. 1, p. 24; September 12, 1973, pt. 1, p. 2; "The California Occupational Safety and Health Act of 1973," 914–18; Dallek, *Ronald Reagan,* 39–50; Bean, *California: An Interpretive History,* 559–74.

23. "California Occupational Safety and Health Act," 918–35; Hunter, "California Occupational Safety and Health Act," 304–12; *OSHR* 4, no. 35 (January 30, 1975): 1048, 1079–80; 4, no. 37 (February 13, 1975): 1231–32; 4, no. 48 (May 1, 1975): 1579–81; 5, no. 3 (June 19, 1975): 86; Mintz, *OSHA,* 659–63.

24. Mintz, *OSHA,* 618; Ashford, *Crisis in the Workplace,* 227; *OSHR* 5, no. 6 (July 10, 1975): 182–83.

25. Bardach and Kagan, *Going by the Book,* 79–89, 102–8.

26. Mintz, *OSHA,* 617–18, 627, 635–41, 644–63; Ashford, *Crisis in the Workplace,* 214–25.

27. Kagan, *Adversarial Legalism,* 30–54; Kelman, *Regulating America,* 182, 191; C. Noble, *Liberalism at Work,* 97–99, 179–81, 196–98; Schaller, *Right Turn,* 198; Mintz, *OSHA,* 339–40; Szasz, "Industrial Resistance," 112, 114; Curran, *Dead Law for Dead Men,* 159–65. Stories about overstretched OSHA staff abound. See, for instance, *Chicago Tribune,* May 6, 1973, sec. 2, p. 4.

28. *Los Angeles Times,* May 10, 1973, pt. 7, p. 2; Mintz, *OSHA,* 345–48; Mendeloff, *Regulating Safety,* 102–16; C. Noble, *Liberalism at Work,* 201–5.

Bibliography

Not listed here are court cases, newspaper articles, and interviews with the author, which receive full citations in notes. The Index mentions all court cases discussed in the text. Not listed here also are specific articles and speeches drawn from safety conferences, organizational proceedings, and government magazines, reports, and bulletins. Such articles and speeches are cited in the notes, with reference to their respective original sources listed below.

Archives and Organizational Publications

American Association for Labor Legislation. Papers. Labor-Management Documentation Center. Martin B. Catherwood Library, Cornell University, Ithaca, New York.

Beck, Joseph D., Papers. Archives Division, State Historical Society of Wisconsin, Madison.

Budget Report to Governor. 1925. In *Clippings: Labor Regulation and Regulatory Agencies in Wisconsin.* Legislative Reference Library of Wisconsin, Madison.

Building Officials Conference. *[Third Annual] Proceedings.* Madison: Democrat Printing, 1917.

California Bar Association. *The State Bar Journal.* 1931–1938.

California State Federation of Labor. *Proceedings.* 1915–1936.

Chicago Bar Association Record. 1938–1941.

Crownhart, Charles H., Family Papers. Archives Division, State Historical Society of Wisconsin, Madison.

Factory Inspector Reports of the Bureau of Labor and Industrial Statistics (July–October 1904). Archives Division, State Historical Society of Wisconsin, Madison.

Flower, Frank A., Papers. Archives Division, State Historical Society of Wisconsin, Madison.

Industrial Commission of Wisconsin. Government Documents Division, State Historical Society of Wisconsin, Madison.

———. Numerical Subject "C" Files. Government Documents Division, State Historical Society of Wisconsin, Madison.

International Association of Industrial Accident Boards and Commissions. *Proceedings*. U.S. Bureau of Labor Statistics. *Bulletins*. 1917–1932. U.S. Division of Labor Standards. *Bulletin No. 2* (1935).

La Follette, Robert M., Sr., Papers. The La Follette Family Collection. Manuscript Division, Library of Congress, Washington, D.C.

McCarthy, Charles H., Papers. Archives Division, State Historical Society of Wisconsin, Madison.

Merchants and Manufacturers Association of Milwaukee. *Civics and Commerce*. 1909–1918.

Milwaukee Federated Trades Council. *Proceedings*. 1907–1910.

New York State Federation of Labor. *Proceedings*. 1913–1924.

Ohio State Bar Association Report. 1929–1943.

Ohio State Federation of Labor. *Proceedings*. 1917–1937.

State Bar Association of Wisconsin. *Bulletins*. 1930–1940.

Wausau Insurance Companies Archives. Wausau, Wisconsin.

Wisconsin Legislative Reference Library. *Testimony, Proceedings, and Report of the Wisconsin Legislature's Special Joint Committee on Industrial Insurance (1909–1911)*.

Wisconsin Manufacturers' Association Papers. Archives Division, State Historical Society of Wisconsin, Madison.

Wisconsin State Federation of Labor. *Proceedings*. 1905–1941.

Wisconsin State Medical Society. *Wisconsin Medical Journal* 39–40 (1940–1941).

Public Records

ALABAMA

General Laws of Alabama. 1889–1975.

CALIFORNIA

Industrial Accident Commission of California. *California Safety News*. 1917–1925.

———. *Reports*. 1916–1920.

Statutes and Amendments of California. 1889–1947.

ILLINOIS

Illinois Department of Labor. *Annual Reports*. 1918–1930.

Laws of Illinois. 1911–1975.

MASSACHUSETTS

Laws of Massachusetts. 1913.

NEW YORK

Benjamin, Robert M. *Administrative Adjudications in the State of New York.* Vol. 5, *Supplemental Report on the Board of Standards and Appeals.* Albany: New York State, 1942.

Laws of New York. 1913–1975.

New York State Department of Labor. *Annual Reports of Commissioner of Labor.* 1902–1910.

———. *Annual Reports of the Industrial Commission.* 1915–1919.

———. *Annual Reports of the Industrial Commissioner.* 1921–1925.

———. *Proceedings of Industrial Safety Congresses.* 1920–1928.

———. *Special Bulletins.* 1920–1927.

New York State Industrial Commission. *Proceedings of Industrial Safety Congresses.* 1916–1919.

———. *Special Bulletin No. 75.* March 1916.

OHIO

Acts of Ohio. 1884–1943.

Constitution of the State of Ohio.

Department of Industrial Relations. *Annual Reports.* 1921–1933.

Industrial Commission of Ohio. *Annual Reports.* 1913–1918.

———. *Bulletins.* 1915.

———. "Inspection of Workshops, Factories, and Public Buildings in Ohio." *Report No. 26.* 1915.

———. *The Ohio State Insurance Manual.* 1915.

———. "Physical Examination of Wage Earners in Ohio in 1914." *Report No. 18.* 1915.

———. Division of Safety and Hygiene. *Annual Statistical Reports.* 1934–1940.

PENNSYLVANIA

Laws of Pennsylvania. 1913.

WISCONSIN

Assembly Bills. 1878–1937.

Assembly Journal. 1883–1933.

Blue Book. 1911.

Bureau of Labor and Industrial Statistics. *Biennial Reports.* 1884–1911.

Civil Service Commission. *State Civil Service: Manual of Competitive Examinations.* 1913.

Compensation Insurance Board. *Insurance Experience under Compensation Act.* January 3, 1919.

———. *1930 Report of the Compensation Insurance Board.*

Industrial Commission of Wisconsin. *Biennial Reports.* 1918–1942.

———. *Building Code.* 1914.

———. *Bulletin: Workmen's Compensation Insurance.* 1915.

———. *Bulletins.* 1911–1915.

———. *Code of Boiler Rules.* 1914.

———. *First Aid: A Handbook for Use in Shops.* 1915.

———. *General Orders on Dusts, Fumes, Vapors, and Gases.* 1932.

———. *General Orders on Safety.* 1912–1914.

———. *General Orders on Safety in Construction.* 1933.

———. *General Orders on Sanitation.* 1921.

———. *General Orders on Spray Coating.* 1924, 1925, 1929, 1939.

———. *Hearing on Factory Rules.* January 2, 1912.

———. *Heating and Ventilation Code.* 1925.

———. *Labor Laws of the State of Wisconsin and Orders of the Industrial Commission.* 1938.

———. *Minutes.* Department of Industry, Labor, and Human Relations. Madison, Wisconsin.

———. *News Releases, 1917–1940.*

———. *Physical Examinations of Industrial Workers: Report of the Joint Advisory Committee.* 1939.

———. *Proceedings of Conference Concerning Effects of Dust upon the Respiratory System, November 1932.* 1933.

———. *Progress of Work Report.* January 1922.

———. *Report on Allied Functions.* 1912–1918.

———. *State Electrical Code.* 1917, 1922, 1924, 1930, 1934.

———. *Wisconsin Safety Review.* 1918–1923.

———. *Workmen's Compensation: Annual Reports.* 1913–1940.

Laws of Wisconsin. 1874–1943.

Legislative Journals. 1937 (for the 1937 special session).

Revised Statutes. 1878.

Sanborn, Arthur L., and John B. Sanborn, eds., and annotators. *Supplement to the Wisconsin Statutes of 1898.* Chicago: Callaghan, 1906.

Senate Bills. 1911–1937.

Senate Journal. 1911–1939.

Wisconsin Labor Statistics. 1921–1932.

UNITED STATES

Barnako, Frank R. "Enforcing Job Safety: A Managerial View." *Monthly Labor Review* 98 (March 1975): 36–39.

Falasz, John M. "Factory Inspection and Safety." *Bulletin No. 690* (U.S. Bureau of Labor Statistics) (1941): 129–44.

Greenburg, Leonard. "Industrial Hygiene and Worker's Health." *Bulletin No. 678* (U.S. Bureau of Labor Statistics) (1940): 24–42.

Hoffman, Frederick. "Industrial Accidents." *Bulletin of the Bureau of Labor* (U.S. Department of Commerce and Labor) 78 (1908): 417–65.

"Industrial Accidents: Industrial Injuries in the United States during 1939." *Monthly Labor Review* 51 (1941): 86–108.

"Industrial Accidents in Manufacturing Industries, 1926–1932." *Monthly Labor Review* 37 (1933): 1388–1398.

MacLaury, Judson. "The Job Safety Law of 1970: Its Passage Was Perilous." *Monthly Labor Review* 104 (March 1981): 18–24.

"Occupational Safety and Health Act, 1970." *Hearings before the Subcommittee of the Committee on Labor and Public Welfare, United States Senate, Ninety-first Congress.* Washington, D.C., 1970.

Seymour, Sally. "The Federal Role in Job Safety and Health: Forging a Partnership with the States." *Monthly Labor Review* 96 (August 1973): 28–55.

Sheehan, John J. "OSHA and State Job Safety Plans." *Monthly Labor Review* 97 (April 1974): 44–46.

Smith, Florence P. "Chronological Development of Labor Legislation for Women in the United States." *Bulletin of the Women's Bureau* (U.S. Department of Labor), no. 66–II (1932).

U.S. Bureau of Labor Statistics. "Federal and State Agencies Concerned with Problems of Industrial Health." *Bulletin No. 694* (1942): 351.

———. "Labor Laws and Their Administration." *Bulletin No. 678* (1940).

———. "Labor Laws and Their Administration." *Bulletin No. 690* (1941).

———. "Problems of Workmen's Compensation Legislation in the United States and Canada." *Bulletin No. 672* (1940).

———. "Safety Codes and Standard Safety Practices." *Bulletin No. 616* (May 1936): 306–9.

———. "Statistics of Industrial Accidents in the United States to the End of 1927." *Bulletin No. 490* (August 1929).

———. "Workmen's Compensation Legislation of the United States and Canada as of January 1, 1929." *Bulletin No. 496* (1929).

U.S. Bureau of the Census. *United States Census.* Washington, D.C., 1880–1940.

U.S. Department of Labor, Division of Labor Statistics. *National Conference on Silicosis.* Washington, 1937.

U.S. Department of Labor, Occupational Safety and Health Administration. *Occupational Safety and Health Reporter.* 1975–1976.

U.S. Division of Labor Standards. *Bulletins.* 1935.

Willoughby, W. F. "Inspection of Factories and Workshops in the United States." *Bulletin of the Department of Labor,* no. 12 (1897): 549–68.

Other Sources

Aikens, Andrew J., and Lewis A. Proctor. *Men of Progress: Wisconsin.* Milwaukee: Evening Wisconsin, 1897.

Alchon, Guy. *The Invisible Hand of Planning: Capitalism, Social Science, and the State in the 1920s.* Princeton: Princeton University Press, 1985.

Aldrich, Mark. *Death Rode the Rails: American Railroad Accidents and Safety, 1828–1965*. Baltimore: Johns Hopkins University Press, 2006.

———. *Safety First: Technology, Labor, and Business in the Building of American Work Safety, 1870–1939*. Baltimore: Johns Hopkins University Press, 1997.

Altmeyer, Arthur J. *The Industrial Commission of Wisconsin: A Case Study in Labor Law Administration*. Madison: University of Wisconsin Press, 1932.

Andrews, Irene Osgood. "The New Spirit in Factory Inspection." *Survey* 29 (December 21, 1912): 355–59.

Andrews, John B. *Administrative Labor Legislation: A Study of American Experience in the Delegation of Legislative Power*. New York: Harper and Brothers, 1936.

———. *Labor Laws in Action*. New York: Harper and Brothers, 1938.

Andrews, John, and Irene Andrews. "Scientific Standards in Labor Legislation." *American Labor Legislation Review* 1 (December 1911): 123–34.

"Another State Gives 'Run Around' to Silicosis Victims." *American Labor Legislation Review* 30 (June 1940): 64.

"Are Accidents Increasing?" *American Labor Legislation Review* 18 (1928): 239–41.

Asher, Robert. "Business and Workers' Welfare in the Progressive Era: Workmen's Compensation Reform in Massachusetts, 1880–1911." *Business History Review* 43 (1969): 452–75.

———. "Connecticut's First Workmen's Compensation Law." *Connecticut History* (1991): 25–50.

———. "Industrial Safety and Labor Relations in the United States, 1865–1917." In *Life and Labor: Dimensions in American Working-Class History*, edited by Charles Stephenson and Robert Asher, 115–30. Albany: State University of New York Press, 1986.

———. "The 1911 Wisconsin Workmen's Compensation Law: A Study in Conservative Labor Reform." *Wisconsin Magazine of History* 57 (Winter 1973–1974): 123–40.

———. "Radicalism and Reform: State Insurance of Workmen's Compensation in Minnesota, 1910–1933." *Labor History* 14 (1973): 19–41.

Ashford, Nicholas A. *Crisis in the Workplace: Occupational Disease and Injury*. Cambridge: MIT Press, 1976.

Ashford, Nicholas A., and Charles C. Caldart. *Technology, Law, and the Working Environment*. New York: Van Nostrand Reinhold, 1991.

Auerbach, Jerold S. *Unequal Justice: Lawyers and Social Change in Modern America*. New York: Oxford University Press, 1976.

Baer, Judith A. *The Chains of Protection: The Judicial Response to Women's Labor Legislation*. Westport, Conn.: Greenwood Press, 1978.

Bailey, Mark Warren. *Guardians of the Moral Order: The Legal Philosophy of the Supreme Court*. DeKalb: Northern Illinois University Press, 2004.

Baker, Tom. "Risk, Insurance, and the Social Construction of Responsibility." In *Embracing Risk: The Changing Culture of Insurance and Responsibility*, edited by Tom Baker and Jonathan Simon, 33–45. Chicago: University of Chicago Press, 2002.

Baker, Tom, and Jonathan Simon, eds. *Embracing Risk: The Changing Culture of Insurance and Responsibility.* Chicago: University of Chicago Press, 2002.

Bale, Anthony. "Assuming the Risks: Occupational Disease in the Years before Workers' Compensation." *American Journal of Industrial Medicine* 13 (1988): 499–513.

———. "Compensation Crisis: The Value and Meaning of Work-Related Injuries and Illnesses in the United States, 1842–1932." Ph.D. diss., Brandeis University, 1986.

Bardach, Eugene, and Robert A. Kagan. *Going by the Book: The Problem of Regulatory Unreasonableness.* Philadelphia: Temple University Press, 1982.

Batlon, Felice. "A Reevaluation of the New York Court of Appeals: The Home, the Market, and Labor, 1885–1905." *Law and Social Inquiry* 27 (2002): 489–528.

Bean, Walton. *California: An Interpretative History.* New York: McGraw-Hill Book, 1973.

Beckner, Earl R. *A History of Labor Legislation in Illinois.* Chicago: University of Chicago Press, 1929.

Benedict, Michael Les. "Law and Regulation in the Gilded Age and Progressive Era." In *Law as Culture and Culture as Law: Essays in Honor of John Phillip Reid,* edited by Hendrik Hartog and William E. Nelson, 225–63. Madison: Madison House Publishers, 2000.

Bennett, Dianne, and William Graebner. "Safety First: Slogan and Symbol of the Industrial Safety Movement." *Journal of the Illinois State Historical Society* 68 (June 1975): 243–56.

Bergstrom, Randolph E. *Courting Danger: Injury and Law in New York City, 1870–1910.* Ithaca: Cornell University Press, 1992.

Berkowitz, Edward D. *Something Happened: A Political and Cultural Overview of the Seventies.* New York: Columbia University Press, 2006.

Berman, Daniel M. *Death on the Job.* New York: Monthly Review Press, 1978.

Berman, William C. *America's Right Turn: From Nixon to Bush.* Baltimore: Johns Hopkins University Press, 1994.

Bernstein, Irving. *The Lean Years: A History of the American Worker, 1920–1933.* Boston: Houghton-Mifflin, 1960.

Bernstein, Marver. *Regulating Business by Independent Commission.* Princeton: Princeton University Press, 1955.

Bird, Francis H. "The Advisory Representative Labor Council: A Study of the Relation of Employers, Employees, and the General Public to Labor Legislation and Administration with Especial Reference to the Belgian Superior Council of Labor." Ph.D. diss., University of Wisconsin, 1917.

———. "Proposed Wisconsin Industrial Commission." *Survey* 26 (April 28, 1911): 151–52.

———. "Standardization of Safety." *Survey* 25 (1911): 1021–23.

Block, Fred. *Revising State Theory: Essays in Politics and Postindustrialism.* Philadelphia: Temple University Press, 1987.

Brandeis, Elizabeth. "The Administration of Labor Laws." In *History of Labor in the*

United States, 1896–1932, edited by John R. Commons et al., 3:625–58. New York: Macmillan, 1935.

———. "Minimum Wage Legislation." In *History of Labor in the United States, 1896–1932,* edited by John R. Commons et al., 3:501–39. New York: Macmillan, 1935.

———. "Women's Hours Legislation." In *History of Labor in the United States, 1896–1932,* edited by John R. Commons et al., 3:457–500. New York: Macmillan, 1935.

Breyer, Stephen G., and Richard B. Stewart. *Administrative Law and Regulatory Policy.* Boston: Little, Brown, 1979.

Brody, David. *Workers in Industrial America: Essays on the Twentieth-Century Struggle.* New York: Oxford University Press, 1980.

Bronstein, Jamie L. *Caught in the Machinery: Workplace Accidents and Injured Workers in Nineteenth-Century Britain.* Stanford: Sanford University Press, 2008.

Brownlee, Eliot W. *Progressivism and Economic Growth: The Wisconsin Income Tax, 1911–1929.* Port Washington, N.Y.: Kennikat Press, 1974.

Buenker, John. *The History of Wisconsin.* Vol. 4, *The Progressive Era, 1893–1914.* Madison: State Historical Society of Wisconsin, 1998.

Burgess, John W. *Political Science and Comparative Constitutional Law.* 2 vols. Boston: Ginn, 1890.

Burke, Thomas F. *Lawyers, Lawsuits, and Legal Rights: The Battle over Litigation in American Society.* Berkeley and Los Angeles: University of California Press, 2002.

Caine, Stanley P. *The Myth of Progressive Reform: Railroad Regulation in Wisconsin, 1903–1910.* Madison: State Historical Society of Wisconsin, 1970.

"The California Occupational Safety and Health Act of 1973." *Loyola University Law Review* 9 (1976): 905–60.

Carpenter, Daniel P. *The Forging of Bureaucratic Autonomy: Reputation, Networks, and the Policy Innovations in Executive Agencies, 1862–1928.* Princeton: Princeton University Press, 2001.

Cayton, Andrew R. L. *Ohio: A History of a People.* Columbus: Ohio State University Press, 2002.

Cebula, James E. *James M. Cox: Journalist and Politician.* New York: Garland Publishing, 1985.

Chambers, John Whiteclay. *The Tyranny of Change: America in the Progressive Era, 1900–1917.* New York: St. Martin's Press, 1980.

Chandler, Alfred D., Jr. *The Visible Hand: The Managerial Revolution in American Business.* Cambridge: Harvard University Press, Belknap Press, 1977.

Chase, William G. *The American Law School and the Rise of Administrative Government.* Madison: University of Wisconsin Press, 1982.

Cherniack, Martin. *The Hawk's Nest Incident: America's Worst Industrial Disaster.* New Haven: Yale University Press, 1986.

Chomsky, Carol. "Progressive Judges in a Progressive Age: Regulatory Legislation in the Minnesota Supreme Court, 1880–1925." *Law and History Review* 11 (1993): 383–440.

Church, Robert L. "Economists as Experts: The Rise of an Academic Profession in the United States, 1870–1920." In *The University in Society,* edited by Lawrence Stone, 2:596–626. Princeton: Princeton University Press, 1974.

Clark, Claudia. "Managing an Epidemic: Industrial Radium Poisoning and the Waterbury Clock Company." *Connecticut History* 38 (Fall 1997–Spring 1999): 12–27.

———. *Radium Girls: Women and Industrial Health Reform, 1910–1935.* Chapel Hill: University of North Carolina Press, 1997.

Clark, Thomas Ralph. *Defending Rights: Law, Labor, Politics, and the State in California, 1890–1925.* Detroit: Wayne State University Press, 2002.

Clemens, Elisabeth S. *The People's Lobby: Organizational Innovation and the Rise of Interest Group Politics in the United States, 1890–1925.* Chicago: University of Chicago Press, 1997.

Clemens, Eli Winston. *Economics and Public Utilities.* New York: Appleton-Century-Crofts, 1950.

Cohen, Lizabeth. *Making a New Deal: Industrial Workers in Chicago, 1919–1939.* New York: Cambridge University Press, 1990.

The Columbian Biographical Dictionary and Portrait Gallery of the Representative Men of the United States: Wisconsin Volume. Chicago: Lewis Publishing, 1895.

"Comment—the Creation of a Common-Law Rule: The Fellow Servant Rule, 1837–1860." *University of Pennsylvania Law Review* 132 (1984): 579–620.

Commons, John R. "The Industrial Commission of Wisconsin." *American Labor Legislation Review* 1 (1911): 1–8.

———. *The Industrial Commission of Wisconsin: Its Organization and Methods.* Pamphlet reprinted by the Industrial Commission of Wisconsin, ca. 1913.

———. *Labor and Administration.* New York: Macmillan, 1913.

———. *Myself.* New York: Macmillan, 1934.

———. *Proportional Representation.* New York: Thomas Y. Cromwell, 1896.

———. "The Wisconsin Public Utilities Law." *Review of Reviews* 36 (1907): 221–24.

Commons, John R., and John B. Andrews. *Principles of Labor Legislation.* New York: Harper and Brothers, 1916.

Cooley, Thomas M. *A Treatise on the Law of Torts; or, The Wrongs Which Arise Independent of Contract.* Chicago: Callaghan, 1888.

Crandall, William "Rick," and Richard E. Crandall. "Revisiting the Hawks Nest Tunnel Incident: Lessons Learned from an American Tragedy." *Journal of Appalachian Studies* 8 (2002): 261–83.

Crownhart, Charles H. *The Workmen's Compensation Act.* Pamphlet reprinted by the Industrial Commission of Wisconsin, ca. February 1913.

Curran, Daniel J. *Dead Laws for Dead Men: The Politics of Federal Coal Mine Health and Safety Legislation.* Pittsburgh: University of Pittsburgh Press, 1993.

Cushman, Robert E. *The Independent Regulatory Agencies.* New York: Oxford University, 1941.

Dallek, Robert. *Ronald Reagan: The Politics of Symbolism.* Cambridge: Harvard University Press, 1984.

Davis, Kenneth Culp. *Administrative Law Text.* St. Paul: West Publishing, 1972.

Derickson, Alan. *Black Lung: Anatomy of a Public Health Disaster.* Ithaca: Cornell University Press, 1998.

———. *Workers' Health, Workers' Democracy: The Western Miners' Struggle, 1891–1925.* Ithaca: Cornell University Press, 1988.

"Detroit Convention in Industry's Forward March." *Foundry* (May 1936): 76.

Dicey, A. V. *Introduction to the Study of the Law of the Constitution* 8th ed. London: Macmillan, 1927.

Dickinson, John. *Administrative Justice and the Supremacy of Law in the United States.* New York: Russell and Russell, 1927.

Dubofsky, Melvyn. *Industrialism and the American Worker, 1865–1920.* Arlington Heights, Ill.: AHM Publishing, 1975.

———. *State and Labor in Modern America.* Chapel Hill: University of North Carolina Press, 1994.

Dulles, Foster Rhea, and Melvyn Dubofsky. *Labor in America: A History.* 4th ed. Arlington Heights, Ill.: Harlan Davidson, 1984.

Eisner, Marc Allen. "Discovering Patterns in Regulatory History: Continuity, Change, and Regulatory Regimes." *Journal of Policy History* 6 (1994): 157–87.

Ellis, David M., et al. *A History of New York State.* Ithaca: Cornell University Press, 1967.

Ely, James W., Jr. *Railroads and American Law.* Lawrence: University Press of Kansas, 2001.

Erickson, Nancy S. "*Muller v. Oregon* Reconsidered: The Origins of a Sex-Based Doctrine of Liberty of Contract." *Labor History* 30 (Spring 1989): 228–49.

Ernst, Daniel R. "Common Laborers? Industrial Pluralists, Legal Realists, and the Law of Industrial Disputes." *Law and History Review* 11 (Spring 1993): 59–100.

Fetner, Gerald L. *Ordered Liberty: Legal Reform in the Twentieth Century.* New York: Alfred A. Knopf, 1983.

Fine, Sidney. *Laissez-Faire and the General Welfare State: A Study in Conflict in American Thought, 1865–1901.* Ann Arbor: University of Michigan Press, 1964.

Fink, Leon. "'Intellectuals' versus 'Workers': Academic Requirements and the Creation of Labor History." *American Historical Review* 96 (April 1991): 395–421.

———. "Labor, Liberty, and the Law: Trade Unionism and the Problem of the American Constitutional Order." *Journal of American History* 74 (1987): 904–25.

———. *Progressive Intellectuals and the Dilemmas of Democratic Commitment.* Cambridge: Harvard University Press, 1997.

———. *Workingmen's Democracy: The Knights of Labor in American Politics.* Urbana: University of Illinois Press, 1983.

Fishback, Price V., and Shawn Everett Kantor. *A Prelude to the Welfare State: The Origins of Workers' Compensation.* Chicago: University of Chicago Press, 2000.

Forbath, William E. *Law and the Shaping of the American Labor Movement.* Cambridge: Harvard University Press, 1991.

Freund, Ernst. *The Police Power: Public Policy and Constitutional Rights.* Chicago: Callaghan, 1904.

Freyer, Tony A. *Producers versus Capitalists: Constitutional Conflict in Antebellum America.* Charlottesville: University Press of Virginia, 1994.

Friedman, Lawrence M. *A History of American Law.* 2d ed. New York: Simon and Schuster, 1985.

Friedman, Lawrence M., and Jack Ladinsky. "Social Change and the Law of Industrial Accidents." *Columbia Law Review* 67 (1967): 50–82.

Furner, Mary O. *Advocacy and Objectivity: A Crisis in the Professionalization of American Social Science, 1865–1905.* Lexington: University Press of Kentucky, 1975.

———. "Knowing Capitalism: Public Investigation on the Labor Question in the Progressive Era." In *The State and Economic Knowledge: The American and British Experiences,* edited by Mary O. Furner and Barry Supple, 241–86. New York: Cambridge University Press, 1990.

Galambos, Louis. "The Emerging Organizational Synthesis in Modern American History." *Business History Review* 44 (1970): 279–90.

Galie, Peter J. *The New York State Constitution: A Reference Guide.* New York: Greenwood Press, 1991.

———. *Ordered Liberty: A Constitutional History of New York.* New York: Fordham University Press, 1996.

Gavett, Thomas William. "The Development of the Labor Movement in Milwaukee." Ph.D. diss., University of Wisconsin, 1957.

———. *Development of the Labor Movement in Milwaukee.* Madison: University of Wisconsin Press, 1965.

Gillman, Howard. *The Constitution Besieged: The Rise and Demise of Lochner-Era Police Powers Jurisprudence.* Durham: Duke University Press, 1993.

Glad, Paul W. *The History of Wisconsin: War, a New Era, and Depression, 1914–1940.* Madison: State Historical Society of Wisconsin, 1990.

Glaeser, Martin G. *Outline of Public Utilities Economics.* New York: Macmillan, 1929.

Glenn, Brian J. "Risk, Insurance, and the Changing Nature of Mutual Obligation." *Law and Social Inquiry* 28 (2003): 295–314.

"Good Housecleaning at Falks." Pts. 1 and 2. *Foundry* (February 1933): 10–11; (March 1933): 16–18.

Goodnow, Frank J. *Comparative Administrative Law: An Analysis of the Administrative Systems, National and Local, of the United States, England, France, and Germany.* New York: G. P. Putnam's Sons, 1893.

———. *Politics and Administration: A Study in Government.* New York: Macmillan, 1900.

Gordon, Colin. "Does the Ruling Class Rule?" *Reviews in American History* 25 (June 1997): 288–93.

———. *New Deals, Business, Labor, and Politics in America, 1920–1935.* New York: Cambridge University Press, 1994.

Graebner, William. *Coal-Mining Safety in the Progressive Period: The Political Economy of Reform*. Lexington: University Press of Kentucky, 1976.

———. "Federalism in the Progressive Era: A Structural Interpretation of Reform." *Journal of American History* 64 (1977): 331–57.

———. "Private Power, Private Knowledge, and Public Health: Science, Engineering, and Lead Poisoning, 1900–1970." In *The Health and Safety of Workers: Case Studies in the Politics of Professional Responsibility*, edited by Ronald Bayer, 15–71. New York: Oxford University Press, 1988.

———. "Radium Girls, Corporate Boys." *Reviews in American History* 26 (September 1998): 587–92.

Grant, H. Roger. *Insurance Reform: Consumer Action in the Progressive Era*. Ames: Iowa State University Press, 1979.

Grantham, Dewey W. *Southern Progressivism: The Reconciliation of Progress and Tradition*. Knoxville: University of Tennessee Press, 1983.

Green, Marguerite. *The National Civic Federation and the American Labor Movement, 1900–1925*. Washington, D.C.: Catholic University of America Press, 1956.

Haber, Samuel. *Efficiency and Uplift: Scientific Management in the Progressive Era, 1890–1920*. Chicago: University of Chicago Press, 1964.

Hackney, James R. "The Intellectual Origins of American Strict Products Liability: A Case Study in American Pragmatic Instrumentalism." *American Journal of Legal History* 39 (1995): 443–509.

Hackney, Sheldon. *Populism to Progressivism in Alabama*. Princeton: Princeton University Press, 1969.

Hagglund, George. "Some Factors Contributing to Wisconsin Occupational Injuries." Ph.D. diss., University of Wisconsin, 1966.

Hamby, Alonzo L. *For the Survival of Democracy: FDR and the World Crisis of the 1930s*. New York: Free Press, 2004.

———. *Liberalism and Its Challengers, FDR to Reagan*. New York: Oxford University Press, 1985.

Hamilton, Alice. *Exploring the Dangerous Trades*. Boston: Little, Brown, 1942.

Hamm, Richard F. *Shaping the Eighteenth Amendment: Temperance Reform, Legal Culture, and the Polity, 1880–1920*. Chapel Hill: University of North Carolina Press, 1995.

Hard, William. "The Law of the Killed and the Wounded." *Everybody's Magazine* 19 (September 1908): 361–71.

———. "Making Steel and Killing Men." *Everybody's Magazine* 17 (November 1907): 579–91.

Harter, Lafayette G., Jr. *John R. Commons: His Assault on Laissez-Faire*. Corvallis: Oregon State University Press, 1962.

Hatch, Leonard W. "The Prevention of Accidents." *American Labor Legislation Review* 1 (June 1911): 107–9.

———. "What the Accident Record Shows." *American Labor Legislation Review* 16 (1926): 167–73.

Hawley, Ellis W. "The Discovery and Study of a 'Corporate Liberalism.'" *Business History Review* 52 (1978): 309–20.

———. *The Great War and the Search for a Modern Order: A History of the American People and Their Institutions, 1917–1933.* New York: St. Martin's Press, 1979.

———. "Herbert Hoover, the Commerce Secretariat, and the Vision of an 'Associative State,' 1921–1928." *Journal of American History* 61 (1974): 116–40.

Hayhurst, Emery R. "Occupational Disease Considerations." *Ohio Bar Association Report* 16 (1943): 87–93.

Hays, Samuel P. "The New Organizational Society." In *American Political History as Social Analysis.* Knoxville: University of Tennessee Press, 1980.

———. "Political Choice in Regulation." In *Regulation in Perspective: Historical Essays,* edited by Thomas K. McCraw, 124–54. Cambridge: Harvard University Press, 1981.

Heimer, Carol A. "Insuring More, Ensuring Less: The Costs and Benefits of Private Regulation through Insurance." In *Embracing Risk: The Changing Culture of Insurance and Responsibility,* edited by Tom Baker and Jonathan Simon, 116–39. Chicago: University of Chicago Press, 2002.

Hepler, Allison L. *Women in Labor: Mothers, Medicine, and Occupational Health in the United States, 1890–1980.* Columbus: Ohio State University Press, 2000.

Herbst, Jurgen. *The German Historical School in American Scholarship: A Study in the Transfer of Culture.* Ithaca: Cornell University Press, 1965.

Herron, Belva M. "Factory Inspection in the United States." *American Journal of Sociology* 12 (1907): 487–99.

Higgens-Evenson, R. Rudy. "From Industrial Police to Workmen's Compensation: Public Policy and Industrial Accidents in New York, 1880–1910." *Labor History* 39 (November 1998): 365–80.

Hobson, Wayne K. "Professionals, Progressives, Bureaucratization: A Reassessment." *Historian* 39 (1977): 639–58.

Hoeveler, J. David. "The University and the Social Gospel: The Intellectual Origins of the 'Wisconsin Idea.'" *Wisconsin Magazine of History* 59 (1976): 289–98.

Horne, Roger D. "Practical Idealism at the Zenith of Progressivism: John R. Commons as an Industrial Reformer." *Mid-America* 76 (1994): 53–70.

Horwitz, Morton J. *The Transformation of American Law, 1780–1860.* Cambridge: Harvard University Press, 1977.

———. *The Transformation of American Law, 1870–1960: The Crisis of Legal Orthodoxy.* New York: Oxford University Press, 1992.

Hovenkamp, Herbert. *Enterprise and American Law, 1836–1937.* Cambridge: Harvard University Press, 1991.

Howard, Robert P. *Illinois: A History of the Prairie State.* Grand Rapids: William B. Eerdmans Publishing, 1972.

Howe, Frederic C. *Wisconsin: An Experiment in Democracy.* New York: Charles Scribner Sons, 1912.

Hoyt, Ralph M. "The Wisconsin Administrative Procedure Act." *Wisconsin Law Review* 1944 (1944): 214–39.

Hunter, Carol. "The California Occupational Safety and Health Act: An Overview." *Los Angeles Bar Bulletin* 50 (1975): 303–12.

Hurst, James Willard. *Law and Economic Growth: The Legal History of the Lumber Industry in Wisconsin, 1836–1915.* Cambridge: Harvard University Press, Belknap Press, 1964.

———. *Law and the Conditions of Freedom in the Nineteenth-Century United States.* Madison: University of Wisconsin Press, 1950.

"Industrial Hygiene Bill." *American Labor Legislation Review* 30 (June 1940): 72.

Jacoby, Sanford M. *Employing Bureaucracy: Managers, Unions, and the Transformation of Work in American Industry, 1900–1945.* New York: Columbia University Press, 1985.

———. *Modern Manors: Welfare Capitalism since the New Deal.* Princeton: Princeton University Press, 1997.

Jones, Ray R. "Silicosis: An Occupational Hazard." *Labor Information Bulletin* (March 1936).

Kaczorowski, Robert J. "The Common-Law Background of Nineteenth-Century Tort Law." *Ohio State Law Journal* 51 (1990): 1127–99.

Kagan, Robert A. *Adversarial Legalism: The American Way of Law.* Cambridge: Harvard University Press, 2001.

———. "On Regulatory Inspectorates and Police." In *Enforcing Regulation,* edited by Keith Hawkins and John M. Thomas, 39–54. Boston: Kluwer-Nijhoff Publishing, 1984.

Karsten, Peter. "Explaining the Fight over the Attractive Nuisance Doctrine: A Kinder, Gentler Instrumentalism in the Age of Formalism." *Law and History Review* 10 (1992): 45–92.

———. *Heart versus Head: Judge-Made Law in Nineteenth-Century America.* Chapel Hill: University of North Carolina Press, 1997.

Kasparek, Jonathan. *Fighting Son: A Biography of Philip F. La Follette.* Madison: Wisconsin Historical Society Press, 2006.

Keller, Morton. *Affairs of State: Public Life in Nineteenth-Century America.* Cambridge: Harvard University Press, 1977.

———. *Regulating a New Economy: Public Policy and Economic Change in America, 1900–1933.* Cambridge: Harvard University Press, 1990.

———. *Regulating a New Society: Public Policy and Social Change in America, 1900–1933.* Cambridge: Harvard University Press, 1994.

Kelly, Brian. *Race, Class, and Power in the Alabama Coal Fields, 1908–1921.* Urbana: University of Illinois Press, 2001.

Kelman, Steven. "Occupational Safety and Health Administration." In *The Politics of Regulation,* edited by James Q. Wilson, 236–66. New York: Basic Books, 1980.

———. *Regulating America, Regulating Sweden: A Comparative Study of Occupational Safety and Health Policy.* Cambridge: MIT Press, 1981.

Kens, Paul. *"Lochner v. New York": Economic Regulation on Trial.* Lawrence: University Press of Kansas, 1998.

———. "The Source of a Myth: Police Powers of the States and Laissez-Faire Constitutionalism, 1900–1937." *American Journal of Legal History* 35 (1991): 70–98.

Kessler-Harris, Alice. *Out to Work: A History of Wage-Earning Women in the United States.* New York: Oxford University Press, 1982.

Kimball, Spencer L. *Insurance and Public Policy: A Study in the Legal Implementation of Social and Economic Public Policy, Based on Wisconsin Records, 1835–1959.* Madison: University of Wisconsin Press, 1960.

Knepper, George W. *Ohio and Its People.* Kent: Kent State University Press, 1989.

Kolko, Gabriel. *The Triumph of Conservatism: A Reinterpretation of American History, 1900–1916.* New York: Quadrangle Books, 1963.

Konesky, Alfred S. "As Best to Subserve Their Own Interests: Lemuel Shaw, Labor Conspiracy, and Fellow Servants." *Law and History Review* 7 (1989): 219–39.

Korman, Gerd. *Industrialization, Immigrants, and Americanizers: The View from Milwaukee.* Madison: State Historical Society of Wisconsin, 1967.

Kuechle, B. E. "Occupational Disease Liabilities—Financial and Humanitarian." *Journal of American Insurance* 22 (May 1945): 17–18.

Landis, James M. *The Administrative State.* New Haven: Yale University Press, 1938.

Larson, Margali S. "The Production of Expertise and the Constitution of Expert Power." In *The Authority of Experts,* edited by Thomas L. Haskell, 28–80. Bloomington: Indiana University Press, 1984.

———. *The Rise of Professionalism: A Sociological Analysis.* Berkeley and Los Angeles: University of California Press, 1977.

"Legislation for Women in Industry." *American Labor Legislation Review* 16 (1916): 356–410.

Leuchtenburg, William E. *The Supreme Court Reborn: The Constitutional Revolution in the Age of Roosevelt.* New York: Oxford University Press, 1995.

Levy, Leonard W. *The Law of the Commonwealth and Chief Justice Shaw.* New York: Oxford University Press, 1957.

Link, Arthur S., and Richard L. McCormick. *Progressivism.* Arlington Heights, Ill.: Harlan Davidson, 1983.

Lippmann, Walter. *Drift and Mastery: An Attempt to Diagnose the Current Unrest.* New York: Mitchell Kennerley, 1914.

Lower, Richard Coke. *A Bloc of One: The Political Career of Hiram Johnson.* Stanford: Stanford University Press, 1993.

Lubove, Roy. *The Struggle for Social Security, 1900–1935.* Pittsburgh: University of Pittsburgh Press, 1986.

Madden, Daniel Richard. "Factory Safety in Wisconsin, 1878–1911." Master's thesis, University of Wisconsin, 1968.

Mahon, Thomas J. "Conserving Human Life." *La Follette's Weekly Magazine* 2 (June 3, 1911): 7–8.

Mandell, Nikki. *The Corporation as Family: The Gendering of Corporate Welfare, 1890–1930.* Chapel Hill: University of North Carolina Press, 2002.

Margulies, Herbert. *The Decline of the Progressive Movement in Wisconsin.* Madison: State Historical Society of Wisconsin, 1968.

Markowitz, Gerald, and David Rosner. *Deceit and Denial: The Deadly Politics of Industrial Pollution.* Berkeley and Los Angeles: University of California Press, 2002.

Mashaw, Jerry L. *Due Process in the Administrative State.* New Haven: Yale University Press, 1985.

Mason, Mary Ann. "Neither Friends nor Foes: Organized Labor and the California Progressives." In *California Progressives Revisited,* edited by William Deverell and Tom Sitton, 57–70. Berkeley and Los Angeles: University of California Press, 1994.

McCarthy, Charles. *The Wisconsin Idea.* New York: Macmillan, 1912.

McClay, Wilfred M. "John W. Burgess and the Search for Cohesion in American Political Thought." *Polity* 26 (1993): 51–73.

McCormick, Richard L. *From Realignment to Reform: Political Change in New York, 1893–1910.* Ithaca: Cornell University Press, 1981.

———. "The Party Period and Public Policy: An Exploratory Hypothesis." *Journal of American History* 66 (1979): 279–98.

McCraw, Thomas K. *Prophets of Regulation.* Cambridge: Harvard University Press, 1984.

———. "Regulation in America: A Review Article." *Business History Review* 49 (Summer 1975): 159–83.

McDowell, Banks. *Deregulation and Competition in the Insurance Industry.* New York: Quorum Books, 1989.

McEvoy, Arthur F. "Freedom of Contract, Labor, and the Administrative State." In *The State and Freedom of Contract,* ed. Harry N. Scheiber, 198–235. Stanford: Stanford University Press, 1998.

———. "The Triangle Shirtwaist Factory Fire of 1911: Social Change, Industrial Accidents, and the Evolution of Common-Sense Causality." *Law and Social Inquiry* 20 (Spring 1995): 621–51.

McGill, J. D., Jr. "Alabama Administrative Regulatory Agencies: Development and Powers." *Alabama Lawyer* 4 (1943): 75–90.

———. "The Alabama Supreme Court and Administrative Regulation, 1920–1940." *Alabama Lawyer* 4 (1943): 158–84.

McNeill, Joseph H. "Massachusetts Board of Boiler Rules." *American Labor Legislation Review* 1 (December 1911): 70–80.

McNulty, Paul J. *The Origins and Development of Labor Economics: A Chapter in the History of Social Thought.* Cambridge: MIT Press, 1980.

Mendeloff, John. *Regulating Safety: An Economic and Political Analysis of Occupational Safety and Health Policy.* Cambridge: MIT Press, 1979.

Mettler, Suzanne. *Dividing Citizens: Gender and Feminism in New Deal Public Policy.* Ithaca: Cornell University Press, 1998.

Mintz, Benjamin. *OSHA: History Law and Policy.* Washington, D.C.: Bureau of National Affairs, 1984.

Montgomery, David. *The Fall of the House of Labor: The Workplace, the State, and American Labor Activism, 1865–1925.* New York: Cambridge University Press, 1987.

———. *Workers' Control in America.* New York: Cambridge University Press, 1979.

Morey, Richard. "The General Duty Clause of the Occupational Safety and Health Act of 1970." *Harvard Law Review* 86 (April 1973): 988–1005.

Mosher, William E., and Finla G. Crawford. *Public Utility Regulation.* New York: Harper and Brothers, 1933.

Moss, David A. *Socializing Security: Progressive-Era Economists and the Origins of American Social Policy.* Cambridge: Harvard University Press, 1996.

Mowry, George E. *The California Progressives.* Berkeley and Los Angeles: University of California Press, 1951.

Nash, Gerald D. "The Influence of Labor on State Policy, 1860–1920: The Experience of California." *California Historical Society Quarterly* 42 (1963): 241–57.

Nelson, Daniel. *Frederick Taylor and the Rise of Scientific Management.* Madison: University of Wisconsin Press, 1980.

———. *Managers and Workers: Origins of the New Factory System in the United States, 1880–1920.* Madison: University of Wisconsin Press, 1975.

Nesbit, Robert C. *Wisconsin: A History.* 2d ed. Madison: University of Wisconsin Press, 1989.

Noble, Charles. *Liberalism at Work: The Rise and Fall of OSHA.* Philadelphia: Temple University Press, 1986.

Noble, David F. *America by Design: Science, Technology, and the Rise of Corporate Capitalism.* New York: Oxford University Press, 1977.

Nonet, Philippe. *Administrative Justice: Advocacy and Change in a Government Agency.* New York: Russell Sage Foundation, 1969.

Novak, William J. "The Legal Origins of the Modern American State." In *Looking Back at Law's Century,* edited by Austin Sarat, Bryant Garth, and Robert A. Kagan, 249–83. Ithaca: Cornell University Press, 2002.

Novkov, Julie. *Constituting Workers, Protecting Women: Gender, Law, and Labor in the Progressive Era and New Deal Years.* Ann Arbor: University of Michigan Press, 2001.

Nugent, Angela. "Fit for Work: The Introduction of Physical Examination in Industry." *Bulletin of the History of Medicine* 57 (1983): 578–95.

———. "Organizing Trade Unions to Combat Disease: The Workers' Health Bureau, 1921–1928." *Labor History* 26 (1985): 423–46.

Orren, Karen. *Belated Feudalism: Labor, the Law, and Liberal Development in the United States.* New York: Cambridge University Press, 1991.

Ozanne, Robert W. *The Labor Movement in Wisconsin: A History.* Madison: State Historical Society of Wisconsin, 1984.

Patterson, James T. *The New Deal and the States: Federalism in Transition.* Princeton: Princeton University Press, 1969.

Paul, Arnold M. *Conservative Crisis and the Rule of Law: Attitudes of Bar and Bench, 1887–1895.* New York: Harper and Row Publishers, 1960.

Pegram, Thomas R. *Partisans and Progressives: Private Interest and Public Policy in Illinois, 1870–1922.* Urbana: University of Illinois Press, 1992.

Pierce, Michael. "Organized Labor and the Law in Ohio." In *The History of Ohio Law,* edited by Michael Les Benedict and John F. Winkler, 883–915. Athens: Ohio University Press, 2004.

Quaife, Milo Milton. *Wisconsin: Its History and Its People, 1634–1924.* 4 vols. Chicago: S. J. Clarke Publishing, 1924.

Rabin, Robert L. "The Historical Development of the Fault Principle." *Georgia Law Review* 15 (1981): 931–61.

Ramirez, Bruno. *When Workers Fight: The Politics of Industrial Relations in the Progressive Era, 1898–1916.* Westport, Conn.: Greenwood Press, 1978.

Raney, William Francis. *Wisconsin: A Story of Progress.* Appleton, Wisc.: Perrin Press, 1940.

Reagan, Patrick D. "The Ideology of Social Harmony and Efficiency: Workmen's Compensation in Ohio, 1904–1919." *Ohio History* 90 (1981): 317–31.

Reeve, Arthur B. "The Death Roll of Industry." *Charities and the Commons* 17 (February 1907): 791–807.

———. "Our Industrial Juggernaut." *Everybody's Magazine* 16 (February 1907): 147–52.

Reuss, Henry S. "Thirty Years of the Safe Place Statute." *Wisconsin Law Review* (1940): 335–61.

Ricci, David M. *The Tragedy of Political Science: Politics, Scholarship, and Democracy.* New Haven: Yale University Press, 1984.

Rodgers, Daniel T. *Atlantic Crossings: Social Politics in a Progressive Age.* Cambridge: Harvard University Press, Belknap Press, 1998.

———. *Contested Truths: Key Words in American Politics since Independence.* Cambridge: Harvard University Press, 1987.

———. "In Search of Progressivism." *Reviews in American History* 10 (December 1982): 113–32.

Rogers, William Warren, et al. *Alabama: The History of a Deep South State.* Tuscaloosa: University of Alabama Press, 1994.

Rose, Patricia Terpack. "Design and Expediency: The Ohio State Federation of Labor as a Legislative Lobby, 1883–1935." Ph.D. diss., Ohio State University, 1975.

Rosner, David, and Gerald Markowitz. *Deadly Dust: Silicosis and the Politics of Occupational Disease in Twentieth-Century America.* Princeton: Princeton University Press, 1991.

———, eds. *Dying for Work: Workers' Safety and Health in Twentieth-Century America.* Bloomington: Indiana University Press, 1987.

Ross, Dorothy. *The Origins of American Social Science.* New York: Cambridge University Press, 1991.

Ross, William G. *A Muted Fury: Populists, Progressives, and Labor Unions Confront the Courts, 1890–1937.* Princeton: Princeton University Press, 1994.

Schaller, Michael. *Right Turn: American Life in the Reagan-Bush Era, 1980–1992.* New York: Oxford University Press, 2007.

Schatz, Ronald W. "From Commons to Dunlap: Rethinking the Field and Theory of Industrial Relations." In *Industrial Democracy in America: The Ambiguous Promise,* edited by Nelson Lichtenstein and Howell John Harris, 87–112. New York: Cambridge University Press, 1993.

Schlanger, Margo. "Injured Women before Common Law Courts, 1860–1930." *Harvard Women's Law Journal* 21 (1998): 79–140.

Schmidt, Gertrude. "History of Labor Legislation in Wisconsin." Ph.D. diss., University of Wisconsin, 1933.

Schmidt, James D. "'Restless Movements Characteristic of Childhood': The Legal Construction of Child Labor in Nineteenth-Century Massachusetts." *Law and History Review* 23 (2005): 315–50.

Schmitter, Phillippe C. "Still the Century of Corporatism." In *Trends toward Corporatist Intermediation,* edited by Phillippe C. Schmitter and Gerhard Lehmbruch, 8–41. Beverly Hills: Sage Publications, 1979.

Schwartz, Bernard. "The Administrative Agency in Historical Perspective." *Indiana Law Journal* 36 (1961): 263–81.

———. *Administrative Law.* Boston: Little, Brown, 1976.

Schwartz, Gary. "The Character of Early American Tort Law." *UCLA Law Review* 36 (1989): 641–718.

———. "Tort Law and the Economy in Nineteenth-Century America: A Reinterpretation." *Yale Law Journal* 90 (1981): 1717–75.

Schweber, Howard. *The Creation of American Common Law, 1850–1880: Technology, Politics, and the Construction of Citizenship.* New York: Cambridge University Press, 2004.

Seager, David R. "Barre, Vermont, Granite Workers, and the Struggle against Silicosis, 1890–1960." *Labor History* 42 (2001): 61–79.

Sellers, Christopher. *Hazards of the Job: From Industrial Disease to Environmental Health Science.* Chapel Hill: University of North Carolina Press, 1997.

Shamir, Ronan. *Managing Legal Uncertainty: Elite Lawyers in the New Deal.* Durham: Duke University Press, 1995.

Shapiro, Martin. "The Supreme Court's 'Return' to Economic Regulation." *Studies in American Political Development* 1 (1986): 91–141.

Sharfman, Leo I. "Commission Regulation of Public Utilities: A Survey of Legislation." *Annals of the American Academic of Political and Social Science* 53 (May 1914): 1–18.

Shaw, James. "Practice and Procedure in Occupational Disease Claims." *Ohio Bar Association Report* 11 (1941): 184–91.

Shepherd, George B. "Fierce Compromise: The Administrative Procedure Act Emerges from New Deal Politics." *Northwestern University Law Review* 90 (1996): 1557–1683.

"Silicosis Compensation Blunder Disgraces New York." *American Labor Legislation Review* 29 (1939): 166–68.

"Silicosis Disgrace Continues in New York." *American Labor Legislation Review* 28 (1938): 172–75.

Simon, Jonathan. "For the Government of Its Servants: Law and Disciplinary Power in the Work Place, 1870–1906." *Studies in Law, Politics, and Society* 13 (1993): 105–36.

Simpson, Brian A. W. *Leading Cases in the Common Law.* New York: Oxford University Press, 1995.

Simson, Gary. "The Occupational Safety and Health Act of 1970: State Plans and the General Duty Clause." *Ohio State Law Journal* 34 (1973): 599–627.

Sklar, Kathryn Kish. *Florence Kelley and the Nation's Work: The Rise of Women's Political Culture, 1830–1900.* New Haven: Yale University Press, 1995.

Sklar, Martin J. *The Corporate Reconstruction of American Capitalism, 1890–1916: The Market, the Law, and Politics.* New York: Cambridge University Press, 1988.

Skocpol, Theda. *Protecting Soldiers and Mothers: The Political Origins of Social Policy in the United States.* Cambridge: Harvard University Press, 1992.

Skowronek, Stephen. *Building a New American State: The Expansion of National Administrative Capacities, 1877–1920.* New York: Cambridge University Press, 1982.

Starr, Kevin. *Inventing the Dream: California through the Progressive Era.* New York: Oxford University Press, 1985.

Stern, Marc J. "Industrial Structure and Occupational Health: The American Pottery Industry, 1897–1929." *Business History Review* 27 (2003): 417–45.

Stewart, Ethelbert. "Are Accidents Increasing?" *American Labor Legislation Review* 16 (1926): 161–65.

Stewart, Richard B. "The Reformation of American Administrative Law." *Harvard Law Review* 88 (June 1975): 1669–1813.

Stigler, George J. "The Theory of Economic Regulation." *Bell Journal of Economics* 2 (1971): 3–21.

Stone, Katherine Van Wezel. "The Post-war Paradigm in American Labor Law." *Yale Law Journal* 90 (1981): 1509–80.

Szasz, Andrew. "Industrial Resistance to Occupational Safety and Health Legislation, 1971–1981." *Social Problems* 32 (1984): 103–16.

Taft, Philip. *Labor Politics American Style: The California State Federation of Labor.* Cambridge: Harvard University Press, 1968.

Teaford, Jon C. *The Rise of the States: Evolution of American State Government.* Baltimore: Johns Hopkins University Press, 2002.

Thelen, David P. *Robert M. La Follette and the Insurgent Spirit.* Boston: Little, Brown, 1976.

Tindall, George B. *The Emergence of the New South, 1913–1945.* Baton Rouge: Louisiana State University Press, 1967.

Tomlins, Christopher L. *Law, Labor, and Ideology in the Early American Republic.* New York: Cambridge University Press, 1993.

———. "Law and Power in the Employment Relationship." In *Labor Law in America,* edited by Christopher L. Tomlins and Andrew J. King, 71–98. Baltimore: Johns Hopkins University Press, 1992.

———. "The New Deal, Collective Bargaining, and the Triumph of Industrial Pluralism." *Industrial and Labor Relations Review* 39 (1985): 19–34.

———. *The State and the Unions: Labor Relations, Law, and the Organized Labor Movement in America, 1880–1960.* New York: Cambridge University Press, 1985.

Tone, Andrea. *The Business of Benevolence: Industrial Paternalism in Progressive America.* Ithaca: Cornell University Press, 1997.

Tripp, Joseph F. "An Instance of Labor and Business Cooperation: Workmen's Compensation in Washington State (1911)." *Labor History* 17 (1976): 530–50.

———. "Law and Social Control: Historians' Views of Progressive-Era Labor Legislation." *Labor History* 28 (1987): 447–83.

Unger, Nancy C. *Fighting Bob La Follette: The Righteous Reformer.* Chapel Hill: University of North Carolina Press, 2000.

Urofsky, Melvin I. "State Courts and Protective Labor Legislation during the Progressive Era: A Reevaluation." *Journal of American History* 72 (1985): 63–91.

Verkuil, Paul R. "The Emerging Concept of Administrative Procedure." *Columbia Law Review* 78 (1978): 258–329.

Vogel, David. "The 'New' Social Regulation in Historical and Comparative Perspective." In *Regulation in Perspective,* edited by Thomas K. McCraw, 155–85. Cambridge: Harvard University Press, 1981.

Wahl, Ana-Maria, and Steven E. Gunkel. "Due Process, Resource Mobilization, and the Occupational Safety and Health Administration, 1971–1996: The Politics of Social Regulation in Historical Perspective." *Social Problems* 46 (1999): 591–616.

Walker, Claire Brandler. "A History of Factory Legislation and Inspection in New York State, 1886–1911." Ph.D. diss., Columbia University, 1969.

Warner, Hoyt Landon. *Progressivism in Ohio, 1897–1917.* Columbus: Ohio State University Press, 1964.

Warren, Christian. *Brush with Death: A Social History of Lead Poisoning.* Baltimore: Johns Hopkins University Press, 2000.

Watkins, Bari. "Review Essay." *History and Theory* 15 (1976): 57–66.

Weinstein, James. *The Corporate Ideal in the Liberal State, 1900–1918.* Boston: Beacon Press, 1968.

Weir, Margaret, Ann Shola Orloff, and Theda Skocpol, eds. *The Politics of Social Policy in the United States.* Princeton: Princeton University Press, 1988.

Welke, Barbara Y. "Unreasonable Women: Gender and the Law of Accidental Injury, 1870–1920." *Law and Social Inquiry* 19 (1994): 369–403.

Wesser, Robert F. *Charles Evans Hughes: Politics and Reform in New York, 1905–1910.* Ithaca: Cornell University Press, 1967.

———. "Conflict and Compromise: The Workmen's Compensation Movement in New York State, 1890s–1913." *Labor History* 12 (1971): 345–72.

———. *A Response to Progressivism: The Democratic Party and New York Politics, 1902–1918.* New York: New York University Press, 1986.

Wharton, Francis. *A Treatise on the Law of Negligence.* Philadelphia: Kay and Brother, 1874.

White, G. Edward. "Allocating Power between Agencies and Courts: The Legacy of Justice Brandeis." In *Patterns of American Legal Thought,* 227–87. New York: Bobbs-Merrill, 1978.

———. *The Constitution and the New Deal.* Cambridge: Harvard University Press, 2000.

———. "From Sociological Jurisprudence to Realism: Jurisprudence and Social Change in Early Twentieth-Century America." In *Patterns of American Legal Thought,* 99–135. New York: Bobbs-Merrill, 1978.

———. *Tort Law in America: An Intellectual History.* New York: Oxford University Press, 2003.

Whiteside, James. *Regulating Danger: The Struggle for Mine Safety in the Rocky Mountain Coal Industry.* Lincoln: University of Nebraska Press, 1990.

Wiebe, Robert H. *The Search for Order, 1877–1920.* New York: Hill and Wang, 1967.

Wiecek, William M. *The Lost World of Classical Legal Thought: Law and Ideology in America, 1886–1937.* New York: Oxford University Press, 1998.

Wilson, James Q., ed. *The Politics of Regulation.* New York: Basic Books, 1980.

Wilson, Woodrow. "The Study of Administration." *Political Science Quarterly* 2 (June 1887): 197–222.

Wisconsin Manufacturers Association. *Classified Directory of Wisconsin Manufacturers.* Madison: Wisconsin Manufacturers Association, 1927.

Witt, John Fabian. *The Accidental Republic: Crippled Workingmen, Destitute Widows, and the Remaking of American Law.* Cambridge: Harvard University Press, 2004.

———. *Patriots and Cosmopolitans: Hidden Histories of American Law.* Cambridge: Harvard University Press, 2007.

———. "Speedy Fred Taylor and the Ironies of Enterprise Liability." *Columbia Law Review* 103 (2003): 1–49.

———. "Toward a New History of American Accident Law." *Harvard Law Review* 114 (2001): 690–841.

———. "The Transformation of Work and the Law of Workplace Accidents, 1842–1910." *Yale Law Review* 107 (1998): 1467–1502.

Woloch, Nancy. *"Muller v. Oregon": A Brief History with Documents.* Boston: Bedford/St. Martin's Press, 1996.

Woodward, C. Vann. *Origins of the New South, 1877–1913.* Baton Rouge: Louisiana State University Press, 1966.

Wunderlin, Clarence E. *Visions of a New Industrial Order: Social Science and Labor Theory in America's Progressive Era.* New York: Columbia University Press, 1992.

Yellowitz, Irwin. *Labor and the Progressive Movement in New York State, 1897–1916.* Ithaca: Cornell University Press, 1965.

Zainaldin, Jamil. *Law in Antebellum Society: Legal Change and Economic Expansion.* New York: Alfred A. Knopf, 1983.

Zeppos, Nicholas S. "The Legal Profession and the Development of Administrative Law." *Chicago-Kent Law Review* 72 (1997): 1119–57.

Index

DONALD W. ROGERS is an instructor in history at
Central Connecticut State University and Housatonic
Community College.

The University of Illinois Press
is a founding member of the
Association of American University Presses.

Composed in 10.5/13 Adobe Minion Pro
with Meta display
by Jim Proefrock
at the University of Illinois Press
Manufactured by Thomson-Shore, Inc.

University of Illinois Press
1325 South Oak Street
Champaign, IL 61820-6903
www.press.uillinois.edu